INDUSTRIAL POWER DISTRIBUTION

IEEE Press
445 Hoes Lane
Piscataway, NJ 08854

IEEE Press Editorial Board
Tariq Samad, *Editor in Chief*

George W. Arnold   Vladimir Lumelsky   Linda Shafer
Dmitry Goldgof   Pui-In Mak   Zidong Wang
Ekram Hossain   Jeffrey Nanzer   MengChu Zhou
Mary Lanzerotti   Ray Perez   George Zobrist

Kenneth Moore, *Director of IEEE Book and Information Services (BIS)*

Technical Reviewer

Om Malik, *University of Calgary*

SECOND EDITION

INDUSTRIAL POWER DISTRIBUTION

RALPH E. FEHR, III

IEEE PRESS

WILEY

Copyright © 2016 by The Institute of Electrical and Electronics Engineers, Inc.

Published by John Wiley & Sons, Inc., Hoboken, New Jersey. All rights reserved.
Published simultaneously in Canada.

No part of this publication may be reproduced, stored in a retrieval system, or transmitted in any form or by any means, electronic, mechanical, photocopying, recording, scanning, or otherwise, except as permitted under Section 107 or 108 of the 1976 United States Copyright Act, without either the prior written permission of the Publisher, or authorization through payment of the appropriate per-copy fee to the Copyright Clearance Center, Inc., 222 Rosewood Drive, Danvers, MA 01923, (978) 750-8400, fax (978) 750-4470, or on the web at www.copyright.com. Requests to the Publisher for permission should be addressed to the Permissions Department, John Wiley & Sons, Inc., 111 River Street, Hoboken, NJ 07030, (201) 748-6011, fax (201) 748-6008, or online at http://www.wiley.com/go/permission.

Limit of Liability/Disclaimer of Warranty: While the publisher and author have used their best efforts in preparing this book, they make no representations or warranties with respect to the accuracy or completeness of the contents of this book and specifically disclaim any implied warranties of merchantability or fitness for a particular purpose. No warranty may be created or extended by sales representatives or written sales materials. The advice and strategies contained herein may not be suitable for your situation. You should consult with a professional where appropriate. Neither the publisher nor author shall be liable for any loss of profit or any other commercial damages, including but not limited to special, incidental, consequential, or other damages.

For general information on our other products and services or for technical support, please contact our Customer Care Department within the United States at (800) 762-2974, outside the United States at (317) 572-3993 or fax (317) 572-4002.

Wiley also publishes its books in a variety of electronic formats. Some content that appears in print may not be available in electronic formats. For more information about Wiley products, visit our web site at www.wiley.com.

Library of Congress Cataloging-in-Publication Data is available.

ISBN: 978-1-119-06334-6

Printed in the United States of America

10 9 8 7 6 5 4 3 2 1

CONTENTS

PREFACE	xi
PREFACE TO THE FIRST EDITION	xiii
ACKNOWLEDGMENTS	xv
ABOUT THE AUTHOR	xvii

CHAPTER 1   UTILITY SOURCE 1

1.1   Electrical Safety   1
1.2   Delivery Voltage   3
1.3   One-Line Diagrams   4
1.4   Zones of Protection   5
1.5   Source Configuration   6
1.6   The Per-Unit System   14
1.7   Power in AC Systems   18
1.8   Voltage Drop Calculations   20
1.9   Short-Circuit Availability   22
1.10 Conductor Sizing   23
1.11 Transformer Sizing   26
1.12 Liquid-Immersed Transformer kVA Ratings   30
       Summary   32
       For Further Reading   33
       Questions   33
       Problems   34

CHAPTER 2   INSTRUMENT TRANSFORMERS AND METERING   37

2.1   Definitions   37
2.2   Instrument Transformers   39
       2.2.1   Fundamentals   39
       2.2.2   Correction Factors   46
       2.2.3   Burden Calculations   47
       2.2.4   ANSI Accuracy Classes   49
2.3   Metering Fundamentals   49
2.4   Watthour Metering   50
       2.4.1   Single-Stator Watthour Metering   50
       2.4.2   Multi-Stator Watthour Metering   52
2.5   Demand Metering   52
       2.5.1   Kilowatt Demand   53
       2.5.2   Kilovar and kVA Demand   53
2.6   Pulse-Operated Meters   54

v

2.7 Time-of-Use Meters 54
2.8 Special Metering 55
    2.8.1 Voltage and Current Metering 55
    2.8.2 Var and Q Metering 57
    2.8.3 Compensating Metering 59
    2.8.4 Totalizing Metering 60
    2.8.5 Pulse Recorders 60
2.9 Digital Metering 61
2.10 Smart Meters 61
Summary 62
For Further Reading 63
Questions 63
Problems 64

CHAPTER 3 TRANSFORMER CONNECTIONS 65

3.1 Voltage Selection 65
3.2 Ideal Transformer Model 66
3.3 Transformer Fundamentals 68
3.4 Transformer Circuit Model 71
3.5 Single-Phase Transformer Connections 71
3.6 Three-Phase Transformer Connections 73
    3.6.1 Delta–Delta 74
    3.6.2 Wye–Wye 76
    3.6.3 Delta–Wye 78
    3.6.4 Wye–Delta 82
    3.6.5 Open Delta–Open Delta 82
    3.6.6 Open Wye–Open Delta 86
3.7 Two-Phase Transformer Connections 88
    3.7.1 T-Connection (Scott Connection) 89
3.8 Six-Phase Transformer Connections 92
3.9 Transformer Phase Shifts 93
3.10 Grounding Transformers 95
    3.10.1 Wye–Delta 96
    3.10.2 Zig–Zag Connection 96
3.11 Ferroresonance 97
Summary 98
For Further Reading 98
Questions 99
Problems 99

CHAPTER 4 FAULT CALCULATIONS 101

4.1 Overview 101
4.2 Types of Faults 102
4.3 Data Preparation 103
4.4 First-Cycle Symmetrical Current Calculations 105
4.5 Contact-Parting Symmetrical Current Calculations 112
4.6 Analyzing Unbalanced Systems 113
4.7 Physical Example of Vector Components 114

4.8   Application of Symmetrical Components to a Three-Phase Power System   **116**
4.9   Electrical Characteristics of the Sequence Currents   **121**
4.10 Sequence Networks   **124**
4.11 Short-Circuit Faults   **134**
    4.11.1 Three-Phase Fault   **134**
    4.11.2 Line-to-Ground Fault   **136**
    4.11.3 Double Line-to-Ground Fault   **138**
    4.11.4 Line-to-Line Fault   **141**
4.12 Open-Circuit Faults   **143**
    4.12.1 One-Line-Open Fault   **143**
    4.12.2 Two-Lines-Open Fault   **147**
    Summary   **150**
    For Further Reading   **150**
    Questions   **151**
    Problems   **152**

CHAPTER 5 PROTECTIVE DEVICE SELECTION AND COORDINATION   155

5.1 Overview   **155**
5.2 Power Circuit Breaker Selection   **158**
5.3 Fused Low-Voltage Circuit Breaker Selection   **160**
5.4 Molded-Case Circuit Breaker Selection   **162**
5.5 Medium-Voltage Fuse Selection   **163**
5.6 Current-Limiting Fuse Selection   **166**
5.7 Low-Voltage Fuse Selection   **168**
5.8 Overcurrent Device Coordination   **169**
5.9 Summary   **174**
    For Further Reading   **175**
    Questions   **175**
    Problems   **176**

CHAPTER 6 RACEWAY DESIGN   179

6.1 Overview   **179**
6.2 Conduit and Duct Systems   **181**
    6.2.1 Pulling Tension   **187**
    6.2.2 Sidewall Pressure   **188**
    6.2.3 Design Examples   **189**
6.3 Cable Tray Systems   **194**
    6.3.1 Design Example   **202**
    Summary   **203**
    For Further Reading   **203**
    Questions   **204**
    Problems   **204**

CHAPTER 7 SWITCHGEAR AND MOTOR CONTROL CENTERS   207

7.1 Overview   **207**
7.2 NEMA Enclosures   **208**

viii CONTENTS

- 7.3 Switchgear 208
  - 7.3.1 Source Transfer 213
  - 7.3.2 Configuration 214
  - 7.3.3 Ratings 215
  - 7.3.4 Circuit Breakers 217
- 7.4 Motor Control Centers 222
  - 7.4.1 Configuration 223
  - 7.4.2 Ratings 223
  - 7.4.3 Starters 223
  - 7.4.4 Protection 225
- 7.5 ARC Flash Hazard 226
  - Summary 231
  - For Further Reading 232
  - Questions 233
  - Problems 233

CHAPTER 8   LADDER LOGIC   235

- 8.1 Fundamentals 235
- 8.2 Considerations When Designing Logic 236
- 8.3 Logic Implementation 239
- 8.4 Seal-In Circuits 240
- 8.5 Interlocks 243
- 8.6 Remote Control and Indication 245
- 8.7 Reversing Starters 246
- 8.8 Jogging 248
- 8.9 Plugging 250
  - Summary 251
  - For Further Reading 251
  - Questions 251
  - Problems 252

CHAPTER 9   MOTOR APPLICATION   255

- 9.1 Fundamentals 255
- 9.2 Energy Conversion and Losses 259
- 9.3 Speed–Torque Curves 260
- 9.4 Motor Starting Time 263
- 9.5 Cable Sizing 264
- 9.6 Motor Protection 265
- 9.7 Circuit Protection 266
- 9.8 Winding Protection 266
- 9.9 Motor Starting Methods 267
  - 9.9.1 Across-the-Line 267
  - 9.9.2 Reduced Voltage Starting 267
  - 9.9.3 Wye–Delta Starting 276
  - 9.9.4 Part-Winding Starting 278
  - 9.9.5 Solid-State Starting Options 278
  - Summary 283
  - For Further Reading 283

Questions 283
Problems 284

CHAPTER 10 *LIGHTING SYSTEMS* 287

10.1 Fundamentals 287
10.2 Lighting Technologies 288
    10.2.1 Incandescent 288
    10.2.2 Low-Pressure Discharge 290
    10.2.3 High-Intensity Discharge 294
    10.2.4 Light-Emitting Diode (LED) Lighting 297
10.3 Luminaire Designs 299
10.4 Electrical Requirements 301
10.5 Lighting System Design Examples 303
    10.5.1 Parking Lot Lighting 303
    10.5.2 Interior Lighting 311
    Summary 315
    For Further Reading 316
    Questions 316
    Problems 317

CHAPTER 11 *POWER FACTOR CORRECTION* 319

11.1 Overview 319
11.2 Configuration 321
    11.2.1 Delta 321
    11.2.2 Wye 322
    11.2.3 Grounded Wye 322
11.3 Sizing and Placement 323
11.4 Capacitor Switching 324
11.5 Harmonics 329
11.6 Resonance 330
11.7 Protection 330
    Summary 331
    For Further Reading 332
    Questions 332
    Problems 332

CHAPTER 12 *POWER QUALITY* 335

12.1 Overview 335
12.2 Historical Perspective 335
12.3 Quantifying Power Quality 336
12.4 Continuity of Service 338
12.5 Voltage Requirements 340
12.6 Transients 341
12.7 Harmonics 341
    12.7.1 Fourier Analysis 343
    12.7.2 Effects of Harmonics 346
    12.7.3 Harmonic Filters 349

12.8 Power Factor **352**
Summary **353**
For Further Reading **354**
Questions **355**
Problems **355**

APPENDIX A: UNITS OF MEASUREMENT — **357**

APPENDIX B: CIRCUIT ANALYSIS TECHNIQUES — **361**

APPENDIX C: PHASORS AND COMPLEX NUMBER MATHEMATICS — **369**

APPENDIX D: IMPEDANCE DATA — **373**

APPENDIX E: AMPACITY DATA — **381**

APPENDIX F: CONDUIT DATA — **401**

INDEX — **405**

PREFACE

SINCE THE RELEASE of the first edition in October 2001, I have received many comments from readers saying how helpful the text was while preparing for and taking the Professional Engineering (PE) examination. Truth be told, I never really thought about the PE exam while preparing the first edition manuscript. My objective was simply to produce a book useful to both students and practitioners. When the National Council of Examiners for Engineering and Surveying (NCEES) restructured the Electrical Engineering PE exam in April 2009 creating a power-specific exam, I saw the relevance of aligning this text more with the new exam format.

Several new topics, including arc flash hazard and lighting systems, have been added to this edition, keeping current with the latest trends and practices in the power field. More depth and some derivations have also been added to other sections, such as an in-depth exploration of the per-unit system, additional material on motor application, a thorough analysis of AC power in both the time and phasor domains, and an intuitive development of three-phase symmetrical components. I have never believed in deriving an equation simply for the sake of deriving it (that is a great exercise for mathematicians, but engineers should focus their efforts on application of the formula), but AC power and symmetrical components are exceptions to my tenet on derivations since these are so widely used by power engineers, and complete comprehension, in my opinion, requires going through the step-by-step development. Enhancements such as these should add to the level of understanding of critical power engineering concepts for many students.

Practitioners in the electric power industry will also find this book useful both as a reference and as a means of filling in gaps in their understanding of key concepts. Based largely on practices in the United States, this book references primarily US standards and codes. Often times, similar standards and codes have been developed and adopted abroad, and readers outside the United States are encouraged to investigate local standards and codes. Similarly, customary US units of measurements, as are commonly encountered in the United States, are used throughout the book.

The material covered in this book will be a valuable reference for the practicing power engineer, but will be just as helpful to the engineering student pursuing a career in the power industry. Few US universities have comprehensive programs in electric power that align with industry requirements. Narrowing the gap between what topics are taught at the university and what skills are needed by industry should be made a very high priority, both in academia and in industry. A close academia–industry partnership is needed to accomplish such a lofty goal. It is my hope that this book will help narrow the gap and forge the partnership. Developing new courses based on *Industrial Power Distribution* and incorporating the book into existing courses will facilitate the alignment of the academic curriculum with industry needs.

Over 370 students at the University of South Florida and many more through short courses conducted worldwide have studied the contents of the first edition, and as a result, many constructive comments were made. Between those comments and the expertise of the manuscript reviewers, I believe the second edition of *Industrial Power Distribution* will be a valuable resource for engineering students and practicing power engineers alike, and will help create an academic environment better able to address the needs of the electric power industry.

Ralph E. Fehr, III

PREFACE TO THE FIRST EDITION

ELECTRICITY has been an essential part of our lives since the late nineteenth century. During the early twentieth century, electricity routinely began replacing steam as the primary power source in industrial plants. As factories became larger and more complex, so did their power requirements. As such, engineers were faced with a new challenge: how to distribute electricity safely, reliably, and economically within the industrial facility. Although this challenge has been addressed for over a century, the theories, requirements, and procedures for safe, reliable, and economical industrial power distribution have not been presented in textbook form. This work attempts to do so by drawing on more than a hundred years of lessons learned and refinements in the electrical power distribution field.

The book begins by analyzing the source of the electricity—the utility system. Requirements including the delivery voltage and the topology of the source are explored, along with their impact on the operation and reliability of the industrial facility. Next, a powerful calculation method using the per-unit system is reviewed, and this method is applied extensively throughout the text. Other power source issues such as short circuit availability and transformer sizing are discussed, followed by a brief presentation on metering methods.

The next topic covered is medium-voltage distribution systems within the industrial facility. Guidelines for selecting the optimum voltage are presented. Various transformer connections are analyzed to determine their effects on the behavior of the electrical system. Ferroresonance and methods of eliminating its undesirable effects are introduced. Methods of conductor sizing consistent with the National Electrical Code (NEC) also are presented.

Two chapters are devoted to the calculation of fault currents and the sizing of protective equipment such as circuit breakers and fuses. The method of symmetrical components is used to analyze unbalanced fault conditions.

Switchgear and motor control centers, two types of equipment used extensively to distribute electricity within industrial facilities, are examined in detail including specification and selection requirements. Raceway design including conduit, duct banks, and cable tray is covered consistent with NEC specifications.

Methods of motor starting and motor control are discussed, and a detailed tutorial on ladder logic is included. The application of shunt capacitors for power factor correction rounds out the text.

Useful engineering data is presented in the appendices. This data, including units of measurement, circuit analysis techniques, impedance data, ampacity data reprinted from the NEC, and conduit data, provides a valuable reference source for the engineer.

The material in this book serves not only as an informative textbook, but also as a concise reference book. Objectives are clearly stated at the beginning of each chapter, and a succinct summary is presented at the end of each chapter. Questions to test the reader's comprehension of important concepts and problems to check the reader's ability to apply theory to solve practical problems are provided for each chapter.

This text is designed for use at either the upper-division undergraduate or graduate level. An understanding of basic power system analysis is the only prerequisite necessary to comprehend fully the material presented in this book.

Ralph E. Fehr, III

ACKNOWLEDGMENTS

THE AUTHOR wishes to thank the many people who contributed their expertise and insight to this book, particularly the technical reviewers of the first edition: Ralph D. Painter, the late Charles Concordia, and the late Joseph P. Skala.

Mr. Painter has been a resident of Florida for over 50 years. He earned BSEE and ME degrees from the University of South Florida, and a JD degree from Stetson University College of Law. He is a senior member of the IEEE, a registered professional engineer in Florida, a Master Electrician in the City of Tampa, and a member of the Florida Bar. Mr. Painter has been employed by Tampa Electric Company for over 40 years in various engineering positions related to power plant electrical system operation, maintenance, and design and regulatory compliance. He is currently Manager—FERC Compliance.

Dr. Concordia needs no introduction to the members of the power engineering community. He worked for over 40 years for the General Electric Company in Schenectady and later as a consultant, making countless contributions to the areas of system analysis and machine design. He was awarded the prestigious IEEE Medal of Honor in 1999 and has worked in the electric power industry for over 70 years. His accomplishments in the area of electric power are astonishing, and his willingness to share his expertise and experience are deeply appreciated.

Mr. Skala had been a dear friend and colleague since I arrived in Florida in 1992. His interest in teaching encouraged me to become more involved in academia. He retired as a power systems engineer with a 30-year background in the utility industry, including transmission and substation design, transmission and distribution maintenance, system planning, and research and development. His extensive experience allowed him to build a tremendous base of expertise, from which I have received benefits many times. Mr. Skala was recognized for his creativity and innovation, both as an engineer and as an educator. For 20 years, he has taught electrical engineering and mathematics courses in the State of Florida University System. As a teacher, he established the graduate power systems program at the University of South Florida and co-founded the DiNapoli-Skala Families Scholarship at St. Petersburg Junior College, which provides assistance to students with a demonstrated financial need. Mr. Skala's contributions to the power industry and academia were many, and through them, he will live on.

I would also like to thank two additional electrical engineers who were instrumental in the development of this book. My father, Ralph E. Fehr, II, worked as an electrical power engineer for over 33 years. He was a registered professional engineer in Pennsylvania, was active as a member and section/committee chairman of the IEEE, and served on numerous technical committees of the Pennsylvania Electric Association. My uncle, Arthur R. Hill, taught electrical engineering and electrical engineering technology courses at the Berks Campus of the Pennsylvania State University for over 36 years. He also prepared electrical and electronics training and testing programs for several industries. In addition to reviewing the manuscript of this book, my father and uncle provided a supportive academic environment and guided me into the field of engineering.

Key contributors to the second edition include Serge Beauzile (Lakeland Electric Co.), Marcel Bertran (GE Energy Services), Thomas Blair (Tampa Electric Co.), Dave Darden (Tampa Electric Co.), Craig Kalhoefer (Harold Hart & Associates, Inc.), Michael Milbert (Engineering Consultant), John Raksany (University of Wisconsin—Madison), Titipong Samakpong (Provincial Electricity Authority of Thailand), and Harianto Suryo (Lakeland Electric Co.).

And a special thank you to Karen Fehr and Margaret McMullen, who read every page of this second edition manuscript and helped to make this book the best it could be.

ABOUT THE AUTHOR

RALPH E. FEHR III, earned a Bachelor of Science degree in Electrical Engineering from the Pennsylvania State University in 1983, a Master of Engineering degree in Electrical Engineering (Power) from the University of Colorado at Boulder in 1987, and a Doctor of Philosophy in Electrical Engineering from the University of South Florida in 2005. He has worked in the generation engineering field, designing power distribution and control systems for nuclear and fossil-fired power plants. Dr. Fehr also has worked for electric utilities for more than 15 years in the operations, planning, and design areas including transmission, distribution, and substation engineering.

Teaching always has been an important aspect of Dr. Fehr's career. On an adjunct basis, he has taught courses ranging from computer operating systems to mathematics to power system analysis for several institutions including the University of New Mexico at Albuquerque, St. Petersburg (Florida) Junior College, and the University of South Florida at Tampa. He has also taught short courses in the power engineering area domestically and abroad for the Pennsylvania State University and the University of Wisconsin–Madison. Since 1996, he also has taught a review course for professional engineer examination candidates. Dr. Fehr has expanded the power engineering program at the University of South Florida with the goal of producing a greater number of proficient engineering graduates for the power industry.

Dr. Fehr is a senior member of the Institute of Electrical and Electronics Engineers (IEEE) and is a registered professional engineer in New Mexico and Florida. His biography was published in *Who's Who in Science and Engineering*. He also received the 2010 Joseph Biedenbach Award from IEEE Region 3 as Outstanding Engineering Educator from the southeast United States. He currently serves on the Electrical Engineering Faculty at the University of South Florida, where he oversees the school's power engineering program.

CHAPTER 1

UTILITY SOURCE

OBJECTIVES

- Be aware of the need to always exercise safety while working around electricity
- Understand the significance of the voltage level at the point of delivery to the industrial customer
- Be proficient in the use of one-line diagrams to represent a three-phase electrical system
- Understand the concept of protection zones
- Recognize various source configurations and know the advantages and disadvantages of each
- Calculate per-unit quantities, convert between actual and per-unit quantities, and be able to apply the per-unit system to do electrical calculations
- Understand the components of power in an AC system
- Be able to calculate voltage drop in a balanced three-phase system
- Understand the significance of short circuit availability, both from a fault interrupting standpoint and from a motor starting standpoint 2
- Comprehend the importance of properly sizing conductors and supply transformers

1.1 ELECTRICAL SAFETY

A 6-W night light bulb draws 50 mA of current. Even this small magnitude of current can kill a person. Humans can perceive an electric current as small as 0.5 mA. As the current magnitude rises to the 1–5 mA range, muscles will contract. Currents in the 3–10 mA range cause pain. Currents in the 10–40 mA range fall in the "let go" threshold, meaning they will cause muscles to contract so tightly that "letting go" of a grasped wire becomes impossible. Respiratory paralysis can occur in the 30–75 mA range. The heart will be affected with ventricular fibrillation occurring in the 75–100 mA range and heart paralysis occurring in the 250–300 mA range. Note that we have not yet reached one-third of an ampere! In the range of 5–6 A, organ burns will occur. These typical values are summarized in Figure 1.1.

Industrial Power Distribution, Second Edition. Ralph E. Fehr, III.
© 2016 The Institute of Electrical and Electronics Engineers, Inc. Published 2016 by John Wiley & Sons, Inc.

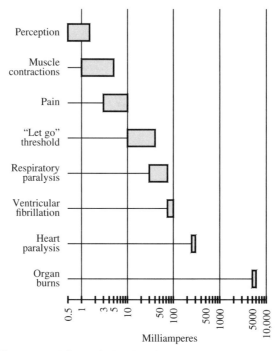

Figure 1.1   Effects of Electrical Current on the Human Body

Although electrocution is a major concern when working around electricity, most electrical injuries are the result of burns occurring when large amounts of thermal energy are released during arcing faults. *Electric arc flash* presents a hazard that is distinct from electric shock. The heat and blast released during a short circuit fault can injure or kill a person located many feet from the fault location and who never come in contact with an energized conductor. The arc flash hazard will be discussed in Section 7.5.

Because of the shock and burn hazards posed by electricity, safety must *always* be exercised and considered as the *highest* priority when working around electrical equipment. Nothing supersedes safety. Carelessness for even a moment could result in a serious injury—or worse.

Much of what is considered safe working practices falls soundly into the category of *common sense*. For example, do not assume a circuit is de-energized—test it and ground it to make sure. Follow all grounding and bonding requirements when performing maintenance of electrical equipment. Stringently adhere to all switching and tagging practices required by the owner of the facility in which you are working. Be aware of working clearances and approach boundaries (these topics will be discussed in Section 7.5). Be sure that all tools and equipment including personal protective equipment (PPE) are in good condition, are being used properly, and have current testing or certification credentials, if applicable. Adhere to all safety procedures, even if the procedure does not seem important to you. *It probably is important, but even if it is not, better safe than sorry.* And be sure to complete all required training, including refresher courses, required for your job.

Electricity need not be feared, but certainly does command respect. Safe working practices need to be standard operating procedure and never be compromised. Safety must also be considered when designing electrical systems. Engineering additional safety features beyond what is minimally required for a system is a hallmark of excellent design. If an unsafe condition is discovered, it needs to be reported and made safe as quickly as possible. Take safety seriously—your life depends on it!

1.2 DELIVERY VOLTAGE

Electricity for industrial facilities can be supplied by the local utility either at distribution voltages (2.4–34.5 kV) or at subtransmission or transmission voltages (46–230 kV) depending on the size of the load, the topology of the utility system, and the tariffs and rate schedules under which the service is supplied. Since the highest utilization voltage in the plant may be lower than the delivery voltage, a voltage transformation is often necessary at the point of delivery.

If the industrial facility is small enough to be fed from a single distribution transformer, the utility may serve the facility at *utilization voltage,* which is defined as the voltage at the line terminals of utilization equipment. This voltage is often 480Y/277 V, meaning that a four-wire wye-connected service is provided with a line-to-line voltage of 480 V and a line-to-neutral voltage of 277 V, as shown in Figure 1.2.

Three-phase 480 V and single-phase 240-V service also can be provided by a *high-leg delta* system as shown in Figure 1.3. One of the delta legs is center tapped and grounded to form the system neutral. Care must be taken not to connect load from the high leg phase of the delta to neutral, as this voltage (416 V) is not a normally utilized voltage.

Single-phase 120 V and three- or single-phase 208 V can be provided using dry-type transformers rated 480 – 208Y/120 V.

As the size of the industrial facility increases, it becomes necessary to supply the facility at a higher voltage to limit excessive voltage drop and to reduce losses.

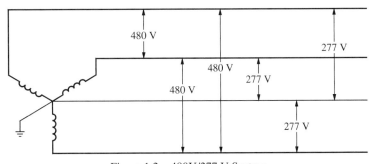

Figure 1.2 480Y/277 V System

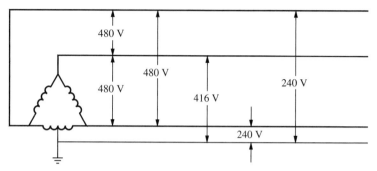

Figure 1.3   High-Leg Delta System

Loss reduction results in a reduced operating cost. The lower currents resulting from operating the system at a higher voltage also allow smaller conductor to be used, thereby reducing the system construction cost. Many times, a three-phase tap is built from a distribution feeder to the facility. At the facility, distribution transformers and their associated high- and low-voltage switching equipment are called *unit substations*. The unit substations are fed from the distribution tap, and in turn, feed the facility. Distribution feeders are typically in the 2.4–15 kV range, with a trend toward the higher part of that range. Some utilities are using distribution voltages higher than 15 kV, particularly if the load density is high or the feeders are very long. Equipment in the 25 kV class can be used on 22 and 24.9 kV systems, while 35 kV class equipment can be used on 33 and 34.5 kV systems.

1.3   ONE-LINE DIAGRAMS

The one-line or single-line diagram is an important type of drawing used by power engineers to convey topological information about a power system. Most industrial power systems are three-phase systems, where each of the three phases is very similar to if not identical to the other two. This symmetry can be exploited by showing only one of the three phases, thus the term *one-line* diagram. Any asymmetries can be noted on the diagram.

Special forms of one-line diagrams can be produced to convey special information such as protection schemes or switching procedures. One-line diagrams frequently reference other drawings which provide details not shown on the one-line. Although not a necessity, one-line diagrams are usually drawn in such a way that they closely resemble the physical system they represent. Sometimes, a north arrow is shown to establish a physical orientation.

Similar to a schematic diagram, the one-line diagram uses fairly standardized symbols to represent system components. Although much standardization in one-line symbols has taken place over the years, differences will be found from drawing to drawing. Some of the more common symbols are shown in Table 1.1.

TABLE 1.1 Common One-Line Diagram Symbols

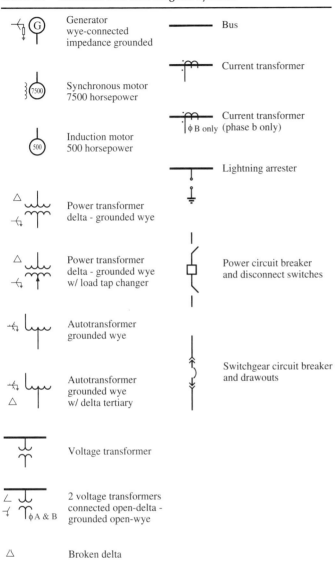

1.4 ZONES OF PROTECTION

A short circuit fault is an unintentional connection of a phase to another phase or to ground through a low impedance, resulting in very high current magnitudes. The high current poses a serious compromise to safety and can cause severe equipment damage. The ability to quickly and accurately detect a short circuit fault and the means to reliably remove, or clear, that fault from the system are primary considerations

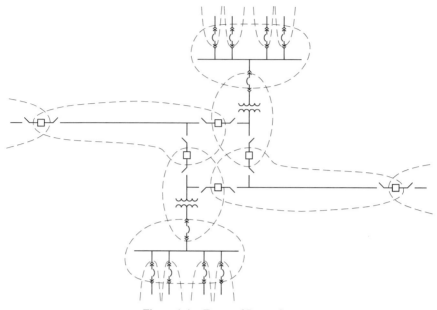

Figure 1.4 Zones of Protection

when designing any power distribution system. Fault detection is done using a variety of protective devices, such as protective relays and fuses. Some protective devices are able to detect faults within a certain proximity of the device. This limited range is necessary to prevent multiple devices from operating simultaneously. The range within which a device can detect a fault is known as the device's *zone of protection*.

When designing a power distribution system, it is mandatory that every portion of the system lies within a zone of protection; otherwise, that portion of the system would be *unprotected*. A fault occurring in an unprotected zone would not be cleared. It is also highly desirable to have *overlapping zones*, as in Figure 1.4, so that a fault occurring in a portion of the system included in multiple zones would be detected by multiple devices. By varying the time characteristics of the protective devices, it can be assured that only one device would operate. But by having multiple devices aware of the fault, backup protection exists in the event that the primary protection device is unable to clear the fault.

1.5 SOURCE CONFIGURATION

Early in the power system design process, a decision must be made regarding the configuration of the utility source. This decision is always a compromise between reliability and economics. The simplest and least expensive option is a single radial source as shown in Figure 1.5.

This option can be implemented at any voltage level. A single feed is provided to the industrial plant by the utility. Depending on the delivery voltage, the feed can

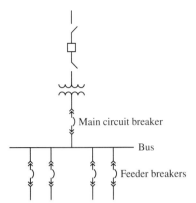

Figure 1.5   Radial System

either be transformed to a lower voltage or used to supply switchgear for subsequent distribution. Radial buses require one circuit breaker per element connected to the bus.

When a fault occurs on the radial bus, every breaker must *trip*, or open, to isolate the fault. This results in the complete de-energization of the bus, which is highly objectionable for many applications. Note that the main circuit breaker (shown between the transformer and the bus in Figure 1.5) is optional in a radial bus scheme. Its presence simplifies switching, but does so at the cost of an additional breaker.

When a fault occurs on a feeder leaving the radial bus, the feeder breaker should trip to clear the fault without interrupting any other loads. But if the feeder breaker fails to clear the fault, every breaker must open to clear the fault. This condition, known as *breaker failure*, also de-energizes the entire bus. With two fairly probable scenarios (bus fault and feeder breaker failure) resulting in complete de-energization of the bus, the radial bus should not be implemented when reliability is a concern, unless an alternate source and source transfer scheme is provided, as is discussed in Chapter 7.

The poor reliability of the radial bus can be improved greatly by adding a second source configured in a *ring bus* as shown in Figure 1.6.

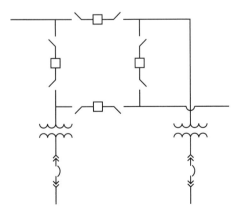

Figure 1.6   Ring Bus

8 CHAPTER 1 UTILITY SOURCE

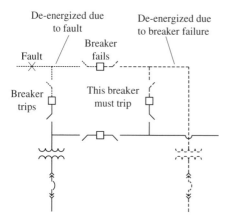

Figure 1.7 Ring Bus with Breaker Failure

The ring bus, like the radial bus, requires one circuit breaker per element connected to the bus. When a fault occurs on the ring bus, each bus section is protected by the protection for one of the elements connected to the bus. For example, if a fault occurs on the bus between the breakers at the 9 o'clock and 12 o'clock positions in Figure 1.6, the protection for the circuit leaving the upper left corner of the ring would detect the bus fault. That line would be de-energized, but the remainder of the bus would remain in service. Note that two breakers must trip to clear a fault in a ring bus topology. Also note that after two breakers open to clear a fault, the ring is no longer intact—the bus essentially becomes a radial configuration with what amounts to tie breakers in the bus.

In the event of breaker failure, the breaker in the next position in the ring must trip to clear the fault. This results in the de-energization of one circuit in addition to the faulted circuit, as seen in Figure 1.7.

When one breaker in the ring is opened, all circuits leaving the ring bus remain energized but the reliability of the ring is compromised. With one breaker open, losing another breaker will either de-energize a circuit or split the substation into two separate subsystems, as shown in Figure 1.8. Both of these scenarios are undesirable.

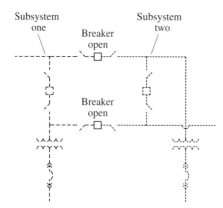

Figure 1.8 Split Ring Bus

1.5 SOURCE CONFIGURATION 9

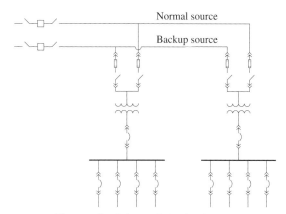

Figure 1.9 Primary Selective System

Because of this characteristic, care must be exercised when designing a ring bus. To minimize the risk of losing multiple sources or multiple loads in the event that a circuit breaker fails during a fault, source and load connections should alternate. Connecting two sources to adjacent ring bus positions would result in the loss of both sources if a fault occurred on one of the sources and the circuit breaker between the sources failed to open to clear the fault. Also, circuits terminated on the ring bus should have disconnected the switches installed which can be used to isolate the circuit so the ring breakers can be reclosed. Disconnect switches cannot break current; they can only be opened after circuit breakers de-energize the circuit.

The ring bus configuration allows multiple sources to feed multiple loads with automatic fault isolation and a great deal of flexibility. All circuit breakers in a ring bus are normally closed. Since ring buses increase short circuit availability (SCA) by paralleling sources, care must be taken so device short circuit capabilities are not exceeded.

Reliability also can be improved by adding a second source configured in a *primary selective* scheme as shown in Figure 1.9. Here, the primary selective topology is supplying two radial buses. One source is designated as the *normal* source and the second serves as a *backup* source. When the normal source fails, the backup source can be energized either by manual or automatic switching.

Referring to Figure 1.9, assume that the transformer on the left is normally fed from the top feeder (the bottom feeder is the backup source). Conversely, the transformer on the right is normally fed from the bottom feeder, with the top feeder serving as a backup source. This normal configuration is clarified in Figure 1.10.

When a fault occurs on the top feeder, the transformer on the left experiences an interruption in service when the top feeder breaker trips to clear the fault, while the transformer on the right is unaffected. To restore service to the transformer on the left, the normally closed switch just upstream from the transformer is opened, and the accompanying normally open switch is closed. This backup configuration is shown in Figure 1.11.

10 CHAPTER 1 UTILITY SOURCE

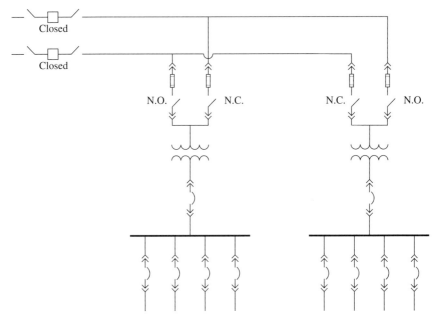

Figure 1.10 Primary Selective System Normal Configuration

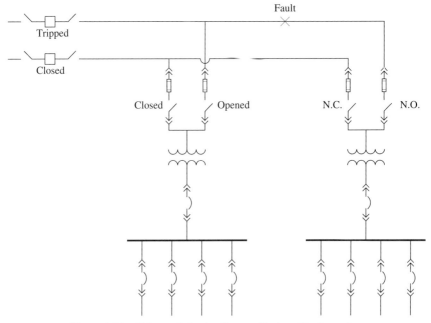

Figure 1.11 Primary Selective System Backup Configuration

The system can be designed such that the switches upstream from the transformers toggle automatically when necessary. This feature is called *automatic source transfer*. Automatic source transfer can be accomplished by a *fast transfer* or by an *open* or *closed transfer*. A *fast transfer* involves opening the first source breaker then closing the second source breaker very quickly. A brief de-energized period exists during the transfer. If most of the motor load is driving high-inertia loads such as fans, completing the transfer in less than six cycles (0.1 seconds) should prevent running motors from slowing to the point where an out-of-phase condition at re-energization produces dangerously high torques that could damage the driven loads. But if many of the motors are driving low-inertia loads such as centrifugal pumps, the transfer must be completed even faster to prevent damage due to transient torques. It is good practice to use specialized relays designed for motor bus transfers to assure that the transition does not produce dangerous torques.

If a fast transfer cannot be achieved, an *open transfer* can be used, where the first source breaker is opened and the running motors are allowed to slow down. When the residual voltage is 25% or less of the nominal system voltage, the second source breaker is closed. Torques generated during re-energization are tolerable because of the decayed residual voltage. A *closed transfer* briefly parallels the two sources before opening the first source breaker. Due to the reduced source impedance while the two sources are paralleled, interrupting requirements may be very high. A decision may be made to accept the risk of exceeding equipment-interrupting ratings during the transfer period, because the probability of a fault occurring during this brief period is very small. Closed transfers frequently are used to switch sources for maintenance purposes. Unless a closed transfer method is used, there will be an interruption of service until the backup source can be connected.

Because of the extra equipment required (cable, fuses, and switchgear), the primary selective scheme is more costly than the radial system. Depending on the voltage and the method of transfer, the primary selective scheme may be more or less costly than the ring bus. A detailed economic analysis is required to determine the lowest cost option.

Another method of increasing reliability is to implement a *primary loop* system. One variation of the primary loop system is shown in Figure 1.12. In this example, the primary loop supplies three radial bus substations. Circuit breakers supply both ends of the primary loop, and switches are installed in the loop for isolation purposes.

While similar to the primary selective system, this scheme utilizes two independent sources that are operated in parallel. This mode of operation greatly increases reliability and flexibility and also increases the complexity of the protection systems required. The reliability is high, and in many applications, justifies the cost.

If a fault occurs in or downstream from one of the transformers, the power fuses on the primary transformer will blow to clear the fault. If a fault occurs in the loop itself, both circuit breakers will trip, thus de-energizing the entire system. Next, the fault must be located, and a pair of switches must be opened to isolate the fault. Then, both breakers can be reclosed, energizing the unfaulted portions of the system.

Reliability, both a reduction in amount of load interrupted and a substantial reduction in the restoration time for the interrupted load, can be improved by taking one or more measures. Designating one of the isolation switches as a normally open

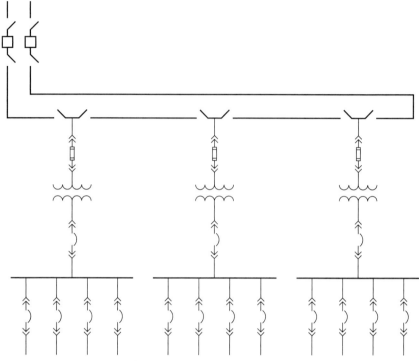

Figure 1.12 Primary Loop System

point, as in Figure 1.13, would result in dropping only half the load when a loop fault occurs. Theoretically, the normally open switch should be located near the null current point—the point on the loop where the current would equal zero if the loop were closed. Additionally, the isolation switches can be automated for automatic sectionalizing of the system after a fault. This measure reduces interruption time dramatically, possibly to less than a minute.

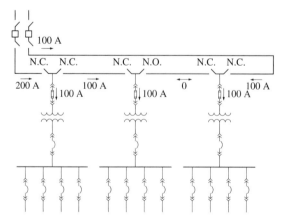

Figure 1.13 Primary Loop System with Normally Open Point

1.5 SOURCE CONFIGURATION

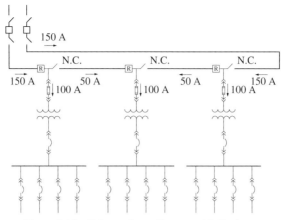

Figure 1.14 Primary Loop System with Reclosers

Another option would be to operate the loop in closed mode (with no normally open point) and install isolation devices capable of interrupting fault current, as in Figure 1.14. A recloser would be an example of an isolation device capable of interrupting fault current. Intelligent reclosers capable of communicating with other reclosers, relays, and motor-operated switches can be used to develop a highly reliable and fully automated system.

When a voltage transformation is made at the delivery point, reliability also can be increased by tying the secondaries of the source transformers together in either a *secondary selective* arrangement, a *secondary spot network*, or by utilizing *tie breakers* and/or *sparing transformers*.

The *secondary selective system*, shown in Figure 1.15, is essentially a normally open bus tie between two secondary buses which is closed either manually or automatically when one source fails.

When a transformer fault occurs, the high-side and low-side breakers trip to isolate the transformer. After the low-side bus is de-energized, the normally open tie breaker can close to re-energize the bus normally fed by the failed transformer.

With this system, both transformers must be sized such that each is capable of serving the entire secondary load for a predetermined period of time. Many

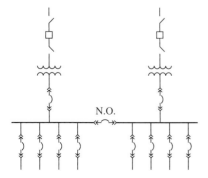

Figure 1.15 Secondary Selective System

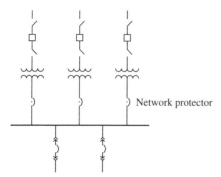

Figure 1.16 Secondary Spot Network

times, a degree of overloading is allowed and the resulting loss of transformer life is accepted.

The tie breaker can be operated normally closed, but this parallels the transformers which greatly increases the SCA. The increase in fault current magnitude usually requires more costly equipment with higher short circuit ratings.

A *secondary spot network*, shown in Figure 1.16, parallels multiple sources.

To prevent backfeeding from the secondary to the source, *network protectors* are used between the transformers and the secondary bus. A network protector is essentially a circuit breaker with a reverse-power relaying package incorporated. If a power flow from secondary to primary is sensed, the network protector opens. While this scheme is feasible when the load density is very high, the extra cost of the network protectors and redundant transformer capacity make this scheme quite expensive.

A *sparing transformer* can be used to protect against the loss of a source. In addition to the expense of the sparing transformer, which is normally not loaded, considerable cabling, circuit breakers, and a tie bus are all required. These items often make the sparing transformer scheme economically unattractive, but the increase in reliability afforded by a spare transformer installed and ready to supply load may offset the economics. The sparing transformer configuration is shown in Figure 1.17.

1.6 THE PER-UNIT SYSTEM

Due primarily to the abundance of transformers in power systems, many power system problems can be tedious to solve using electrical units. This is because, the transformer turns ratio changes the electrical quantities of voltage, current, and impedance differently. If the transformer turns ratio is n, as one moves from the high-voltage side of the transformer to the low-voltage side, the voltage changes by $1/n$, the current changes by n, and the impedance changes by $1/n^2$. These different factors can make even simple calculations rather complicated.

The complications introduced by the transformer turns ratios can be avoided by applying the *per-unit system*. The per-unit system uses dimensionless quantities instead of electrical units (volts, amps, ohms, watts, etc.). The per-unit system, when properly applied, also changes all transformer turns ratios to 1. This way, as one

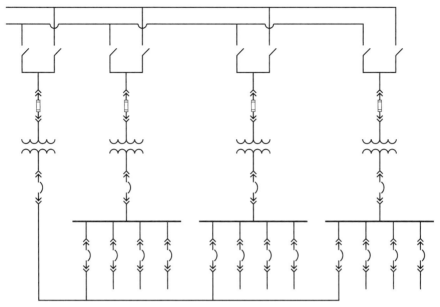

Figure 1.17 Sparing Transformer

moves from the high-voltage side of the transformer to the low-voltage side, the voltage, current, and impedance are all unaffected.

The per-unit system relies on the establishment of four *base quantities*. The required base quantities are *base power*, *base voltage*, *base current*, and *base impedance*. The base quantities are selected by the person doing the problem. The numeric values of these base quantities are arbitrary. This is because the per-unit system is a linear transformation. The problem to be solved is transformed into the per-unit system, solved, and then transformed back to electrical quantities. Since the transformation back to electrical quantities is the inverse of the transformation into the per-unit system, the base quantities which define the transformations can assume any numeric value, except zero.

Of the four base quantities, two are mathematically independent. The other two are then defined by the first two. For example, if voltage and current are assumed to be independent, power can be thought of as the product of voltage and current, and impedance as the quotient of voltage and current. The typical way to apply the per-unit system is to arbitrarily assign the power and voltage bases, then using the mathematical relationships between the electrical quantities to determine the current and impedance bases.

Typically, the base power (kVA or MVA base) is selected arbitrarily, often as 10 or 100 MVA. Or, the power base can be selected to match the kVA rating of a particular piece of equipment, such as a transformer. The power base is constant through the entire system. The base voltage (kV base) is arbitrary, but is frequently assigned as the nominal operating voltage at a given bus in the system. At every voltage transformation, the base voltage is adjusted by the transformer turns ratio.

Therefore, many different base voltages may exist throughout the system. The proper selection of voltage bases throughout the system found by multiplying the base voltage in one circuit by the turns ratio of the transformer connecting that circuit to another circuit effectively makes the transformer turns ratios equal to one and merges the two electric circuits into one.

After the power and voltage bases are chosen, the other two base quantities can be calculated from the established bases by using the formulas

$$\text{Base Current} = \frac{\text{Base kVA}_{3\Phi}}{\sqrt{3} \times \text{Base kV}_{L-L}} \quad (1.1)$$

and

$$\text{Base Impedance} = \frac{(\text{Base kV}_{L-L})^2}{\text{Base MVA}_{3\Phi}} \quad (1.2)$$

where the subscripts 3Φ and L–L respectively denote three-phase power and line-to-line voltage.

Actual electrical quantities are converted to dimensionless per-unit quantities using the formula

$$\text{Per-Unit Quantity} = \frac{\text{Actual Quantity}}{\text{Base Quantity}} \quad (1.3)$$

Often, a per-unit quantity must be converted from a particular base to a new base. Toward that end, we use the relationship

$$\text{Per-Unit Quantity}_{\text{New}} = \text{Per-Unit Quantity}_{\text{Old}} \times \left(\frac{\text{kV Base}_{\text{Old}}}{\text{kV Base}_{\text{New}}}\right)^2 \times \left(\frac{\text{kVA Base}_{\text{New}}}{\text{kVA Base}_{\text{Old}}}\right). \quad (1.4)$$

Examples

1. A generator has an impedance of 2.65 Ω. What is its impedance in per-unit, using bases of 500 MVA and 22 kV?

   Solution:

   Calculate the base impedance using Eq. (1.2).

   $$\text{Base Impedance} = \frac{(\text{Base kV}_{L-L})^2}{\text{Base MVA}_{3\Phi}} = \frac{22^2}{500} = 0.968 \, \Omega$$

   Calculate per-unit impedance using Eq. (1.3).

   $$\text{Per-unit impedance} = \frac{\text{Actual impedance}}{\text{Base impedance}} = \frac{2.65 \, \Omega}{0.968 \, \Omega} = 2.738 \text{ p.u.}$$

2. A transformer rated 12/16/20 MVA, 44/13.2 kV has an impedance of 6.25%. What is its percent impedance on 100 MVA and 46 kV bases?

   Solution:

   For transformers with multiple stages of cooling, the *self-cooled* (lowest) MVA rating (12 in this case) is always used.

Convert to new per-unit bases using Eq. (1.4).

Per-Unit Quantity$_{New}$ = Per-Unit Quantity$_{Old}$

$$\times \left(\frac{\text{kV Base}_{Old}}{\text{kV Base}_{New}}\right)^2 \times \left(\frac{\text{kVA Base}_{New}}{\text{kVA Base}_{Old}}\right)$$

$$= 6.25\% \times \left(\frac{44 \text{ kV}}{46 \text{ kV}}\right)^2 \times \left(\frac{100 \text{ MVA}}{12 \text{ MVA}}\right) = 47.65\%$$

3. Consider the power system shown in Figure 1.18:

   Rated Voltages:
   - Utility Source: 12.47 kV
   - Generator: 460 V

   Transformers
   - T1: 13.2 kV/4.16 kV
   - T2: 460 V/4 kV
   - T3: 4.16 kV/480 V

   Motors
   - M1: 4000 V
   - M2: 460 V

   Let the base voltage at Bus 1 be 4.16 kV. Find the base voltage at
   a. the utility connection point;
   b. the generator terminals; and
   c. Bus 2.

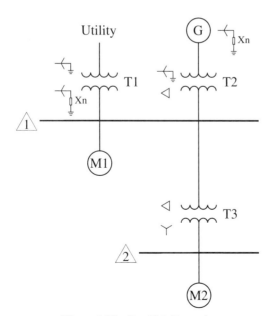

Figure 1.18 Per-Unit Example

Solution:

a. Starting with the given base voltage at Bus 1 and applying the turns ratio of transformer T1,

$$V_{\text{base (Utility)}} = 4.16 \text{ kV} \left(\frac{13.2 \text{ kV}}{4.16 \text{ kV}}\right) = 13.2 \text{ kV}$$

b. Starting with the given base voltage at Bus 1 and applying the turns ratio of transformer T2,

$$V_{\text{base (Generator)}} = 4.16 \text{ kV} \left(\frac{460 \text{ V}}{4000 \text{ V}}\right) = 478.4 \text{ V}$$

c. Starting with the given base voltage at Bus 1 and applying the turns ratio of transformer T3,

$$V_{\text{base (Bus 2)}} = 4.16 \text{ kV} \left(\frac{480 \text{ V}}{4160 \text{ V}}\right) = 480 \text{ V}$$

Notice that the base voltage is independent of rated voltage, nominal voltage, actual voltage, or any other voltage quantity. It is simply a number to use in a mathematical transformation—nothing more. Since the same numeric value is used when performing the inverse transformation (per-unit back to electrical units), its value is arbitrary. It is essential, however, that after a base voltage is arbitrarily assigned at one bus, the other voltage bases are determined based on that voltage using transformer turns ratios, as in this example.

1.7 POWER IN AC SYSTEMS

Power in AC systems is calculated like in DC systems as the product of voltage and current. But since voltage and current each vary sinusoidally with time, this seemingly simple multiplication is worthy of some analysis. Begin in the time domain by letting

$$v(t) = V_m \cos(\omega t + \theta_v) \tag{1.5}$$

and

$$i(t) = I_m \cos(\omega t + \theta_i) \tag{1.6}$$

Then the total (or *complex*) power $s(t)$ is

$$s(t) = v(t) \cdot i(t) = V_m I_m \cos(\omega t + \theta_v) \cos(\omega t + \theta_i) \tag{1.7}$$

Applying the product of cosines property,

$$s(t) = \frac{V_m I_m}{2}[\cos(\theta_v - \theta_i) + \cos(2\omega t + \theta_v + \theta_i)]. \tag{1.8}$$

Notice that the first cosine term in Eq. (1.8) is not a sinusoid but a constant, and the second cosine term is a sinusoid that oscillates at twice the frequency of the voltage and the current. Noting that the 2 in the denominator can be factored into

$\sqrt{2} \times \sqrt{2}$ and rearranging the argument terms in the second cosine function to force a $(\theta_v - \theta_i)$ term as in the first cosine function,

$$s(t) = \frac{V_m}{\sqrt{2}} \cdot \frac{I_m}{\sqrt{2}} \cdot \{\cos(\theta_v - \theta_i) + \cos[2(\omega t + \theta_v) - (\theta_v - \theta_i)]\}.$$

(1.9)

Writing the magnitudes as RMS values and applying the cosine of a difference property,

$$s(t) = V_{rms} I_{rms} \langle \cos(\theta_v - \theta_i) + \{\cos[2(\omega t + \theta_v)] \cos(\theta_v - \theta_i) \\ + \sin[2(\omega t + \theta_v)] \sin(\theta_v - \theta_i)\} \rangle \quad (1.10)$$

The first term $\{V_{rms} I_{rms} \cos(\theta_v - \theta_i)\}$ represents the constant component of the real power (P), a DC offset with a magnitude equal to half the peak-to-peak magnitude of the second term. This DC offset keeps the real power sinusoid positive or zero at all times.

The second term $\{V_{rms} I_{rms} \cos(\theta_v - \theta_i)\cos[2(\omega t + \theta_v)]\}$ represents the oscillating component of the real power. When the second term is added to the first term, the resulting expression for the real power component oscillates at twice the frequency of the voltage and current sinusoids and is positive or zero at all times.

The third term $\{V_{rms} I_{rms} \sin(\theta_v - \theta_i)\sin[2(\omega t + \theta_v)]\}$ represents the reactive power (Q). The reactive power oscillates at twice the frequency of the voltage and current sinusoids, is centered about the x-axis, and lags the real power sinusoid by 90°.

Therefore, $s(t) = P + P \cos[2(\omega t + \theta_v)] + Q \sin[2(\omega t + \theta_v)]$. The total power $s(t)$ and its real and reactive components are shown in Figure 1.19 for $V_m = 10$ V, $I_m = 7$ A, and $\theta_i = -26°$. The voltage $v(\omega t)$ and the current $i(\omega t)$ are shown as a reference.

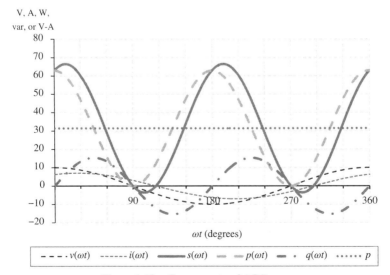

Figure 1.19 Components of AC Power

The total power curve $S(t)$ dips below the x-axis as the power factor drops from unity. This is significant, as the area under the power curve represents energy and any area below the x-axis deducts from the total energy.

Expressing $s(t)$ as a phasor, the DC offset is ignored since it does not vary with time. Using the second and third terms of the total power expression and realizing that the $\cos[2(\omega t + \theta_v)]$ and $\sin[2(\omega t + \theta_v)]$ terms merely provide time variation, the phasor expression for total power can be written as

$$\tilde{S} = \tilde{P} + j\tilde{Q} = V_{rms}I_{rms}\cos(\theta_v - \theta_i) + jV_{rms}I_{rms}\sin(\theta_v - \theta_i)$$
$$= V_{rms}I_{rms}[\cos(\theta_v - \theta_i) + j\sin(\theta_v - \theta_i)] \quad (1.11)$$

Applying Euler's formula,

$$\tilde{S} = V_{rms}I_{rms}e^{j(\theta_v - \theta_i)} = V_{rms}e^{j\theta_v} \cdot I_{rms}e^{-j\theta_i} = V_{rms}\underline{/\theta_v} \cdot I_{rms}\underline{/-\theta_i}$$
$$= \tilde{V}_{rms} \cdot \tilde{I}^*_{rms} \quad (1.12)$$

Note that it is the current phasor that is conjugated while the voltage phasor is used directly.

1.8 VOLTAGE DROP CALCULATIONS

Voltage drop in an AC power system occurs as predicted by Ohm's law. As a current flows through an impedance, a voltage drop occurs:

$$V_{drop} = I \times Z \quad (1.13)$$

Recalling that the current I is a phasor and the impedance Z consists of a real component (resistance) and an imaginary component (reactance), Eq. (1.13) can be restated as

$$V_{drop} = I \times Z = (I_R + jI_X) \times (R + jX)$$
$$= (I_R \times R) + (I_R \times jX) + (jI_X \times R) + (jI_X \times jX) \quad (1.14)$$
$$= (I_R R - I_X X) + j(I_R X + I_X R)$$

Since the imaginary term is in quadrature with the real component of the voltage, its effect on the overall voltage magnitude is much smaller than that of the first term. So a reasonable approximation for voltage drop is to consider only the real part of the total voltage drop:

$$V_{drop} \approx \text{Re}\{V_{drop}\} = I_R \times R - I_X \times X = \text{Re}\{I\} \times R - \text{Im}\{I\} \times X \quad (1.15)$$

Note that the real component of the current, or the component of the current in-phase with the voltage, produces a voltage drop across the resistance, and the reactive component of the current, or the component of the current in quadrature with the voltage, produces a voltage drop across the reactance. These fundamental concepts are important to remember when doing many types of calculations, power factor correction to name one.

When calculating voltage drop in power systems, the per-unit system is very advantageous, particularly when transformers are present. As current flows through a

1.8 VOLTAGE DROP CALCULATIONS

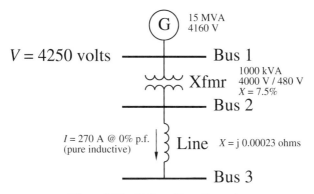

Figure 1.20   Voltage Drop Example

transformer, it effectively exits one electric circuit and enters another. In addition, the transformer windings, through which the current flows, have impedance (generally $X \gg R$). This means a change in voltage occurs due to both the turns ratio of the transformer and the voltage drop across the transformer impedance. The per-unit system makes capturing both of these voltage drop components very easy. Consider the example shown in Figure 1.20.

To calculate the voltage on Bus 3, the per-unit system can be used as follows. First, choose an MVA base to use for this problem. This selection is arbitrary, so a convenient value of 10 MVA is chosen. Keep in mind that this base MVA applies to the entire system being studied.

Next, a voltage base is selected for one of the buses. This choice, too, is arbitrary, so we will define the nominal voltage of Bus 1 (4160 V) as the base voltage of Bus 1. That voltage base of 4160 V applies to the entire circuit containing Bus 1, which in this case is Bus 1 and the generator.

Moving through the transformer takes us to a second electric circuit. That circuit, consisting of Bus 2, the line, and Bus 3, must also have a base voltage assigned, but this assignment is not arbitrary. In order to essentially make the transformer turns ratio become 1:1, the base voltage of Bus 2 must be a function of both the Bus 1 voltage base and the transformer turns ratio. Specifically,

$$V_{\text{base (Bus 2)}} = V_{\text{base (Bus 1)}} \times \frac{480}{4000} = 4160 \times \frac{480}{4000} = 499.2 \text{ V} \quad (1.16)$$

This base voltage of 499.2 V applies to the entire circuit containing Bus 2.

Next, all impedances must be converted to per-unit values. The transformer reactance is given as 7.5%, or 0.075 p.u., but this per-unit value was determined on the transformer's power and voltage bases (1 MVA and 4000 V/480 V), not the bases we are using in this problem (10 MVA and 4160 V/499.2 V). So we can use Eq. (1.4) to convert the 0.075 p.u. to the appropriate bases:

$$X_{\text{xfmr}} = 0.075 \times \left(\frac{4000}{4160}\right)^2 \times \left(\frac{10}{1}\right) = 0.6934 \text{ p.u.} \quad (1.17)$$

Note that the voltage correction term of Eq. (1.17) could have been the quotient squared of 480 and 499.2 instead of 4000 and 4160, as each represents the same value.

Then, the reactance of the line ($j\ 0.00023\ \Omega$) can be converted to per-unit by dividing it by the base impedance in this circuit. The base impedance is found using Eq. (1.2). The per-unit reactance of the line is

$$X_{line} = \frac{0.00023}{\left(\frac{0.4992^2}{10}\right)} = 0.00023 \times \left(\frac{10}{0.4992^2}\right) = 0.0092\ \text{p.u.} \quad (1.18)$$

Now the Bus 1 voltage can be converted to per-unit by dividing the actual voltage by the base voltage.

$$V_{Bus\ 1} = \frac{4250}{4160} = 1.0216\ \text{p.u.} \quad (1.19)$$

Next, the current can be converted to per-unit by dividing the actual current by the base current:

$$I = \frac{270}{\left(\frac{10,000}{0.4992\sqrt{3}}\right)} = 270 \times \left(\frac{0.4992\sqrt{3}}{10,000}\right) = 0.0233\ \text{p.u.} \quad (1.20)$$

Now that every electrical quantity has been converted to per-unit, the voltage drop calculation can be done easily using Ohm's law:

$$V_{Bus\ 3} = 1.0216 - 0.0233(0.6934 + 0.0092) = 1.0052\ \text{p.u.} \quad (1.21)$$

The final step is to convert the per-unit answer back to electrical units (volts) by multiplying it by the base voltage.

$$V_{Bus\ 3} = 1.0052\ \text{p.u.} \times 0.4992\ \text{V} = 501.8\ \text{V} \quad (1.22)$$

1.9 SHORT-CIRCUIT AVAILABILITY

The major source of short circuit current in an industrial plant is the utility source. Remote utility generators are often located in very large generating stations far from the industrial plant. When a short circuit occurs in the industrial plant, the fault current appears as an additional increment of load current to the remote generators, so the station simply furnishes the extra power requirement.

Some depression of voltage will occur at the utility level when a fault occurs in the industrial plant. To model the "stiffness" of the utility source, a Thévenin equivalent is derived for the source. An ideal voltage source, or *infinite bus*, is modeled in series with a *system impedance*. The system impedance consists of a resistance in series with an inductive reactance. In a typical power system, the reactance component is much larger than the resistance component, so the X to R ratio (X/R) is high—X many times exceeding R by more than an order of magnitude. If the utility X/R is not known, it can be assumed to be infinite ($X = Z$, $R = 0$). This Thévenin equivalent accurately represents the voltage drop seen at the utility level when a fault occurs downstream.

The system impedance varies considerably throughout the interconnected power system. The lower the system impedance, the stronger or *stiffer* the system is at that location. In other words, the voltage at a stiff location in the system will drop less than the voltage at a less stiff location for the same fault. One would expect lower system impedances on higher voltage systems close to strong generation sources than on lower voltage systems far removed from strong generation sources. Starting a motor on a stiff system produces less of a voltage drop than on a weaker system. But fault currents will be higher on a stiff system than on a weaker system.

System impedances frequently are expressed in per-unit values. An arbitrary MVA base is selected, often 10 MVA or 100 MVA. A kV base is chosen equal to the nominal operating voltage at the delivery point. From these two bases, all other per-unit bases are derived.

The system impedance is inversely related to the SCA of the source: the lower the system impedance, the higher the fault current contribution from the source. In fact, when doing per-unit calculations, the per-unit system impedance is the base MVA divided by the SCA in MVA.

$$Z_{p.u.-system} = \frac{MVA_{base}}{SCA_{MVA}} \qquad (1.23)$$

The lower the system impedance, the larger the load can be without excessively depressing the source voltage. This is a key factor when starting large motors since the heavy starting current can cause large voltage drops if the system impedance is too high.

The system impedance represents the Thévenin (effective) impedance located between the remote generator bus and the bus of interest. It is the Thévenin impedance that determines the SCA at any point in the system as well as the voltage drop experienced when a given amount of current flows past any point in the system.

1.10 CONDUCTOR SIZING

Electrical conductors are sized in such a way to produce a specific temperature rise at a particular loading. Heating of a conductor is a function of its resistance and the current passing through it according to the relationship

$$P = I^2 R \qquad (1.24)$$

where

P represents the power dissipated in the form of heat

I is the current passing through the conductor

R is the resistance of the conductor.

Resistance is determined by the equation

$$R = \frac{\rho L}{A} \tag{1.25}$$

where

ρ is the *resistivity* of the conductor,
L represents the conductor's length, and
A represents the conductor's cross-sectional area.

Resistivity depends on the material of which the conductor is made. The resistivity of hard-drawn copper at 20°C is 1.77×10^{-8} Ω-m. This value is lower than that of aluminum at 20°C (2.83×10^{-8} Ω-m), since copper is the better conductor of the two metals. Silver, the best conducting metal known, has a resistivity of 1.59×10^{-8} Ω-m. Because of its very low resistivity, silver plating is often used where resistance must be minimized such as switchgear bus joints and circuit breaker terminals, as long as the environment in which the equipment is installed is chemical friendly toward silver (more on this in Chapter 7).

The resistance of a conductor is directly proportional to its length, but since physical requirements typically determine the required length of a conductor, length usually cannot be varied to control resistance. The cross-sectional area of the conductor is a parameter that can be varied to change the conductor's resistance. By increasing the cross-section of the conductor, the resistance decreases proportionally. This means that for a given temperature rise, a larger conductor can carry more current than a smaller conductor. While this is a rather obvious conclusion, good judgment should be exercised when applying this principle. As conductors become very large, they become unwieldy and difficult to install. Insulated cables require more insulation to cover large conductors, which greatly increases the cost of the cable. Also, the tendency for current density to be higher near the surface of a conductor than closer to the center, or *skin effect*, becomes an important consideration with very large conductors, since the skin depth at 60 Hz is on the order of 1 cm. These concerns suggest that using multiple smaller conductors per phase may be a more practical option than using one large conductor per phase.

Skin depth (δ) is defined as the depth below the surface of the conductor at which the current density falls to 1/e or about 36.8% of its density at the conductor surface. The classical formula for calculating skin depth in meters is

$$\delta = \frac{1}{\sqrt{\pi f \sigma \mu}} \text{ (meters)} \tag{1.26}$$

where

f is the frequency in hertz
σ is the conductivity of the conductor in meters/ohm-meter^2
μ is the permeability of the conductor in henrys/meter.

Equation (1.26) can be manipulated to incorporate more readily available values. Replacing conductivity with the reciprocal of resistivity (ρ),

$$\delta = \frac{1}{\sqrt{\pi f \frac{1}{\rho} \mu}} \qquad (1.27)$$

Rearranging terms,

$$\delta = \frac{1}{\sqrt{\pi}} \sqrt{\frac{\rho}{f \mu}} \qquad (1.28)$$

Expressing the permeability as the product of permeability of free space (μ_o) and the relative permeability of the conductor (μ_r),

$$\delta = \frac{1}{\sqrt{\pi}} \sqrt{\frac{\rho}{f \mu_o \mu_r}} \qquad (1.29)$$

Normalizing the resistivity to the resistivity of copper (ρ_{cu}),

$$\delta = \sqrt{\frac{\rho_{cu}}{\pi}} \sqrt{\frac{\frac{\rho}{\rho_{cu}}}{f \mu_o \mu_r}} \qquad (1.30)$$

Rearranging terms,

$$\delta = \sqrt{\frac{\rho_{cu}}{\pi \mu_o}} \sqrt{\frac{\frac{1}{\mu_r} \cdot \frac{\rho}{\rho_{cu}}}{f}} \qquad (1.31)$$

Substituting numerical values into the first radical of Eq. (1.31) and expressing length units in centimeters, skin depth can be calculated using Eq. (1.32):

$$\delta = 6.6 \sqrt{\frac{\frac{1}{\mu_r} \cdot \frac{\rho}{\rho_{cu}}}{f}} \text{ (centimeters)} \qquad (1.32)$$

where

μ_r is the relative permeability of the conductor (typically 1.0)
ρ is the resistivity of the conductor in ohm-meters
ρ_{cu} is the resistivity of copper (1.77×10^{-8} Ω-m)
f is the frequency in hertz

Using Eq. (1.32), the skin depth of aluminum at 60 Hz is found to be 1.049 cm, while the skin depth of copper at 60 Hz is 0.853 cm. Current density more than 1 cm beneath the surface of the conductor will be very low, so ampacity does not increase linearly with conductor size for very large conductors.

Large conductors can be avoided by paralleling two or more smaller conductors per phase. Using too many paralleled conductors can present installation problems; therefore, sound engineering judgment must be used when determining how many cables of what size should be paralleled in a given situation. Sound judgment is best developed by working closely with field personnel and experiencing firsthand which cable configurations work well and which are problematic.

The National Electrical Code (NEC), published by the National Fire Protection Association as NFPA standard 70, is used to size electrical conductors in industrial, commercial, and residential applications. Article 310 of the NEC stipulates minimum conductor sizes for given ampacities. The type of cable and the type of raceway into which the cable is installed affect the allowable ampacity given by the Code. It should be emphasized that these sizes are minimum values and may have to be increased if substantial cable lengths result in large voltage drops. Tables D.1 through D.24 in Appendix D are reprinted from the 2014 edition of the NEC.

Particular attention must be paid to the type of insulation used in the cable, as this determines the maximum allowable conductor temperature. The maximum conductor temperature can range from 60°C to 250°C depending on the materials used in the cable insulation.

Ambient temperature is also an important factor in determining the ampacity of a cable. The NEC ampacity tables are developed for an ambient temperature of 30°C (86°F) or 40°C (104°F), depending on the installation parameters for which the table was derived. As the ambient temperature increases, the cables must be derated per the correction factors shown in the NEC tables. Conversely, as the ambient temperature falls below the referenced ambient temperature, the ampacities can be increased.

1.11 TRANSFORMER SIZING

Determining the kVA rating of any transformer represents a critical decision in the design process. In the case of a main transformer supplying an industrial facility, many unknowns must be assumed prior to ordering the transformer. If the installation is new, the connected kVA load is known but the *diversity* of that load probably is not known. Diversity represents the percentage of the connected load that will be energized at one given time. For example, a switchgear bus may supply six 250-hp motors, but because of the design and operating procedures of the plant, if no more than four of these motors can run at one time, it would be overly conservative to design the electrical system to feed all six motors concurrently. On the other hand, if plant operating practices change, all motors may have to run simultaneously. This dilemma can be difficult to resolve.

Also uncertain are future changes that will occur both in the plant and in the utility system. Although utilities typically prepare planning documents that project 5–10 years into the future, plans can change suddenly and unexpectedly. If an independent power producer builds a 500-MW generating plant near your industrial plant, the SCA at the plant could increase dramatically. Transmission system enhancements could have a similar effect. In addition, business factors could force plant expansions that were not anticipated. All of these scenarios impact the rating of the plant's supply transformer.

A good approach when sizing a main transformer is to make conservative yet realistic assumptions about system change and then determine an expansion plan that can be executed when the main transformer becomes inadequate.

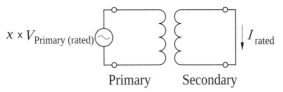
Figure 1.21 Definition of Transformer Impedance

The main transformer at an industrial installation represents substantial impedance in series with the utility (system) impedance. This can be crucial in limiting fault current in the plant to acceptable levels. If the impedance of the main transformer is too high, its voltage regulation will be poor. This will cause unacceptably high voltage drops when starting large motors, or perhaps even during times of heavy load.

A transformer's impedance is an important parameter that helps determine its suitability for a given application. Transformer impedance is expressed in percent based on the self-cooled kVA rating. The impedance is numerically equal to the percentage of rated voltage that would have to be applied across the primary winding to cause rated current to flow in the short-circuited secondary winding. This concept is illustrated in Figure 1.21, where x equals the nameplate value %Z of the transformer.

Transformer impedance varies according to design parameters, particularly kVA rating and basic lightning impulse insulation level (BIL). In general, impedance increases with kVA rating and also increases with BIL. Most distribution class transformers have an impedance in the 2–10% range, with many falling in the 5.5–7.5% range. Specific designs can cause the impedance to lie outside this range. Since transformer impedance is critical for calculating voltage drop and short circuit magnitudes, actual nameplate data should be used whenever possible.

Two basic types of distribution class transformers exist, each with a different type of insulation system. Insulation must be provided to prevent the winding conductor from short circuiting to the transformer core, to the tank or case, or to an adjacent winding.

Liquid immersed transformers, as shown in Figure 1.22, contain the windings in a tank filled with a dielectric liquid, typically mineral oil. Synthetic materials such as polyalpha olefins and silicone compounds also can be used as dielectric fluids, but mineral oil is the least expensive option. Although mineral oil is the most economical dielectric, it is very flammable. Over the years, attempts have been made to reduce its flammability.

Beginning in 1929, polychlorinated biphenyls (PCBs) were added to mineral oil to raise substantially its flash point, greatly reducing the risk of fire. PCBs were sold in the United States under the trade names *Askarel* (Westinghouse Electric Corporation), and *Pyranol* and *Chrorinol* (General Electric Company). The PCBs, manufactured in the United States solely by Monsanto Company, worked well to reduce fire hazard, but were eventually identified as potential carcinogens. Concern over the toxicity and environmental persistence of PCBs led Congress in 1976 to enact Section 6(e) of the Toxic Substances Control Act (TSCA) that included among other things, prohibitions on the manufacture, processing, and distribution in commerce of PCBs.

Figure 1.22 Liquid-Immersed Transformer (*Photo courtesy of Pennsylvania* Transformer Technology, Inc.)

The TSCA required "cradle to grave" management of PCBs in the United States, meaning that the equipment owners were completely responsible for any damages that may result from improper handling of the fluids, even after the equipment is discarded. All materials with PCB levels over 50 parts per million were prohibited under the TSCA, and as a result, all electrical equipment containing dielectric fluids with more than 50 parts per million of PCBs could no longer be used in the United States. Other less-flammable synthetic dielectric fluids do not contain PCBs, but because these fluids are very expensive compared to mineral oil, they are used only where the risk of fire must be kept to an absolute minimum.

An alternative to mineral oil that has been gaining popularity in recent years is biodegradable vegetable oil. Although spills are still an environmental concern and oil containment is still required, spilled vegetable oil is not considered a hazardous waste and can be disposed of by ordinary means. Sunflower and safflower seeds are the chief source for these oils, which because of their high percentage content of mono-unsaturated fatty acids, they tend to be very stable when exposed to oxygen. In addition to excellent dielectric properties, vegetable oils have a flash point of around 330°C, which exceeds the flash point of mineral oil by more than 180°C. Because of their better high-temperature performance than mineral oil, vegetable oils show little thermal decomposition in the vicinity of hot-spot locations and when the transformer is overloaded.

The insulation in a liquid immersed transformer is usually a paper or a similar cellulose-based material. This is why the winding temperature of a transformer is so critical. Cellulose breaks down at high temperatures, producing gases such as carbon monoxide and carbon dioxide. Arcing under oil produces acetylene, and other hydrocarbon gases and hydrogen. An overheated steel core will cause hydrogen,

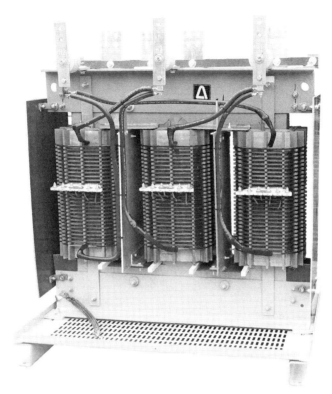

Figure 1.23 Dry-Type Transformer (*Photo courtesy of Alfa Transformer*)

methane, ethane, and ethylene to form in the oil. Low-current sparking under oil forms methane and ethane without increasing acetylene, ethylene, or hydrogen levels. Excessive corona generates hydrogen without increasing hydrocarbon gas levels. By monitoring the levels and types of dissolved gases in transformer oil, the health of the machine can be tracked. It is usually not the amount of a gas, but the rate of increase of a gas level that indicates problems. This is why a regularly scheduled oil analysis program is important. Often, serious problems can be detected before failure occurs, greatly reducing repair cost and equipment downtime.

Dry-type transformers, as shown in Figure 1.23, typically are used in indoor applications, but as environmental restrictions safeguarding against mineral oil leaks become more stringent, dry-type transformers are applied more frequently in environmentally sensitive outdoor locations as well.

Different classifications of dry-type transformers are built including *cast coil, ventilated, enclosed non-ventilated, sealed gas-filled*, and *vacuum pressure impregnated*. Cast coil transformers are more resistant to moisture and airborne dust contamination than other dry types and commonly are used in outdoor or harsh industrial environments. Dry-type transformers typically have lower BIL ratings and are less durable than comparably sized liquid-immersed units, although some manufacturers provide BIL ratings comparable to liquid-immersed units. In spite of these drawbacks, dry-type transformers usually are chosen for indoor and environmentally sensitive outdoor applications, because installing liquid immersed transformers indoors involves very elaborate provisions including fire suppression systems and vault construction.

1.12 LIQUID-IMMERSED TRANSFORMER kVA RATINGS

In theory, many factors limit the amount of power, actually current, that can be handled by a transformer. Current density in the windings and flux density in the core certainly impose limits on power transfer, but these limits are usually academic in practical transformer designs. Long before current density or flux density becomes a problem, thermal issues arise, driving the winding insulation above its design temperature. These thermal issues can be mitigated by improving the heat transfer away from the windings.

Installing radiators as a heat exchanger between the dielectric oil and the air surrounding the transformer enhance heat transfer, thereby increasing the kVA rating of the transformer. As the oil in the main tank heats, it rises and makes its way to the radiators through the upper radiator inlets. When the hot oil enters the radiator, it cools due to enhanced heat transfer with the surrounding air. The cool oil gains density and drops through the radiator tubes to the bottom, where it is drawn back into the main tank by thermosyphonic flow.

The addition of fans on the radiators, as shown in Figure 1.24, and oil pumps, as shown in Figure 1.25, to assist the thermosyphonic flow by forcing the cooled oil back into the transformer tank further enhance heat transfer away from the windings.

Figure 1.24 Radiator with Fans (*Photo courtesy of Progress Energy*)

Figure 1.25   Oil Pump with Flow Gauge (*Photo courtesy of Progress Energy*)

A four-letter designation is used to describe the cooling class of a liquid-immersed transformer. The first letter describes the internal cooling medium in contact with the windings. The second letter denotes the circulation mechanism for the internal cooling medium. The third and fourth letters indicate the external cooling medium and the circulation method for that medium, respectively. Each level of cooling stipulates a different kVA rating. The lowest (self-cooled) kVA rating is always the value used in per-unit calculations. Institute of Electrical and Electronics Engineers (IEEE) standard C57.12.00 specifies liquid-immersed transformer kVA ratings. Prior to the year 2000, a different designation method was used for cooling classification. The older method was sometime ambiguous, as two different cooling mechanisms could have the same designation. The present cooling designations are more descriptive of the cooling mechanisms used and are shown in Table 1.2, while Table 1.3 shows the designations used prior to the year 2000.

TABLE 1.2   Liquid-Immersed Transformer Cooling Class Letter Designations

Position	Letter	Description
1st	O	Mineral oil or synthetic insulating liquid with fire point $\leq 300°C$
	K	Insulating liquid with fire point $>300°C$
	L	Insulating liquid with no measurable fire point
2nd	N	Natural convection flow through cooling equipment and in windings
	F	Forced circulation through cooling equipment (i.e., pumps), natural convection flow in windings (also called *nondirected flow*)
	D	Forced circulation through cooling equipment, directed from the cooling equipment into at least the main windings
3rd	A	Air
	W	Water
4th	N	Natural convection
	F	Forced circulation (fans for air, pumps for water)

TABLE 1.3 Comparison of Cooling Class Designations (IEEE Std. C57.12.00-2000 to C57.12.00-1993 and Before)

IEEE Std. C57.12.00-2000 Designation	Previous Designation
ONAN	OA
ONAF	FA
ONAN/ONAF/ONAF	OA/FA/FA
ONAN/ONAF/OFAF	OA/FA/FOA
ONAN/ODAF	OA/FOA
ONAN/ODAF/ODAF	OA/FOA/FOA
OFAF	FOA
OFWF	FOW
ODAF	FOA
ODWF	FOW

SUMMARY

Electricity for an industrial facility is provided by the local utility at a voltage determined by the size of the load and the topology of the utility's system. In general, if the load is small enough to be fed from a single distribution transformer, the facility will likely be served at *utilization voltage*, which is often 480Y/277 V. As the load becomes larger, utilities opt to serve it at subtransmission or transmission voltages as high as 230 kV.

One-line diagrams are used as a simplified means of describing the topology of a power system. Although not truly a circuit diagram, one-lines are often used as a starting point for constructing circuit diagrams. Commonly used one-line diagram symbols were introduced.

Zones of protection are used to assure that a short circuit fault occurring anywhere on the power system can be detected by at least one, and ideally by multiple protective devices. Protection zones must overlap to assure 100% protection coverage.

The configuration of the utility source has a large bearing on the reliability of the service and the cost of the installation. A thorough economic analysis must be done to determine which configuration option provides the best balance between cost and reliability of service. Configurations discussed were the *radial system*, *ring bus*, *primary selective*, *primary loop*, *secondary selective*, *secondary spot network*, and *sparing transformer* schemes.

The *per-unit system* of calculation is used in power systems where transformers are present. The per-unit system effectively makes the turns ratio of each transformer 1:1, allowing the entire power system to be represented as a single electrical circuit. Typically, *base power* and *base voltage* are arbitrarily assigned, and *base current* and *base impedance* are calculated by the formulas provided.

Power in an AC system is the product of two sinusoidally varying terms—voltage and current. The power sinusoid has a frequency twice that of the voltage or current and has three distinct terms: a *constant component of real power*, an *oscillating component of real power*, and *reactive power*. A phasor representation of power, $\tilde{S} = \tilde{V} \cdot \tilde{I}^*$, was also developed.

Voltage drop was examined and the useful approximation of $V_{drop} \approx \text{Re}\{V_{drop}\} = I_R \times R - I_X \times X = \text{Re}\{I\} \times R - \text{Im}\{I\} \times X$ was derived.

The utility connection is the largest source of short circuit current. A Thévenin equivalent circuit consisting of an ideal voltage source, or *infinite bus*, in series with a resistance and inductive reactance, or *system impedance*, is derived which describes the *stiffness* of the utility source, or how much voltage drop occurs at the point of delivery. The lower the system impedance, the stiffer the source and the higher the SCA. High SCA is desirable when starting large motors, so the resulting voltage drop is tolerable.

A distribution transformer in series with the utility source greatly reduces SCA by adding considerable impedance to the system. Although transformer impedances lie in a relatively narrow range, it is important to select wisely the proper impedance for a transformer to sufficiently limit fault current without producing an excessive voltage drop. In addition to impedance, the kVA rating of the transformer must be adequate for planned electrical system expansions. Liquid-immersed transformers have multiple kVA ratings based on cooling stages. BIL is also an important consideration when specifying a transformer to assure that insulation levels throughout the system are properly coordinated.

FOR FURTHER READING

IEEE Guide for Performing Arc Flash Hazard Calculations, IEEE Standard 1584, 2002.
IEEE Recommended Practice for Electric Power Distribution for Industrial Plants (Red Book), IEEE Standard 141, 1993.
IEEE Standard General Requirements for Liquid-Immersed Distribution, Power, and Regulating Transformers, IEEE Standard C57.12.00, 1993.
IEEE Standard General Requirements for Liquid-Immersed Distribution, Power, and Regulating Transformers, IEEE Standard C57.12.00, 2000.
Joffe, E. B., and Lock, K. *Grounds for Grounding: A Circuit to System Handbook*. Wiley-IEEE Press, 2010. ISBN: 978-0-471-66008-8.
Nabours, R. E. Dalziel revisited. *Industry Applications Magazine, IEEE*, vol. 15, no. 3, pp. 18–21, May–June 2009.
National Electrical Code, National Fire Protection Association, NFPA 70, 2014.
Standard for Electrical Safety in the Workplace, National Fire Protection Association, NFPA 70E, 2015.

QUESTIONS

1. Propose a methodology for determining the configuration of the utility source for a specific industrial facility.
2. What physical characteristic of a generator limits its ability to produce short circuit current?
3. As more generators are connected to a bus, the SCA rises. Eventually, the SCA will exceed the interrupting ratings of the circuit breakers on the bus. How can this be avoided?

4. A small industrial facility is supplied from a single radial feed through a 10 MVA, 69/4.16 kV transformer. The local utility has a second 69 kV circuit nearby. A major expansion is planned which will triple the facility's electrical demand, so two additional transformers, identical to the first, must be added. Because of the expansion, reliability will be more critical in the future than at present. Propose and justify a new source configuration for this facility.

5. Dissolved gas analysis can be a useful diagnostic tool in determining the health of oil-filled equipment. What are some procedures and practices necessary for a good dissolved gas analysis program?

6. Would you expect the dissolved gas levels to be similar in all oil-filled transformers? In an oil-filled transformer versus an oil circuit breaker? Discuss.

7. Why is fault detection and isolation more complicated with a networked distribution system than with a radial distribution system?

8. In the first six decades of the twentieth century, power system calculations were typically done using *percent* values instead of per-unit values. What were the advantages and disadvantages of using percent values, and why do you suppose the change was made to per-unit values?

9. Explain why the sinusoid representing reactive power is centered about the x-axis, but the sinusoid representing real power is offset by a DC component above the x-axis?

10. Why might a vegetable-based dielectric oil be preferred to a mineral-based oil?

11. What concern must be addressed when operating a secondary selective scheme with its tie breaker normally closed?

12. Why is it essential that zones of protection overlap?

13. List some nonpower engineering examples of applications of the per-unit system.

14. What are some advantages of using a one-line diagram to describe a power system over other types of representations?

15. Explain how the configuration of the electrical source affects both the reliability and the cost of the system.

PROBLEMS

1. A 600 A three-phase four-wire 12.47Y/7.2 kV service is fed from a 69 kV utility source having an SCA of 1800 A. The three-phase fault current at the main 12.47 kV switchgear must be limited to 5500 A. What is the minimum impedance for a 14 MVA transformer that will sufficiently limit the fault current?

2. A transformer has the following nameplate data:

    138/13.8 kV

    12/16/20 MVA

    Reactance = 5.42%

    This transformer will be connected to a 138 kV source to supply a 13.8 kV bus. Calculate the per-unit impedance of this transformer using a 10 MVA base.

3. An industrial plant is to be supplied by three identical transformers. A decision must be made whether to purchase a fourth transformer to build a sparing transformer system as shown in Figure 1.16, or to configure the source as a secondary selective system as shown in Figure 1.14. If the secondary selective option is chosen,

   a. how much larger must each transformer in the secondary selective option be compared to each transformer in the sparing transformer option if each transformer is to be able to supply the full plant load in the event that one transformer fails?

   b. how many additional circuit breakers must be purchased to provide the necessary bus tie capability for the secondary selective option compared to the sparing transformer option?

4. Source configurations can be combined to increase operating and maintenance flexibility. Sketch a hybrid primary/secondary selective system by combining Figures 1.8 and 1.14. Describe how the system would operate in the event of

   a. loss of one of the sources and
   b. loss of one of the transformers.

5. A factory currently has six large induction motors connected to 4.16 kV switchgear supplied radially from a single transformer, as in Figure 1.3. A plant expansion is planned, where four more large 4 kV motors must be added. The current switchgear cannot supply all 10 motors, because both the switchgear bus rating and the transformer rating would be exceeded. Propose a 4.16 kV system design to accommodate the new and existing motors.

6. A 1500-ft-long 4.16 kV distribution feeder has an impedance of $(0.02 + j\,0.12)\ \Omega/1000$ ft. Find its per-unit impedance on a 10 MVA base.

7. A three-phase, 480-V, 100 kVA constant-impedance load operates at 85% lagging power factor. Find its per-unit impedance on a 10 MVA base.

8. Refer to Figure 4.5. Assuming a base power of 10 MVA and a base voltage of 13.8 kV at Bus 1, find the base voltage, base current, and base impedance at Buses 2 and 3, and in the high-voltage circuit of transformer T_1.

9. Repeat Problem 8 assuming a base power of 25 MVA and a base voltage of 13.2 kV at Bus 1.

10. If $v(t) = 170 \cos{(377t)}$ and $i(t) = 5 \cos{(377t - 31.8°)}$, find $s(t)$, $\tilde{P}$, $\tilde{Q}$, and $\tilde{S}$. Also sketch the power triangle and calculate the power factor.

11. If $v(t) = 680 \cos{(377t + 10°)}$ and $i(t) = 25 \cos{(377t + 35.8°)}$, find $s(t)$, $\tilde{P}$, $\tilde{Q}$, and $\tilde{S}$. Also sketch the power triangle and calculate the power factor.

12. Calculate the voltage drop across a 50 MVA transformer with an impedance of 5% when the transformer is loaded to 90% of its MVA rating at 90% lagging power factor. Assume an X/R ratio of 25.

13. Repeat Problem 12 if the loading is changed to 120% of the transformer's MVA rating at 70% lagging power factor.

14. Calculate the skin depth at 400 Hz of an alloy with a resistivity of 3.25×10^{-8} Ω-m. Assume a relative permeability of 1.

15. Refer to Figure 4.5. Show the necessary zones of protection.

CHAPTER 2

INSTRUMENT TRANSFORMERS AND METERING

OBJECTIVES

- Be familiar with terminology pertaining to instrument transformers and metering
- Know the criteria necessary to specify instrument transformers, particularly for metering applications
- Understand the fundamental concepts of electricity metering
- Be aware of present metering technologies and emerging trends in electricity metering

2.1 DEFINITIONS

This section defines some of the commonly used terminology in the field of electricity metering.

Automatic Meter Reading (AMR)—The reading of meters from a location remote from where the meter is installed. Telephone circuits, radio transmission, and power line communication are some of the technologies used for AMR.

Burden—The load placed on an instrument transformer secondary circuit by metering (or other) equipment. Burden is expressed in volt-amperes at a specific power factor.

Class—Maximum continuous current the meter is capable of handling. Abbreviated by the letters "CL."

Common Meter Class Ratings:

CL 100	100 A—Self-contained
CL 200	200 A—Self-contained
CL 320	320 A—Self-contained
CL 480	480 A—Self-contained
CL 2	2 A—Transformer-rated
CL 10	10 A—Transformer-rated
CL 20	20 Amps—Transformer-rated

CTR—Current Transformer Ratio

Example:	A 400:5 CT results in a CTR of 80

Elements—A combination of a voltage-sensing unit and a current-sensing unit which provides an output proportional to the power measured.

Form Letter—Letter following the form number that indicates the socket/terminal configuration of the meter. Common forms are:

S	Socket-based
A	Adapter-based (screw terminals for service wiring)

Form Number—Number conforming with the internal connection diagram shown in ANSI C12.10. Meters with the same form number have identical internal connections. Abbreviated as "FM."

K_h—the watt-hour constant, which represents the number of watt-hours represented by each disk revolution on an electromechanical meter, or the number of watt-hours represented by one pulse of serial data.

Light Load—Current in amps that the meter is tested for low current accuracy—typically 10% of TA. Abbreviated by the letters "LL."

Example:	LL = 3 A on a meter with a TA 30 A

Self-contained—Meter is connected directly to the load being measured. Often abbreviated by the letters "SC."

Service Voltage—Line-to-neutral or line-to-line voltage at the meter service, often referred to as meter nameplate voltage.

Example:	Residential meters typically have 120 or 240 V service voltages, while modern solid-state 3-phase meters have 120–480 V service voltages (auto-voltage ranging)

Service Wiring—Number of wires from the distribution transformer and their phase connection. Abbreviated as "#W."

Example: 1 Phase **2W** or **3W**
3 Phase **3W** wye (network) or **3W** delta
3 Phase **4W** wye or **4W** delta

Stators—Old term from electromechanical meters that is synonymous with *elements*.

Test Amps—Current in amps that the meter is tested for "Full Load (FL) accuracy." Abbreviated by the letters "TA" (15% of CL).

TA 0.5	CL2 meter tested at 0.5 A
TA 2.5	CL10 or CL20 meter tested at 2.5 A
TA 15	CL100 meter tested at 15 A
TA 30	CL200/CL320 meter tested at 30 A
TA 50	CL320 meter tested at 50 A

Transformer-Rated—Meter measures scaled down representation of the load. Scaling accomplished by use of external current transformers (CTs) and optional voltage transformers (VTs, or PTs which is an abbreviation for *potential transformer*). Abbreviated by the letters "TR."

VTR – Voltage Transformer Ratio

Example:	A 480:120 VT results in a VTR of 4.

2.2 INSTRUMENT TRANSFORMERS

2.2.1 Fundamentals

When metering services of 480 V or less, a *self-contained meter* can be used if the full-load current does not exceed 600 A. Self-contained meters are connected directly to the service, usually through a plug-in socket. But if the above voltage or current criteria are exceeded, a *transformer-rated meter* must be used. Transformer-rated meters require *instrument transformers* to reduce the voltage and/or current to levels that can be handled by the metering equipment.

Instrument transformers are also used to supply voltages and currents to protective relays. The instrument transformers allow the meters and relays to operate at low voltages and currents, which reduces the cost and complexity of the devices while also reducing the hazards resulting from higher voltage and current applications.

PTs or *VTs* typically have a secondary winding rating of 120 V. As the turns ratio becomes very large, *coupling capacitor voltage transformers* or *CCVTs*, such as shown in Figure 2.1, are advantageous, as they connect the VT across a string of series capacitors from line to ground, forming a voltage divider. This reduces the required turns ratio for the VT. Lower voltage VTs often have integrated fuse protection, as shown in Figure 2.2.

40 CHAPTER 2 INSTRUMENT TRANSFORMERS AND METERING

Figure 2.1   245 kV Capacitive Voltage Transformer. (*Photo courtesy of Trench Group.*)

Figure 2.2   15 kV Voltage Transformer. (*Photo courtesy of GE Power Sensing.*)

Figure 2.3 Voltage Transformer Polarity. (*Photo courtesy of GE Power Sensing.*)

The *polarity* of a VT is designated by terminal numbering. Like a power transformer, the high-voltage terminals are denoted by the letter "H," while the letter "X" indicates the low-voltage terminals, as seen in Figure 2.3. Polarity is defined such that an instantaneous current of I_P flowing *into* terminal H1 produces a current I_S flowing *out of* terminal X1. On drawings, the polarity is shown using the *dot convention*, where the dot on the high voltage winding corresponds to terminal H1 and the dot on the low-voltage winding corresponds to terminal X1, as shown in Figure 2.4.

Figure 2.4 Voltage Transformer Dot Convention

Figure 2.5 69 kV Current Transformer. (*Photo courtesy of ABB, Inc.*)

CTs usually have a secondary winding rating of 5 A, although 1 A secondary ratings are common in International Electrotechnical Commission (IEC) standards. Metering equipment easily can accommodate these current magnitudes. CTs are often installed on equipment bushings, allowing the use of lower-cost bushing-type CTs. Where bushing installation is not feasible, free-standing CTs, such as the one shown in Figure 2.5, can be used.

It is important to assure that the secondary circuit of a CT never becomes open circuited. CTs have a very small turns ratio (a 1200:5 CT has a turns ratio of 1/240), meaning that the voltage across the secondary windings would be increased tremendously (240 times in our example) if there is no load in that circuit. The dangerously high voltage built up across the open-circuited secondary circuit could easily cause an arc and explosion. To prevent dangerously high voltages from building up, a low-impedance load must be designed into the secondary circuit, putting a *burden* on the transformer. If a CT is not needed, its secondary windings must be short circuited and grounded. Test switches in the CT secondary must be shorting-type switches, and selector switches must be of the make-before-break type.

Figure 2.6 Current Transformer Polarity. (*Photo courtesy of GE Power Sensing.*)

The polarity of a CT is shown by marking the primary terminals H1 and H2 and the secondary terminals X1 and X2. Toroidal CTs will mark one face of the toroid H1 and the other H2, as shown in Figure 2.6. On drawings, like with VTs, the polarity is shown using the dot convention. A current flowing into the dotted terminal of the high-voltage winding produces a current flowing out of the dotted low-voltage terminal, as shown in Figure 2.7.

Many CTs are of the toroidal type. Toroidal CTs are slipped over a bushing, cable, or busbar. The high-power circuit serves as a one-turn primary winding. The toroid is a core around which hundreds of secondary turns are wound. Three different types of toroidal CTs are shown in Figure 2.8.

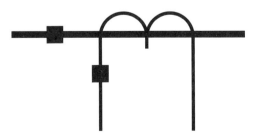

Figure 2.7 Current Transformer Dot Convention

Figure 2.8 Bushing-type CTs Left to Right: Cast Resin (outdoor type), Tape Wound (indoor type), Board Mount (indoor type, typical for generators). (*Photo courtesy of GE Power Sensing*.)

CTs can become magnetized if a DC component is present in the primary current, if the primary circuit is open-circuited under heavy load, or if the secondary circuit is accidently opened with load on the primary circuit (dangerous). Magnetization of the CT can affect accuracy, so demagnetization is recommended before testing a CT. One method of demagnetizing a CT is to provide sufficient current in the primary circuit to produce 5 A of current in the secondary circuit, then gradually increasing the resistance in the secondary circuit to about 50 Ω using a rheostat. Then, the secondary circuit resistance is gradually decreased to zero and the primary current is reduced to zero. Care must be taken throughout this process to assure that the secondary circuit never becomes open circuited.

VTs and CTs can be combined into a single device for convenience. Combination units, or *metering units*, such as those shown in Figures 2.9 and 2.10,

Figure 2.9 34.5 kV Metering Units. (*Photo courtesy of Trench Group*.)

Figure 2.10 230 kV Metering Units. (*Photo courtesy of Trench Group.*)

are frequently used to meter the output of smaller power producers such as cogenerators.

CTs have a *rating factor* (*RF*) that implies the maximum continuous current that can flow through the CT primary without exceeding the thermal capability of the CT windings. Typical RFs range from 1.0 to 4.0. This means that a CT with an RF of 4.0 can produce up to 20 A of secondary current. This is well in excess of the "rated" secondary current of 5 A, so the RF must always be considered when designing metering circuits.

As more impedance is added to the CT secondary circuit, the accuracy of the CT is affected. CTs have a *burden rating*, which defines how much load the CT can serve without deviating from its quoted accuracy. Burden ratings are equal to the maximum external impedance in ohms that can be put in the secondary circuit. Standard CT burden ratings are B0.1, B0.2, B0.5, B1.0, and B2.0, where the numeric values represent the maximum external impedance. The maximum burden in volt-amperes can be calculated for each burden rating by multiplying the rated secondary current squared ($5^2 = 25$) by the burden in ohms. The standard burden ratings in volt-amperes are 2.5, 5.0, 12.5, 25.0, and 50.0.

Different classes of instrument transformers exist for different applications. Those used for revenue billing need a higher accuracy than those used for protective relaying or monitoring. The two main classes of instrument transformers are *metering* and *relaying*.

2.2.2 Correction Factors

Accuracy classes for metering applications are based on the requirement that the transformer correction factor of the instrument transformer shall be within specified limits when the metered load has a power factor between 60% lagging and unity.

Instrument transformers typically have ratios slightly different than what is marked on the nameplate. For example, a CT rated 200:5 should have a 40:1 ratio, but upon testing a ratio of 40.052:1 may be measured. The measured value of 40.052 can be written as a product of the marked ratio and a ratio correction factor, or RCF. In this example, the marked ratio of 40 must be multiplied by an RCF of 1.0013 to give the true ratio (40 × 1.0013 = 40.052).

Both VTs and CTs have an RCF that is defined by IEEE standard C57.13 as

$$\text{RCF} = \frac{\text{True primary quantity}}{\text{True secondary quantity} \times \text{Marked ratio}}. \quad (2.1)$$

Many times, a combined ratio correction factor is used in conjunction with metering circuits. The combined RCF is simply the product of the CT RCF and the VT RCF.

$$\text{RCF}_K = \text{RCF}_{CT} \times \text{RCF}_{VT} \quad (2.2)$$

Similar to the ratio correction factors, another correction factor must be introduced to account for the slight error in phase angle between the primary and secondary instrument transformer circuits. Figure 2.11 shows a wattmeter supplied by instrument transformers.

The true power drawn by the load is the power measured in the primary circuit:

$$P_P = V_P I_P \cos \theta_P \quad (2.3)$$

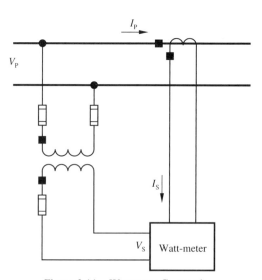

Figure 2.11   Wattmeter Connection

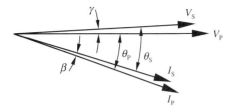

Figure 2.12 Primary Circuit and Secondary Circuit Voltages and Currents

But the meter reads the power determined from the voltage and current in the instrument transformer secondary circuits:

$$P_S = V_S I_S \cos \theta_S \tag{2.4}$$

The phasor diagram in Figure 2.12 shows the relationship between the primary circuit voltages and currents and the secondary circuit voltages and currents.

The combined (for circuits with both CTs and VTs) phase angle correction factor is

$$\text{PACF}_K = \frac{\cos(\theta_S + \beta - \gamma)}{\cos \theta_S} \tag{2.5}$$

Therefore, the true power drawn by the load can be precisely determined by the quantities measured in the instrument transformer secondary circuits and using correction factors, as shown in Eq. (2.6):

$$P_P = V_P I_P \cos \theta_P = (V_S I_S \cos \theta_S)(\text{VTR})(\text{CTR})(\text{RCF}_K)(\text{PACF}_K) \tag{2.6}$$

Figure 2.13 shows the range of acceptable correction factors for various accuracy classes of CTs. Figure 2.14 shows similar information for VTs.

2.2.3 Burden Calculations

IEEE standard C57.13 defines the burden of an instrument transformer as "that property of the circuit connected to the secondary winding that determines the active and reactive power at its secondary terminals." In a VT secondary circuit, the potential coil of the meter has a very high resistance, but a small current still flows in the secondary circuit. This small current places a burden on the VT. VT burden is typically expressed at the total volt-ampere requirement and power factor of the secondary circuit at rated frequency and rated secondary voltage (normally 120 V for VTs with primary voltages below 25 kV and 115 V for VTs with primary voltages above 25 kV.

In a CT circuit, the very low impedance of the current coil of the meter and the resistance of the secondary circuit leads cause a small voltage drop when current flows in the secondary circuit. The CT must develop a small terminal voltage to offset this voltage drop to maintain the magnitude of the secondary current. The impedance

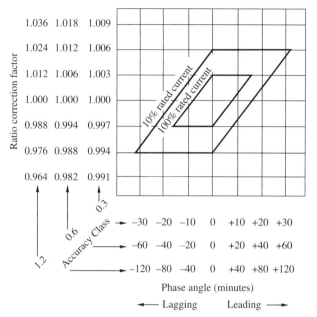

Figure 2.13   CT Correction Factors by Accuracy Class

causing the voltage drop is the CT burden. CT burdens can be expressed like VT burdens (V-A and power factor at rated current and frequency, usually 5 A and 60 Hz), or as ohms of resistance and millihenrys of inductance.

Since VTs have turns ratios greater than one and CTs have turns ratios less than one, the effect of burden impedance is opposite with VTs compared to CTs. Zero burden on a VT is an open circuit, while zero burden on a CT is a short circuit.

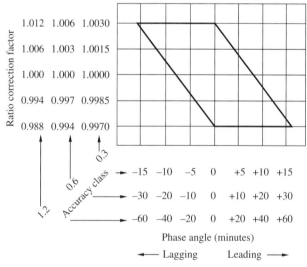

Figure 2.14   VT Correction Factors by Accuracy Class

TABLE 2.1 Voltage Transformer Standard Burdens

			Burden Impedance	
Designation	Volt-Amperes	Power Factor	@ 69.3 V (Ω)	@ 120 V (Ω)
W	12.5	0.10	384	1152
X	25	0.70	192	576
M	35	0.20	137	411
Y	75	0.85	64	192
Z	200	0.85	24	72
ZZ	400	0.85	12	36

2.2.4 ANSI Accuracy Classes

The American National Standards Institute (ANSI) has developed a series of accuracy classes for instrument transformers. These standards are used in the United States and in other countries throughout the world. Based on the calculated burden in an instrument transformer secondary circuit, an appropriate accuracy class can be selected. Since the accuracy of an instrument transformer is a function of burden, a VT may be specified as "1.2 M," meaning that the VT would not affect the meter accuracy by more than ±1.2% at rated current. At light load, or 10% of rated current, the error posed to the meter could double, resulting in a maximum error of ±2.4%.

Similarly, a CT may be specified as "0.3 B-0.9," meaning that the CT would not affect the meter accuracy by more than ±0.3% at 100% rated current. At light load, the maximum error posed to the meter by the CT would be ±0.6%. Tables 2.1 and 2.2 show standard burdens for VTs and CTs per IEEE standard C57.13.

2.3 METERING FUNDAMENTALS

It is common to meter several electrical quantities at the point of connection with the utility. Accurate measurement of energy (kWh) and demand (kW) are necessary for

TABLE 2.2 60 Hz Current Transformer Standard Burdens for Current Transformers with 5 A Secondaries

	Designation	Volt-Amperes @ 5 A	Power Factor	Impedance (Ω)	Resistance (Ω)	Inductance (mH)
Metering	B-0.1	2.5	0.90	0.1	0.09	0.116
	B-0.2	5.0	0.90	0.2	0.18	0.232
	B-0.5	12.5	0.90	0.5	0.45	0.58
	B-0.9	22.5	0.90	0.9	0.81	1.04
	B-1.8	45.0	0.90	1.8	1.62	2.08
Relaying	B-1	25.0	0.50	1.0	0.50	2.3
	B-2	50.0	0.50	2.0	1.00	4.6
	B-4	100.0	0.50	4.0	2.00	9.2
	B-8	200.0	0.50	8.0	4.00	18.4

billing purposes. Consequently, this metering equipment must be sufficiently accurate to satisfy the requirements for the accuracy established in the rate tariffs.

Other metering data could be useful to the industrial customer, such as voltage, current, reactive power, frequency, power factor, so on. Although technologies for measuring these quantities have existed for a century, the past decade has witnessed a tremendous amount of development in electricity metering technology. These new technologies, based on electronics and microprocessors, offer fabulous benefits over the older technologies because of their accuracy, precision, and ability to electronically communicate with other devices in the power system. But the older technologies should not be forgotten. Not only do they remain in use and will undoubtedly do so for years, but also understanding how they work provides valuable insight to the electricity metering process. Before exploring any of the metering technologies, some basic terms must be defined.

2.4 WATTHOUR METERING

Watthour metering is required to record energy usage. Both electromechanical and electronic metering technologies are in widespread use today. Understanding the principles of watthour metering is best accomplished by analyzing the single-stator watthour meter.

2.4.1 Single-Stator Watthour Metering

A watthour meter is basically an induction motor that produces a torque that is proportional to the real power (watts) flowing through it. Many watthour meters are *self-contained*, meaning they need no instrument transformers. Self-contained meters have current coils that are in series with the load, so they must handle full load current. For many years, CL 200 meters, with a continuous current rating of 200 A, were the largest self-contained meters available. Today, self-contained meters as large as CL 480 with a continuous current rating of 480 A can be purchased. When full load current exceeds 600 A, a *transformer-rated* meter, typically CL 20, is used with CTs.

When the watthour mechanism drives a numeric register, integration with respect to time is performed. This integration converts measured power to accumulated energy. The potential and current coils, which act as a stator, are arranged in such a way to produce a torque on the induction disk, which acts as a rotor. The asymmetry of the current coil, which can be seen in the rightmost view of Figure 2.15, provides starting torque, much like in a shaded-pole motor. A permanent magnet serves as a brake to slow the induction disk speed to a usable value. Various adjustments are also provided to increase the accuracy of the meter. A magnetic suspension, made of two like-polarized magnets that repel each other, essentially causes the rotor to float, reducing the rotational friction to almost zero without expensive jewel bearings.

Two important constants, the watthour constant (K_h) and the register ratio (R_r) are printed on the face of the meter. K_h is a function of rated voltage and current and full-load RPM and indicates the number of watthours represented by one revolution

2.4 WATTHOUR METERING

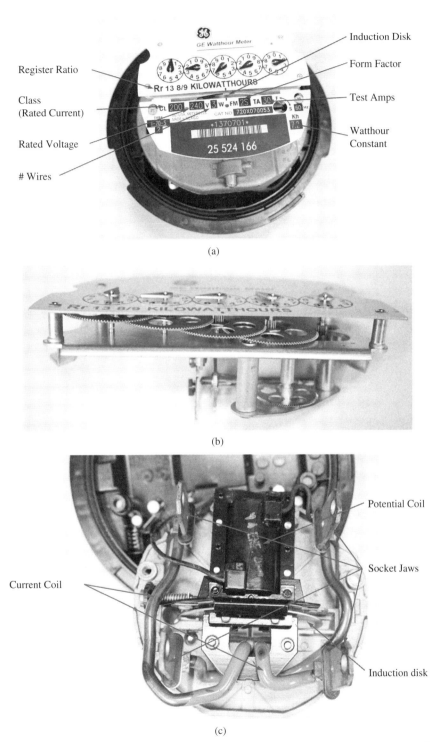

Figure 2.15 Watthour Meter (Front, Side, Rear, and Exploded views). (*Photos courtesy of Michael Celestin.*)

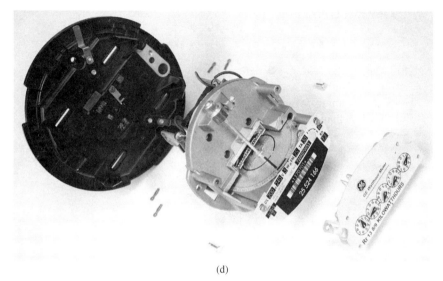

(d)

Figure 2.15 (*Continued*)

of the induction disk. R_r is the number of revolutions of the wheel that meshes with the pinion on the induction disk shaft for one revolution of the first dial pointer. Three views of a single-stator watthour meter are shown in Figure 2.15.

2.4.2 Multi-stator Watthour Metering

An important relationship pertaining to polyphase watt and watthour metering is *Blondel's theorem*, proposed by Andre Blondel in 1893. The theorem states that if energy is supplied to a load by n wires, $n-1$ metering elements are required to measure that energy. Sometimes Blondel's theorem is compromised. For example, two metering elements are sometimes used to meter four-wire systems. A metering scheme that violates Blondel's theorem relies on a balanced load to be accurate. Any significant imbalance will introduce error to the metering. Because of this, much thought must be given to metering schemes that violate Blondel's theorem, and such schemes should never be used for revenue metering.

2.5 DEMAND METERING

Commercial and industrial rate tariffs require the recording of maximum demand, or instantaneous power usage, over a billing period. Usually, billing is not done on the actual peak demand, but on a rolling average basis. Nonetheless, the actual demand must be measured and recorded. Real power demand is always of interest. Some tariffs also require measurement of reactive or total demand. From a metering perspective, kilowatt demand, kilovar demand, and kVA demand can be measured in similar ways. The following sections describe these methods.

2.5.1 Kilowatt Demand

Kilowatt demand is required for billing customers under a commercial or industrial tariff. A kilowatt demand meter must be able to measure demand as defined in the rate tariff. Demand is seldom defined in rate tariffs as a true instantaneous power consumption, but rather as a power consumption sustained for a period of time, usually 15 or 30 minutes. If the instantaneous demand spikes to a high value for a period of time less than the demand interval specified in the tariff, it is ignored.

Electromechanical kilowatt demand meters, as shown in Figure 2.16, employ a drag hand to indicate peak demand. The drag hand is manually reset when the meter is read. Electronic meters, as shown in Figure 2.17, can easily record demand with programmable demand intervals by storing values in registers. These registers can also be cleared when the meter is read.

2.5.2 Kilovar and kVA Demand

Kilovar and kVA demand may be required for particular rate tariffs. If this is the case, electromechanical meters can be used to implement *phase-displaced metering*. Shifting the voltage input by 90° will result in the measurement of vars, while shifting the voltage by the power factor angle [$\cos^{-1}$ (power factor)] will measure volt-amperes.

Figure 2.16 Electromechanical Watthour Meter with Demand Register. (*Photo courtesy of Elster Group GmbH.*)

Figure 2.17 Electronic Watthour Meter with Demand Capability. (*Photo courtesy of Elster Group, GmbH.*)

2.6 PULSE-OPERATED METERS

Pulse-operated meters can be used for both energy and demand measurement. A *pulse initiator* is integrated into the meter, which opens and closes electrical contacts to generate a pulse after a certain amount of energy or power is measured. A register counts the pulses similar to the way a gear-driven register counts the number of revolutions of an electromechanical watthour meter induction disk. The pulses can also be sent to remote locations by way of a communication channel.

2.7 TIME-OF-USE METERS

Time-of-Use (TOU) metering is designed to not only measure how much electricity (demand and energy) was consumed, but *when* it was consumed. Then through specially-developed rate tariffs which price demand and energy at different rates during specified time periods throughout the day, an effective type of load management can be implemented.

At times when utilities would like to reduce their load, such as mid-afternoon, pricing is set higher than at times when there is no desire to reduce demand, such as during off-peak hours. This way, customers can make choices to limit their discretionary use of electricity during prime hours and defer it to the less costly off-peak hours.

TOU tariffs have become increasingly popular in recent years. Microprocessor technology in the meters make the TOU function simple to implement, allowing the implementation of creative rate structures to reduce peak loading while offering customers a financial incentive to help the utility achieve its goal of a flatter load curve.

2.8 SPECIAL METERING

2.8.1 Voltage and Current Metering

Voltage often is measured using a voltmeter and a voltmeter switch, which allows a single meter to measure all three line-to-line voltages or all three line-to-neutral voltages, one at a time. If the voltage exceeds 480 V, VTs are needed to reduce the voltage to a 120-V base for metering.

A VT is a transformer built with a very precise turns ratio. The voltage on the 120 V secondary is measured, and this voltage is multiplied by the turns ratio of the VT to find the primary voltage. VTs should have switches installed in their secondary circuits to disconnect the device for testing. These switches also can be opened when the primary circuit is de-energized to prevent backfeeding from the secondary circuit, which could be very dangerous to maintenance personnel. VT circuits typically are fused on both the primary and secondary sides and the secondaries are grounded for safety reasons. The switch configurations to meter line-to-line and line-to-neutral voltages are shown in Figures 2.18 and 2.19, respectively.

Current often is measured using an ammeter and an ammeter switch, which allows a single meter to measure all three line currents, one at a time. If the current exceeds 600 A, CTs are needed to reduce the current to a 5-A base for metering.

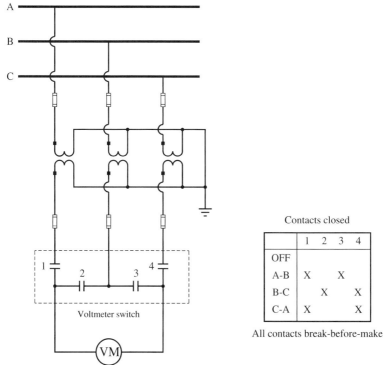

Figure 2.18   Line-to-Line Voltage Metering Using Voltage Transformers

56 CHAPTER 2 INSTRUMENT TRANSFORMERS AND METERING

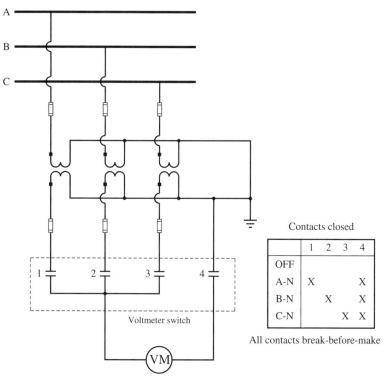

Figure 2.19 Line-to-Neutral Voltage Metering Using Voltage Transformers

Transformer-rated meters are designed for use with CTs and typically have a 20-A rating class, or *CL 20*.

CTs are usually toroidal coils, which form the secondary windings. A high-current conductor passes through the center, or window, of the toroid. This conductor acts as the primary one-turn "winding." CT ratios are stated as the primary current rating followed by a colon then the secondary current rating, which is usually 5 A. A rating of 600:5, read 600 to 5, means that 600 A flowing through the CT window will cause 5 A to flow in the secondary circuit. A designation of "MR" after the CT rating indicates that the CT is of the *multiratio* type, meaning taps are available. CT taps add a great deal of flexibility. If a load grows (or shrinks) considerably, the CT tap can be changed by rewiring terminal blocks in a junction box instead of replacing the CT. Because of their design, CTs easily can be installed around device bushings. Since transformer and power circuit breaker CTs are fit over the bottoms of the bushings inside the tank, they are not easily accessible, and replacing them would be difficult. This adds to the appeal of multiratio CTs. To maximize accuracy, the maximum load in the primary circuit should be close to the CT's primary rating. A 1200:5 CT should not be used in an application where the primary current will never exceed 535 A, but a 1200:600:5 MR CT would work well if set on the 600-A tap.

Dangerously high voltages can develop across the secondary terminals of a CT if the secondary circuit is open circuited. Short-circuiting type (make-before-break)

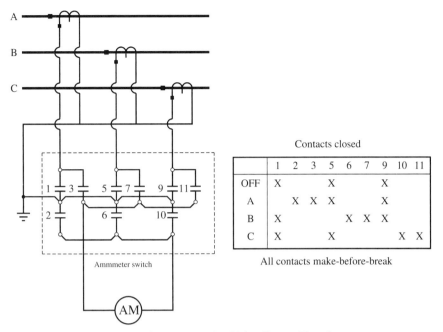

Figure 2.20 Current Metering Using Current Transformers

switches must be used in CT secondary circuits to ensure the circuit is never opened. Protective devices are available to prevent the buildup of very high voltages across CT secondaries, but these devices slightly impact the accuracy of the CT. An ammeter circuit with a selector switch to monitor each of the three phases is shown in Figure 2.20. An unused CT must have its secondary terminals short-circuited with a shorting bar. CT circuits also should be grounded to limit static voltage induction from the high-voltage primary circuit. Since CT circuits must never be open circuited, CTs are never fused.

2.8.2 Var and Q Metering

Reactive power can be measured with a conventional wattmeter and a phase-shifting autotransformer. This transformer, sometimes called a *reactiformer* or *phaseformer*, shifts the phase angle of the potential by 90° (lagging) before supplying it to the wattmeter. Recall that watts = $|V|\,|I|\cos\theta$ and vars = $|V|\,|I|\sin\theta$. Since $\sin\theta = \cos(\theta-90°)$, when the potential phase angle is shifted 90° backwards, the wattmeter actually measures vars instead of watts.

If real power can flow in both directions during differing circumstances, conventional var metering must be used to measure reactive power. But if real power always flows in one direction, which is the case with radial distribution systems, vars can be measured using a specially connected meter called a *Q-meter*. A Q-meter is wired so that the potential fed to the meter lags the actual voltage by 60°. This phase shift can be accomplished without a reactiformer or phaseformer using a

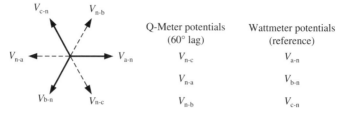

Figure 2.21 Cross-Phasing

technique called *cross-phasing*. The phasor diagram illustrated in Figure 2.21 shows how a three-phase four-wire system is cross-phased to produce a 60° phase shift. Three-wire systems can be cross-phased in a similar fashion.

Next, the operation of the Q-meter must be analyzed. Since the meter is provided a potential that lags the actual voltage by 60°, the torque on the meter will not be proportional to watts or vars, but to a quantity that will be termed Q. The phasor diagram in Figure 2.22 shows the relationship between Q, watts, vars, and volt-amperes.

Phasor SO represents total complex power in volt-amperes. It can be resolved into phasors WO (real power in watts) and VO (reactive power in vars). If point S is projected perpendicularly onto a 60° line drawn through point O, point Q is located. Phasor QO is the quantity measured by the Q-meter.

Since $\triangle OQA$ is a 30°–60°–90° triangle, the magnitude of phasor AO is twice that of phasor QO. Also, $\triangle SWA$ is a 30°–60°–90° triangle, so the distance from point W to point A equals $\sqrt{3}$ times the magnitude of phasor VO.

Because OW, WA, and OA are collinear, $|WA| = |OA| - |OW|$.

Using substitution,

$$\sqrt{3}|VO| = 2|QO| - |OW| \tag{2.7}$$

or

$$\sqrt{3}(\text{vars}) = 2(Q) - \text{watts} \tag{2.8}$$

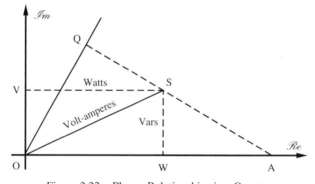

Figure 2.22 Phasor Relationships in a Q-meter

Rearranging terms,

$$\text{vars} = \frac{2Q - \text{watts}}{\sqrt{3}} \qquad (2.9)$$

So, by measuring watts and Q, vars can easily be calculated. However, a physical limitation of the meter must be considered. The torque on the Q-meter is in the forward direction only if the power factor angle lies in the 90° lagging to 30° leading range. Power factor angles outside that range will produce a backward torque on the meter, causing it to run backwards. Since most loads fall in the forward torque range, this limitation is seldom a major issue. If the power factor angle of the load exceeds 30° leading, either the potential must be shifted by less than 60°, or separate leading and lagging meters must be used.

Another issue involving Q metering must be addressed. Since the Q-meter registers positive values for both leading and lagging vars, some means of differentiation must be devised to determine whether the power factor is leading or lagging. A simple comparison of Q and W can make this differentiation. If $Q/W > 0.5$, the power factor is lagging. When $Q/W = 0.5$, the power factor is unity, and $Q/W < 0.5$ indicates a leading power factor.

2.8.3 Compensating Metering

It may not be practical to install metering at the contractual billing location because of physical or economic reasons. When such cases arise, the metering can be installed at a more feasible location, and the meter readings can be compensated for the losses that occur between the metering location and the contractual billing location. This method of metering is called *compensating metering*.

Two cases where compensating metering is frequently used are to account for transformer losses and line losses. Perhaps an industrial customer owns its service transformer, so the losses in that transformer should be borne by the customer. But metering the service on the high-voltage side of the transformer, which would properly account for the losses, would result in the use of very expensive instrument transformers. Instead, the metering can be done on the low-voltage side of the transformer for much less cost, and the losses incurred in the transformer could be calculated by the metering system. Metering the flow on a transmission line at the midpoint instead of at a terminal could be done in a similar fashion.

The calculation of the losses in a transformer or transmission line requires knowledge of the voltage and current at one terminal of the device and a detailed circuit model of the device. The voltage and current are measured by metering equipment, but the circuit model becomes an interesting challenge. Both transformers and transmission lines have classical circuit equivalents, shown in Figures 2.23 and 2.24, but the lumped parameters in these models are represented by numeric values that apply at one specific set of conditions.

The resistance, reactance, conductance, and susceptance values of these models vary with frequency, temperature, and possibly other parameters too. Although the frequency can be considered to be relatively fixed (if not, the voltage and current measurements are subject to error too), temperature can vary greatly. So the question

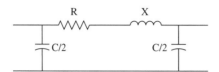

Figure 2.23 Transmission Line Lumped Parameter Circuit Model

becomes: How much variation in the electrical parameters is caused by variation in temperature? This question must be answered for each individual application in order to determine how accurate the loss compensation will be. In general, an acceptable level of accuracy can be achieved by determining the circuit models for the component for which losses are to be calculated, then program the meters to adjust their measured values by the calculated loss.

2.8.4 Totalizing Metering

Totalization is the summation of two or more meters to give a single-metered value. Consider a utility system with five tie lines to neighboring utilities. If watt and var metering was installed on every line, those five meters could be totalized into a single quantity, which would indicate the net interchange from the utility's system in real time. A common method of totalization is to series current inputs to a single meter. Another method utilizes *pulse initiators* which can be accumulated by a *pulse recorder*.

2.8.5 Pulse Recorders

A pulse initiator is a component within a meter which generates an electronic pulse for a discrete amount of a metered quantity. These pulses can be transmitted over a communication channel to a remote device called a *pulse recorder*, which counts the pulses. When using pulse recorders, several issues must be addressed. Since received pulses are being counted, care must be taken to assure that accumulation registers are large enough to hold the maximum number of pulses that could be received during any demand interval. Also, if pulses are being received from multiple sources, two pulses could arrive at the recorder at almost the same time, making the pulses difficult to differentiate, possibly resulting in an undercount. And the amplitude and duration of the pulses sent by the pulse initiators must be compatible with the pulse recorder.

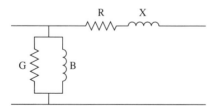

Figure 2.24 Transformer Circuit Model (Based on Per-Unit Values)

2.9 DIGITAL METERING

While traditional analog meters are still in widespread use, solid-state digital metering is gaining increasing popularity, acceptance, and capability. Solid-state meters are very accurate and tend to remain accurately calibrated longer than their analog counterparts. With electronic metering, it is possible to combine many different functions into a single device, saving space on the equipment panel, simplifying installation and wiring, and reducing cost. An example would be a single meter capable of measuring volts, amperes, watts, vars, power factor, watthours, and varhours, packaged in a durable enclosure.

Electronic metering can be interfaced with a computer to allow monitoring and analysis of the measured data, including remote reading and outage notification. Because of various technological developments over the years, it is now common to have metered quantities sent electronically to remote locations via telemetry. There are many means of implementing telemetry, but most require the use of transducers and analog-to-digital converters. A digital signal representing the analog measurements from the metering equipment is sent over a communication circuit, typically telephone line or fiberoptic cable, to a remote location. Other digital data, such as circuit breaker and switch status, also can be telemetered. At large industrial facilities, there can be a considerable distance from the utility service point to the plant control room, so telemetry can be implemented. In more complex installations, a *system control and data acquisition* (SCADA) system can be installed, which allows both monitoring and control of the electrical system. SCADA systems are especially useful in loop and network systems, where complex switching is done routinely.

2.10 SMART METERS

The term *smart meter* is used to describe a metering technology that combines traditional electronic metering functionality with real-time or near-real-time sensors to indicate service interruptions and particular power quality parameters. Smart meters utilize a bidirectional communication interface which allows data exchange between the meter and a central location, usually a utility billing center. The communication is typically done wirelessly, often utilizing IEEE standard 802.15.4, commonly known as ZigBee®, which creates a low-cost robust mesh network using low-power digital radios.

Such data exchange capability can be used for a variety of applications. Better real- or near-real-time load knowledge can lead to a more economical dispatch of generation resources, thereby reducing the cost of operating the power system. When used in conjunction with real-time pricing of electricity, pricing signals can be used to alert customers when electricity costs are substantially higher (or lower) than average, again allowing for customers to potentially reduce their electric bill. When smart meters are interfaced with smart appliances, sophisticated demand-side management methods can be implemented. Such methods could benefit both the customer by lowering bills and the utility by better managing the load profile which will reduce operating expenses.

Bidirectional communication makes the smart meter a valuable addition to a *smart grid*, which is a distribution infrastructure capable of adapting to changing conditions in order to improve performance. Outage restoration, for example, could be vastly improved based on data received from smart meters throughout the system. By specific knowledge of which meters are experiencing a service interruption, utilities can better respond to the service interruption, allowing faster service restoration.

AMR is an older technology that allows data collected by the meter to be sent to the utility using a one-way communication link. While remote meter reading represents a major cost savings for the utility, the benefits afforded by two-way communication far surpass those enabled by one-way communication only. Smart meters, of course, can be remotely read as can AMR-capable meters, but the bidirectional communication ability creates an advanced metering infrastructure (AMI) capable device.

SUMMARY

Instrument transformers are precision devices to reduce high voltages or currents to the levels that can be handled by metering or relaying equipment. *VTs*, or *PTs*, typically have a rated secondary voltage of 120 V, while CTs usually have a secondary rated at 5 A.

Correction factors can be specified to adjust for errors in ratio and phase angle for instrument transformers. The accuracy of an instrument transformer is a function of *burden*, or the load placed in the secondary circuit. *Accuracy classes* quantify the accuracy of an instrument transformer for a given burden.

The ability to accurately meter electrical quantities is necessary to safely and reliably operate a power system. Utilities need to meter certain electrical quantities to properly determine customer bills. A great deal of specialized terminology has developed over the years in the metering area, and knowledge of these key terms is important when working with metering systems.

Watthour metering measures energy usage. Single-stator meters are used to meter single-phase services, while multi-stator meters measure consumption on three-phase services. An electromechanical watthour meter is essentially an induction motor whose torque is proportional to the power passing through it. Counting disk revolutions with a gear-driven register effectively integrates the measured power with respect to time, giving energy usage.

Demand metering records a maximum value since its register was reset. Commonly metered demand quantities include watt demand, var demand, and kVA demand. Commercial and industrial rate tariffs require metering of demand quantities. TOU metering measures how much energy, or demand, occurred during specific time periods. By applying different rates to different time periods, utilities can influence the shape of their load curve while providing customers with a mechanism for reducing their monthly bill.

Metering voltage and current is very important for proper system operation. Instrument transformers are frequently required, since the voltages and currents requiring metering often exceed the physical capabilities of the metering equipment. Voltmeter and ammeter switches allow one meter to monitor all three phases, one at a time. Although modern metering is digital, analog displays offer advantages to operators by showing clearly relative magnitude changes and rates of change. Reactive power can me metered with a wattmeter, provided the voltage provided to the meter is retarded by 90° using a device such as a *reactiformer* or *phaseformer*, which are special types of autotransformers. Reactive power can also be determined by measuring a quantity called "Q" by shifting the input voltage to the meter by 60° using a technique called *cross-phasing*, which eliminates the need for an auxiliary autotransformer.

Compensating metering adjusts for transformer losses, allowing the metering to be installed on the low-voltage side of a transformer for economy, yet be able to meter high-side quantities. *Totalizing metering* sums two or more submeters, providing a single value.

FOR FURTHER READING

Code for Electricity Metering, American National Standards Institute, ANSI Standard C12.1, 2008.
Handbook for Electricity Metering, 11th Edition, Edison Electric Institute 2014, ISBN 978-0-931-03259-2.
IEEE Standard Requirements for Instrument Transformers, IEEE Standard C57.13, 1993.

QUESTIONS

1. Describe the two general types of metering technology, and explain why it is still important to study and understand the older technology.
2. Describe how an electromechanical watthour meter works.
3. State Blondel's theorem.
4. What is the reason for metering kilowatt demand?
5. Three kilowatt demand meters are totalized with a fourth meter. At the end of the month, each submeter reads 2500 kW. The totalizing meter reads 6500 kW. How can this be?
6. Name an advantage of TOU metering.
7. Why are voltmeter switch contacts break-before-make, but ammeter switch contacts are make-before-break?
8. Describe the principle behind compensating metering. Show how the ratio of Q to W in a Q-meter circuit can determine whether the power factor is lagging, leading, or unity.
9. What is the most common voltage rating of a VT secondary and the common current rating of a CT secondary?
10. Why must CT secondary circuits never be open circuited?
11. What measures are taken to prevent open circuits in CT secondary circuits?
12. What happens if a VT secondary circuit becomes short-circuited?
13. What measures are taken to prevent short-circuits in VT secondary circuits?
14. Show how the ratio of Q to W in a Q-meter circuit can determine whether the power factor is lagging, leading, or unity.
15. How can a wattmeter be used to measure quantities other than watts?

PROBLEMS

1. Draw a phasor diagram to show how a three-phase three-wire system can be cross-phased to produce a 60° phase shift.

2. A three-phase four-wire 480Y / 277 V service is sized to carry 375 A per phase. Sketch a metering circuit including instrument transformers to provide line-to-neutral voltage, phase current, and kilowatthour metering.

3. Sketch a Q-meter connection for the service in Problem 2.

4. Design a metering circuit to measure line-to-neutral voltage and line currents on a three-phase, four-wire, 4.16 kV switchgear main circuit breaker rated at 2000 A.

5. Sketch a Q-meter connection to a three-phase, four-wire system including instrument transformer wiring and polarities.

6. A 7200:120 V VT and a 2000:5 A current transformer are used in a metering circuit. They have measured ratios of 59.87:1 and 400.32:1, respectively. Calculate the combined ratio correction factor for the metering circuit.

7. In the same metering circuit described in Problem 6, the phase angle of the VT secondary leads the primary by 12 minutes of angle and the phase angle of the CT secondary lags the primary by 17 minutes of angle. Calculate the combined phase angle correction factor for the metering circuit.

8. Referring to Problems 6 and 7, if the secondary voltage in the metering circuit is 123.65 V and the secondary current is 3.47 A, find the real power delivered by the primary circuit.

9. The induction disk of a CL 200 kilowatthour meter with an R_r of 13 8/9 and a K_h of 7.2 revolves 1200 times. How much energy passed through the meter?

10. The induction disk of a CL 10 kilowatthour meter with an R_r of 41 2/3 and a K_h of 4.8 revolves 1200 times. It is supplied by 200:5 CTs. How much energy passed through the meter?

CHAPTER 3

TRANSFORMER CONNECTIONS

OBJECTIVES

- Understand the rationale used to select industrial system operating voltages
- Know what is meant by polarity and know how to determine the polarity of a single-phase transformer
- Recognize the basic three-phase, two-phase, and six-phase distribution transformer connections, and know the uses and the advantages and disadvantages of each
- Be able to analyze a polyphase transformer connection using phasor analysis
- Identify the causes of ferroresonance and how to prevent it

3.1 VOLTAGE SELECTION

Proper system voltage selection is very important when designing an industrial power system because voltage selection has a direct influence on the cost, versatility, performance, and expandability of the electrical system. The primary factors determining voltage selection are the size of the load, the distance from the source to the load, and the operating voltage ratings of the utilization equipment.

Load magnitude influences the plant's main distribution voltage. Small industrial facilities with demands less than 5–7 MW can be supplied by 480-V transformers connected to the utility distribution system. Facilities with demands more than 7–10 MW are best supplied directly from the utility distribution system that is typically between 12.47 and 14.4 kV. If a utility distribution feeder is not available, a voltage transformation from a transmission voltage may be necessary. Economics typically determine which main distribution voltage to select. The most commonly used main distribution voltages in industrial facilities are 13.8, 6.9, 4.16, and 2.4 kV.

The 13.8 kV option is attractive because the switchgear and cable have a low cost per kVA. However, smaller motors cannot be operated at this voltage so additional transformations are required. These transformers add not only initial cost but also maintenance cost and operating complexity to the system.

The 6.9 kV option is attractive for large- and medium-sized motors because fewer voltage transformations are needed. Losses at 6.9 kV, however, are considerably

---

Industrial Power Distribution, Second Edition. Ralph E. Fehr, III.
© 2016 The Institute of Electrical and Electronics Engineers, Inc. Published 2016 by John Wiley & Sons, Inc.

higher than at 13.8 kV. Additionally, as the voltage levels decrease, the load currents increase proportionately, necessitating larger conductors.

The 4.16 kV option offers the most flexibility for supplying motors, as horsepower ratings from about 150 through 6000 are available in this voltage class. It should be noted that if many large motors are required, the 4.16 kV system could lead to voltage problems during motor starting due to the large starting currents.

The 2.4 kV option was popular years ago and is still utilized in some plants although improved insulation technologies have led to the replacement of many 2.4 kV systems with higher-voltage alternatives.

The distance from the source to the load becomes increasingly important as the operating voltage levels decrease. For example, when the voltage drop on a 240-V circuit becomes excessive, a 480-V circuit may be a good solution. In addition to voltage drop, losses increase quadratically as voltage decreases. When conductor sizes become too large at a given voltage level, an increase in voltage level is necessary. A 480-V system shows many economical advantages over lower-voltage systems. Motors rated at 460 V for use on a 480-V system are readily available in sizes up to about 200 hp. Actually, 600-V systems can show even more financial incentives than 480-V systems, but the 575-V equipment utilized on a 600-V system is not as common as 460-V class equipment, particularly in the United States.

Motors are not the only devices that pose operating voltage limitations. Lighting equipment, for example, can be rated at 120, 208, 240, 277, or 480 V. The cost of the equipment at each operating voltage can vary significantly. Special loads such as furnaces and solid-state drives may have specific voltage requirements. Regardless of the installation, 120-V single phase power is almost always required. Small drytype transformers can provide 120-V single phase from just about any higher voltage source. A detailed economic analysis must be done prior to selecting voltages in an industrial facility to assure economy, versatility, and expandability.

3.2 IDEAL TRANSFORMER MODEL

Transformers are best understood by first considering and comprehending the *ideal transformer*. An ideal transformer is simply two coils of wire that are magnetically coupled. Being "ideal," the magnetic coupling between the coils is perfect. In other words, all the magnetic flux produced by one winding links the other winding, so there is no *stray flux*. The *self-inductance* and the *resistance* of each winding are also disregarded. Only the *mutual inductance* between the windings is considered.

Per Faraday's law, when a time-varying magnetic flux cuts a conductor formed into a coil with N turns, a voltage is induced across the coil:

$$V = -N\frac{d\Phi}{dt} \tag{3.1}$$

Lenz's law determines the polarity of that voltage. The negative sign in Faraday's law indicates that the induced voltage is developed with a polarity that would produce a current (if the coil being considered is part of a closed circuit) that would produce a magnetic flux that would oppose the original flux. Figure 3.1 shows, by the right-hand rule, that the current I produces a flux that opposes the original flux. This stipulation is necessary for the conservation of energy.

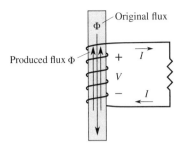

Figure 3.1   Lenz's Law

The time-varying flux is produced by a time-varying current flowing in one of the coils. By convention, the coil across which a voltage is applied is called the *primary* winding, and the coil across which a voltage is induced is called the *secondary* winding. Since it may not be known which winding will have a voltage applied across it, the use of *primary* and *secondary* can be risky. Transformer manufacturers always designate the windings "high voltage" and "low voltage" to avoid this issue.

In an ideal transformer, the ratio of the number of turns on the high-voltage winding (N_H) to the number of turns on the low-voltage winding (N_L) defines the *turns ratio* (n) for the transformer. The turns ratio also describes the ratio of the voltages across the two windings so is also referred to as the *voltage ratio*. If each winding in the ideal transformer is part of a closed circuit, the ratio of the currents that flow through each winding is inversely proportional to the turns ratio. These ratios are shown in Figure 3.2.

Since the ideal transformer is lossless, the power handled by the high-voltage winding must equal the power handled by the low-voltage winding:

$$V_H I_H = V_L I_L \tag{3.2}$$

Also, an impedance put across the low-voltage windings appears to be much higher when viewed from the high-voltage circuit. This can be seen by writing the high-voltage circuit quantities in terms of the low-voltage circuit quantities:

$$V_H = n V_L \tag{3.3}$$

$$I_H = \frac{1}{n} I_L \tag{3.4}$$

Dividing Eq. (3.3) by Eq. (3.4) gives

$$Z_H = n^2 Z_L \tag{3.5}$$

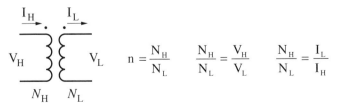

Figure 3.2   The Ideal Transformer

3.3 TRANSFORMER FUNDAMENTALS

Although the ideal transformer model is useful for understanding transformers at a superficial level, the simplifying assumptions made in developing the ideal transformer model must be addressed to fully comprehend the operation of a practical transformer. The windings have resistance which is dependent on the type of material used to make the winding conductor. Both copper and aluminum conductors can be used to wind transformers. It should be noted that copper has significantly more mechanical strength than aluminum so tends to perform better when the windings are subjected to short-circuit forces. Copper, also having a higher conductivity than aluminum, leads to a small conductor cross-section, which can be a manufacturing advantage. The winding conductors, rectangular in cross-section, are typically insulated with paper and immersed in oil. The oil, in addition to enhancing heat transfer from the winding conductors to the outside environment, helps keep oxygen away from the paper, thereby decreasing its rate of oxidation. Oxidation of the paper insulation will eventually lead to a failure of the insulation system. High temperatures accelerate the rate of oxidation, which is the main reason why operating temperature of the transformer is of major concern. Figure 3.3 shows winding conductor that is continuously transposed to expose each conductor to approximately the same magnetic flux (above) and standard single conductor (below).

The resistance of each winding is determined by the resistance per unit length of the winding conductor times the length of the conductor used to form the winding. Each winding also has a self-inductance (L), which is a function of the magnetic permeability (μ), the number of turns in the winding (N), the cross-sectional area of the coil (A), and height of the winding assembly (h):

$$L = \frac{\mu N^2 A}{h} \quad (3.6)$$

The self-inductance of the winding will produce a voltage drop as a time-varying current (I) passes through the coil:

$$V = L \frac{dI}{dt} \quad (3.7)$$

The characteristics of the magnetic core must also be considered. The nonlinear B-H characteristics of the core steel give rise to a hysteresis curve, shown in Figure 3.4, the area inside of which represents a real power loss.

Figure 3.3 Paper-Insulated Transformer Winding Conductor

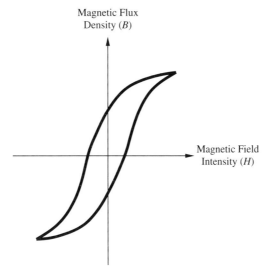

Figure 3.4 Hysteresis Curve for Transformer Core Steel

Electrical currents induced into the core steel by the currents flowing in the windings also produce real power losses. These induced currents, called *eddy currents*, are induced in a direction normal to the magnetic flux. A laminated core reduces eddy currents by placing electrical insulation between the sheets of steel making up the laminated core, thus breaking up the path over which the eddy currents would flow. The eddy currents are limited to a circulation within each lamination sheet. Figure 3.5 shows a three-phase core assembly and Figure 3.6 shows details of the core laminations, with a dime to illustrate the scale of the lamination thickness.

Eddy currents dissipate power in the core steel per Eq. (3.8):

$$P = \frac{\pi^2 B^2 f^2 t^2}{6\rho d} \tag{3.8}$$

where

P = power dissipation in watts/kg of core steel
B = maximum magnetic flux density in teslas
f = frequency in hertz
t = thickness of lamination sheet in meters
ρ = resistivity in ohm-meters
d = specific density of core steel in kg/m³.

Since the permeability of the core is finite, a current is required to maintain magnetic flux in the core. This current lags the induced voltage by 90°. Therefore, a shunt inductive reactance is used to represent core magnetization.

70 CHAPTER 3 TRANSFORMER CONNECTIONS

Figure 3.5 Three-Phase Core Assembly. (*Photo courtesy of Pennsylvania Transformer Technology, Inc.*)

Figure 3.6 Laminated Steel Core

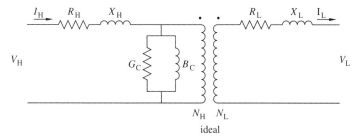

Figure 3.7 Transformer Circuit Model

3.4 TRANSFORMER CIRCUIT MODEL

The characteristics that were disregarded in the ideal transformer model can be accounted for in a lumped parameter circuit model to realistically represent an actual transformer. Figure 3.7 shows a circuit model representing an actual transformer.

The resistances R_H and R_L represent the resistances of the high-voltage and the low-voltage windings, respectively. The inductive reactances X_H and X_L represent the self-inductances of the high-voltage and the low-voltage windings, respectively, multiplied by the radian frequency:

$$X_H = \omega L_H = 2\pi f L_H \qquad (3.9)$$
$$X_L = \omega L_L = 2\pi f L_L \qquad (3.10)$$

G_C paralleled with B_C make up the core branch of the model. G_C is the shunt conductance representing hysteresis and eddy currents. B_C is the shunt susceptance which represents the magnetization current for the core. The total current flowing through both G_C and B_C is the no-load current of the transformer, and can be measured using the open circuit test, which measures I_H with the low-voltage terminals open circuited.

The ideal transformer provides the turns ratio for the model. If the per-unit system is used, the effective turns ratio becomes 1:1, so the ideal transformer can be omitted from the circuit model, resulting in a single circuit instead of two separate electrical circuits.

3.5 SINGLE-PHASE TRANSFORMER CONNECTIONS

An important consideration when connecting transformers is to observe the *polarity* of the transformers. Polarity is a description of the direction that the winding conductor is wound around the core. The right-hand rule shows the direction of the magnetic flux produced when current flows through a coil wound in a particular direction. In a transformer with two windings, two currents can flow, one in each winding. Each of these currents will produce a magnetic flux. If the two fluxes are oriented in the same direction, they are *additive*. If the directions of the fluxes are in opposition, they are

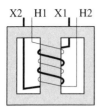

Figure 3.8   Additive Polarity

subtractive. The terms additive and subtractive are used to describe the polarity of the transformer.

Figures 3.8 and 3.9 illustrate additive and subtractive polarities for a single-phase transformer. The leads are labeled corresponding to the bushings on which they terminate.

A simple test for polarity involves shorting bushing H1 to bushing X1, then applying an AC voltage across either winding (either bushings H1 and H2 or X1 and X2). A voltmeter connected across bushings H2 and X2 will read the *sum* of the applied voltage (HV winding voltage) and the applied voltage divided by the turns ratio (LV winding voltage) if the transformer is *additive* polarity, and the *difference* of those two quantities if the transformer is *subtractive* polarity. When performing a polarity test, a sufficiently low voltage must be applied as to not exceed either winding's voltage rating.

A simple means of visually determining additive versus subtractive polarity requires locating the position of bushings H1 and X1. If they are adjacent to other, the transformer is subtractive polarity. If their positions are diagonal with respect to each other, the transformer is additive polarity. Looking down on the transformers from above, the visual test for polarity by identifying bushing locations is illustrated in Figure 3.10.

IEEE standard C57.12.00 stipulates that single-phase power transformers rated 200 kVA and below and having high-voltage winding ratings of 8660 V and below shall have additive polarity, and all others shall have subtractive polarity. Since similarly phased bushings are opposed in a subtractive polarity transformer, the voltage stresses at the winding ends of the bushings inside the transformer tank are minimized. This is a key advantage of a subtractive polarity transformer.

A single-phase transformer with a particular polarity can be replaced with a transformer of opposite polarity as long as each lead is connected to the same *bushing number* on the new transformer as it was on the old transformer. This will result in two of the leads changing relative position as to where they are connected

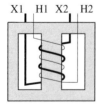

Figure 3.9   Subtractive Polarity

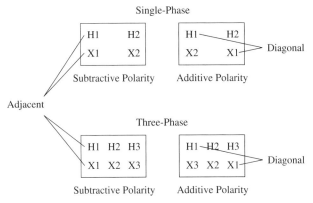

Figure 3.10 Visual Test for Transformer Polarity

to the new transformer. Electrically, both connections shown in Figure 3.11 are equivalent.

3.6 THREE-PHASE TRANSFORMER CONNECTIONS

Three-phase voltage transformations can be accomplished by using *three-phase transformers*, which are single devices with all windings constructed on a single iron core. Three-phase voltage transformations also can be accomplished by using three *single-phase transformers* that are connected externally to form a *three-phase bank*. While three-phase devices are usually the more cost-effective option, the single-phase option provides more versatility and can be attractive from a reliability and maintenance standpoint. If several identical transformers are needed at one location, the single-phase option can include the purchase of a spare unit to reduce outage time in the event of a failure. This practice often is seen with critical autotransformer banks and generator step-up transformers because loss of the transformer for an extended period has very significant impacts. The connections discussed in this chapter will be implemented using single-phase units.

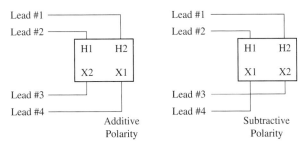

Figure 3.11 Replacing an Additive Polarity Transformer with a Subtractive Polarity Transformer

When connecting single-phase transformers to form a three-phase bank, the winding *polarities* must be carefully observed. Polarity is indicated using a *dot convention*. Current flowing into a dot on the primary winding will induce a current flowing out of the dot on the corresponding secondary winding. Depending on how the windings are connected to the bushings, the polarities can be *additive* or *subtractive*.

The two most commonly used three-phase winding configurations are *delta* and *wye*, named after the Greek and English letter that each resembles. In a delta configuration, the three windings are connected end-to-end to form a closed path. A phase is connected to each corner of the delta. Although delta windings are often operated ungrounded, a leg of the delta can be center tapped and grounded, or a corner of the delta can be grounded. In a wye configuration, one end of each of the three windings is connected to form a *neutral*. A phase is connected to the other end of the three windings. The neutral is usually grounded.

The following sections describe three-phase transformers which utilize the delta and wye connections. Also discussed are three-phase transformers using the open-delta and open-wye connections, where one of the single-phase transformers making up the three-phase bank is omitted. The leg of the transformer with the missing transformer is referred to as the *phantom leg*.

3.6.1 Delta–Delta

Delta–delta transformers, as shown in Figure 3.12, often are used to supply loads that are primarily three phase but may have a small single-phase component.

The three-phase load is typically motor load while the single-phase component is often lighting and low-voltage power. The single-phase load can be fed by grounding a center tap on one of the legs of the delta secondary, then connecting the single-phase load between one of the phases on the grounded leg and this grounded neutral.

Figure 3.13 shows a delta–delta transformer connection. The connection diagram on the left shows how a delta–delta connection can be made, either with three single-phase transformers or with one three-phase transformer. The dashed lines indicate the transformer outlines. The three single-phase transformer implementation can be seen by disregarding the outer dashed outline and the bushing labels shown at that outline, and concentrating on the three smaller (single-phase transformer) outlines. The bushings of the single-phase transformers are connected by external jumpers as

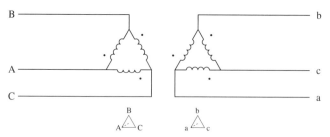

Figure 3.12   Delta–Delta Transformer

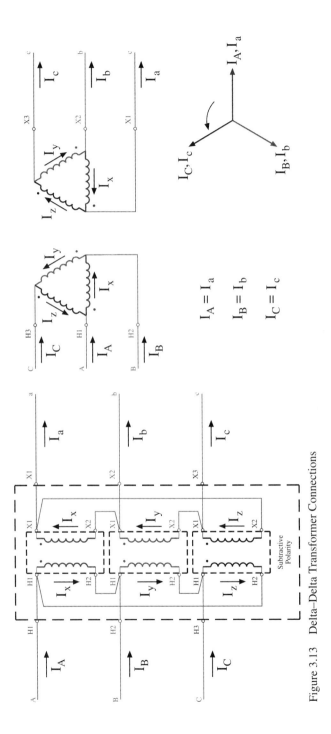

Figure 3.13 Delta–Delta Transformer Connections

shown to accomplish the delta–delta connection. In the case of the one three-phase transformer implementation, the three inner outlines are disregarded, and the jumpers between the windings are made inside the transformer tank. The six bushings on the three-phase transformer outline are available for connection. The schematic diagram at the upper right is perhaps easier to analyze, as the delta connections can clearly be seen. The phasor diagram at the lower right shows the geometric relationships between the high-voltage circuit and low-voltage circuit currents, and the equations at the bottom center show those relationships mathematically.

As the loading on a delta–delta transformer becomes unbalanced, high currents can circulate in the delta windings leading to a voltage imbalance. Balanced loading requires the selection of three transformers with equal voltage ratios and identical impedances. Also, the amount of single-phase load should be kept low because the center-tapped transformer must supply most of the single-phase load. As the single-phase load is increased, the center-tapped transformer will increase its loading more than the other two transformers and will eventually overload.

If one of the single-phase transformers in the delta–delta bank fails, the bank can be operated with only two transformers forming an open delta configuration. The kVA rating of the bank is reduced, but three-phase power is still supplied to the load. See Section 3.6.5 for a thorough discussion of this topic.

3.6.2 Wye–Wye

Wye–wye transformers, as shown in Figure 3.14, can serve both three-phase and single-phase loads. The single-phase load should be distributed as evenly as possible between each of the three phases and neutral.

Figure 3.15 illustrates the wye–wye connection, either as three single-phase transformers or as a single three-phase unit. Both bushing labels and polarity dots are shown.

One problem inherent to wye–wye transformers is the propagation of third-harmonic currents and voltages. These harmonics can cause interference in nearby communication circuits as well as other assorted power quality problems. Another problem is that the possibility exists for resonance to occur between the shunt

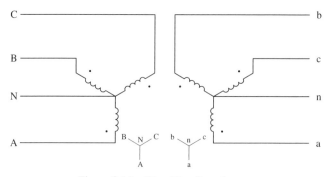

Figure 3.14   Wye–Wye Transformer

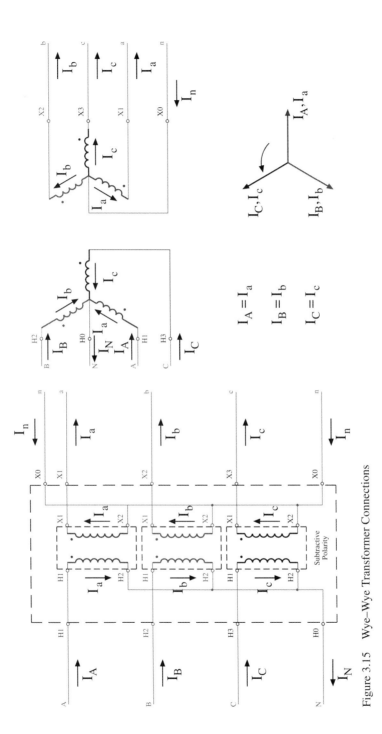

Figure 3.15 Wye–Wye Transformer Connections

capacitance of the circuits connected to the transformer and the magnetizing susceptance of the transformer, especially if the circuits include insulated cable. Because of these problems, wye–wye transformers must be specified and implemented carefully. Adding a third (tertiary) winding connected in delta alleviates many of the concerns mentioned.

3.6.3 Delta–Wye

The delta–wye connection is the most commonly used three-phase transformer connection. The wye-connected secondary allows single-phase load to be distributed among the three phases to neutral instead of being placed all on one winding as with a four-wire delta secondary. This helps keep the phase loading on the transformer balanced and is especially important when the amount of single-phase load becomes large. The stable neutral point also provides a good ground location to allow critical damping of the system to prevent voltage oscillations.

If one of the single-phase transformers in the delta–wye bank fails, the entire bank becomes inoperative. Also, since the delta–wye transformer introduces a 30° phase shift from primary to secondary as can be seen by the phasing symbols in Figure 3.16, it cannot be paralleled with delta–delta and wye–wye transformers that produce no phase shift.

Figure 3.17 illustrates the delta–wye connection, either as three single-phase transformers or as a single three-phase unit. Both bushing labels and polarity dots are shown.

Analyzing the delta–wye transformer illustrates many important concepts regarding the operation of polyphase transformers. The analysis can be done on either a voltage or a current basis. Since voltage (potential difference or the subtraction of two phasor quantities) is rather abstract and difficult to visualize, current (or the flow of charge) will be used as the basis for analysis, since current is easy to conceptualize.

The currents flowing in the windings of a delta–wye transformer are shown in Figure 3.18. Note that the arrows indicate instantaneous directions of the AC current and are consistent with the dot convention.

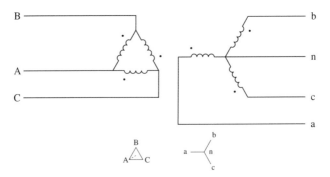

Figure 3.16 Delta–Wye Transformer

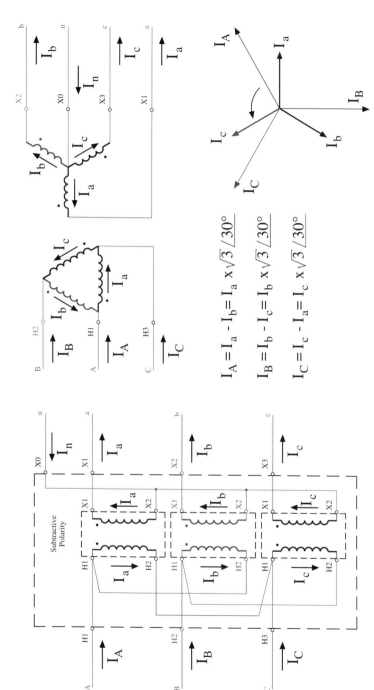

Figure 3.17 Delta–Wye Transformer Connections

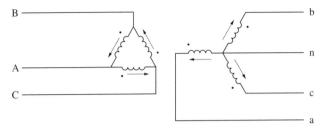

Figure 3.18 Delta and Wye Windings

The analysis must begin in one of the two electric circuits, either the delta-connected high-voltage circuit or the wye-connected low-voltage circuit. Since current is being used as the basis for analysis, the wye-connected circuit is selected as the starting point, since in a wye-connected circuit, the line currents (leaving the transformer) and the phase currents (flowing in the transformer windings) are equal. This relationship between line and phase currents simplifies the analysis.

The analysis starts by labeling all line and phase currents. This is shown in Figure 3.19.

Note that lower-case subscripts indicate line currents in the low-voltage circuit, and upper-case subscripts indicate line currents in the high-voltage circuit. In the low-voltage circuit, the phase currents are identical to the corresponding line currents, so they also are labeled I_a, I_b, and I_c. When the transformer windings are drawn, a particular high-voltage winding corresponds to the low-voltage winding drawn parallel to it. In other words, a high-voltage winding and a low-voltage winding that are drawn parallel to each other constitute a single-phase transformer or two windings on the same leg of the magnetic core of a three-phase transformer.

The high-voltage phase current corresponding to I_a is labeled $I_{a'}$. The direction of $I_{a'}$ relative to that of I_a must honor the dot convention. The magnitude of $I_{a'}$ relative to I_a is the inverse of the transformer turns ratio "n," or

$$I_{a'} = \frac{1}{n} I_a \qquad (3.11)$$

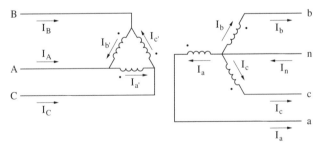

Figure 3.19 Delta–Wye Transformer with Currents Labeled

3.6 THREE-PHASE TRANSFORMER CONNECTIONS

When analyzing a transformer using per-unit, $n = 1$ so Eq. (3.11) becomes

$$I_{a'} = I_a \tag{3.12}$$

So,

$$I_{a'} = I_a(\text{per-unit}) \tag{3.13}$$
$$I_{b'} = I_b(\text{per-unit}) \tag{3.14}$$
$$I_{c'} = I_c(\text{per-unit}) \tag{3.15}$$

Next, Kirchhoff's current law can be applied to each node of the delta:

$$I_A = I_{a'} - I_{b'} = I_a - I_b \tag{3.16}$$
$$I_B = I_{b'} - I_{c'} = I_b - I_c \tag{3.17}$$
$$I_C = I_{c'} - I_{a'} = I_c - I_a \tag{3.18}$$

Equations (3.16) through (3.18) express the high-voltage circuit line currents in terms of the low-voltage circuit line currents. At this point, numerical values can be substituted for I_a, I_b, and I_c. Keeping in mind that I_a, I_b, and I_c represent a balanced set of phasors, arbitrary per-unit values are selected to represent a-b-c phase sequencing:

$$I_a = 1\,\underline{/0°} \tag{3.19}$$
$$I_b = 1\,\underline{/240°} \tag{3.20}$$
$$I_c = 1\,\underline{/120°} \tag{3.21}$$

A positive phase sequencing (a-b-c) must be used, since the IEEE standards for power transformers (the IEEE C57 series) are based on positive phase sequencing. Substituting Eqs. (3.19) through (3.21) into Eqs. (3.16) through (3.18),

$$I_A = 1\,\underline{/0°} - 1\,\underline{/240°} = \sqrt{3}\,\underline{/30°} \tag{3.22}$$
$$I_B = 1\,\underline{/240°} - 1\,\underline{/120°} = \sqrt{3}\,\underline{/270°} \tag{3.23}$$
$$I_C = 1\,\underline{/120°} - 1\,\underline{/0°} = \sqrt{3}\,\underline{/150°} \tag{3.24}$$

Comparing I_a to I_A, a $\sqrt{3}$ magnitude difference and a 30° angular difference are apparent. IEEE Std. C57.12.00 defines the direction in which the phasor angles shall change from one electrical circuit to the other. In a *standard* delta–wye (or wye–delta) transformer, the positive-sequence currents and voltages on the high-voltage side *lead* the positive-sequence currents and voltages on the low-voltage side by 30°. When the high-voltage phasors *lag* the low-voltage phasors, the connection is considered to be *nonstandard*. Sometimes nonstandard connections are necessary to match the phasings on two different systems that must be electrically tied, but normally, standard connections are specified.

Note that the convention to determine a *standard* connection requires that the *high-voltage* phasors lead the *low-voltage* phasors by 30°. No reference is made to *primary* or *secondary*. The primary windings of a transformer are those windings to which voltage is applied. The secondary windings have an induced voltage impressed across them. Usually, the primary windings are the high-voltage windings, but this is not always the case. A good example of an exception is a generator step-up transformer.

82 CHAPTER 3 TRANSFORMER CONNECTIONS

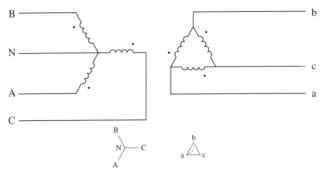

Figure 3.20 Wye–Delta Transformer

3.6.4 Wye–Delta

The wye–delta transformer shown in Figure 3.20 is sometimes used to provide a neutral on a three-wire system (see section 3.10.1) but also can serve load from its secondary.

The primary wye windings are typically grounded. If the secondary is a four-wire delta, the fourth wire originating at a center tap on one of the legs of the delta is grounded.

Figure 3.21 illustrates the wye–delta connection, either as three single-phase transformers or as a single three-phase unit. Both bushing labels and polarity dots are shown.

3.6.5 Open Delta–Open Delta

Two single-phase transformers can be connected open delta–open delta to serve three-phase load, as shown in Figure 3.22.

The connections to configure two single-phase transformers into an open delta–open delta bank are shown in Figure 3.23.

This configuration can be used advantageously if a small load is expected to grow substantially at some point in the future. The initial load can be served by the open delta–open delta bank, and when the load growth occurs, a third transformer can be added converting the bank to delta–delta.

When an open delta–open delta bank is installed, the kVA rating of the bank is not simply the sum of the kVA ratings of the single-phase transformers making up the bank, as is the case with a delta–delta bank. This is because the power factors at which the single-phase transformers operate are substantially different from the power factor of the load.

Consider a delta–delta transformer bank feeding a delta-connected resistive load of $1 \underline{/0°}$ per leg as shown in Figure 3.24.

Assume $V_{an} = 1 \underline{/0°}$ and positive sequence phase sequencing. This leads to $V_{bn} = 1 \underline{/240°}$ and $V_{cn} = 1 \underline{/120°}$. The corresponding line-to-line voltages are $V_{ab} = \sqrt{3} \underline{/30°}$, $V_{bc} = \sqrt{3} \underline{/270°}$, and $V_{ca} = \sqrt{3} \underline{/150°}$. These phasor relationships are illustrated in Figure 3.25.

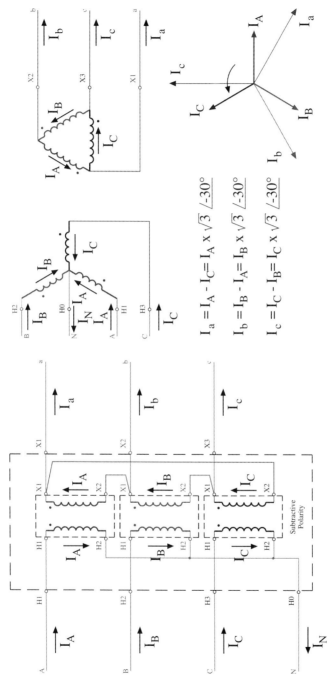

Figure 3.21 Wye–Delta Transformer Connections

84 CHAPTER 3 TRANSFORMER CONNECTIONS

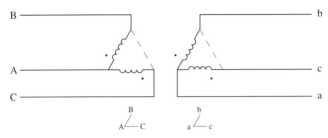

Figure 3.22 Open Delta–Open Delta Transformer

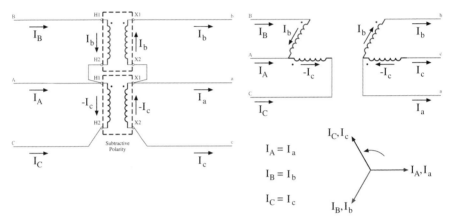

$I_A = I_a$

$I_B = I_b$

$I_C = I_c$

Figure 3.23 Open Delta–Open Delta Transformer Connections

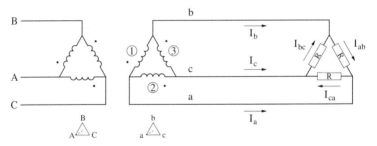

Figure 3.24 Delta–Delta Transformer Serving Delta-Connected Resistive Load

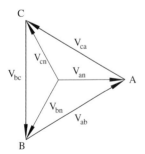

Figure 3.25 Voltage Phasor Relationships

3.6 THREE-PHASE TRANSFORMER CONNECTIONS

The secondary line currents I_a, I_b, and I_c can be calculated from the phase currents in the load by writing node equations for each corner of the load delta:

$$I_a = I_{ca} - I_{ab} = \frac{V_{ca}}{R} - \frac{V_{ab}}{R} = \sqrt{3}\,\underline{/150°} - \sqrt{3}\,\underline{/30°} = 3\,\underline{/180°} \quad (3.25)$$

$$I_b = I_{ab} - I_{bc} = \frac{V_{ab}}{R} - \frac{V_{bc}}{R} = \sqrt{3}\,\underline{/30°} - \sqrt{3}\,\underline{/270°} = 3\,\underline{/60°} \quad (3.26)$$

$$I_c = I_{bc} - I_{ca} = \frac{V_{bc}}{R} - \frac{V_{ca}}{R} = \sqrt{3}\,\underline{/270°} - \sqrt{3}\,\underline{/150°} = 3\,\underline{/300°} \quad (3.27)$$

These secondary line currents are determined by the load and have no dependency on the transformer connection serving the load.

To determine the power furnished by each of the single-phase transformers, the impedance of the secondary winding of transformer 1 is found by writing the node equation of the phase a corner of the delta:

$$I_{ab} - I_{ca} = I_a \quad (3.28)$$

Each of the three transformer secondaries is assumed to have the same impedance, which is represented by Z. Substituting into Eq. (3.29),

$$\frac{V_{ab}}{Z} - \frac{V_{ca}}{Z} = 3\,\underline{/180°} \quad (3.29)$$

Solving for Z,

$$V_{ab} - V_{ca} = 3Z\,\underline{/180°} \quad (3.30)$$

$$\sqrt{3}\,\underline{/30°} - \sqrt{3}\,\underline{/150°} = 3Z\,\underline{/180°} \quad (3.31)$$

$$3\,\underline{/0°} = 3Z\,\underline{/180°} \quad (3.32)$$

$$Z = 1\,\underline{/180°} \quad (3.33)$$

Now the current in the secondary winding of transformer 1 can be found:

$$I_{ab} = \frac{V_{ab}}{1\,\underline{/180°}} = \frac{\sqrt{3}\,\underline{/30°}}{1\,\underline{/180°}} = \sqrt{3}\,\underline{/210°} \quad (3.34)$$

Finally, the power furnished by transformer 1 can be expressed:

$$S_1 = V_{ab} I_{ab}^* = (\sqrt{3}\,\underline{/30°})(\sqrt{3}\,\underline{/150°}) = 3\,\underline{/180°} \quad (3.35)$$

The asterisk indicates complex conjugate, which is necessary for correct var flow direction (see Section 1.7). Because of the symmetry of this problem, S_2 and S_3 are found also to be $3\,\underline{/180°}$.

Next, one of the single-phase transformers is removed to form an open delta–open delta bank as shown in Figure 3.26.

Since the secondary line currents remain the same as calculated in Eqs. (3.26) through (3.28), the power supplied by the two remaining single-phase transformers can be calculated as follows:

$$S_1 = V_{ba} \cdot I_b^* = (\sqrt{3}\,\underline{/210°})(3\,\underline{/-60°}) = 3\sqrt{3}\,\underline{/150°} \quad (3.36)$$

$$S_2 = V_{ca} \cdot I_c^* = (\sqrt{3}\,\underline{/150°})(3\,\underline{/60°}) = 3\sqrt{3}\,\underline{/210°} \quad (3.37)$$

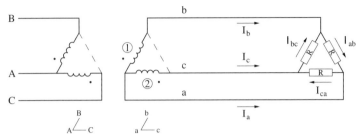

Figure 3.26 Open Delta–Open Delta Transformer Serving Delta-Connected Resistive Load

The total power delivered to the load remains

$$S_{\text{total}} = S_1 + S_2 = (3\sqrt{3}\ \underline{/150°}) + (3\sqrt{3}\ \underline{/210°}) = 9\ \underline{/180°} \quad (3.38)$$

Transformer 1 is operating at a power factor of cos 150° or 86.6% lagging, while transformer 2 is operating at a power factor of cos 210° or 86.6% leading. Compared to the delta–delta configuration, each transformer supplies a power magnitude of $3\sqrt{3}$ instead of 3 as before. This results in a derating to $1/\sqrt{3}$ or 57.7% of the delta–delta bank rating when one single-phase transformer is removed.

3.6.6 Open Wye–Open Delta

The open wye–open delta connection shown in Figure 3.27 is used to supply a three-phase load when only two phases of the three-phase system are available. Because of the 120° phase separation, this is still considered a three-phase connection. It has the peculiar ability to *create* the missing phase in the secondary circuit. This connection is used more commonly in utility distribution systems than in industrial facilities.

The connections to configure two single-phase transformers into an open wye–open delta bank are shown in Figure 3.28.

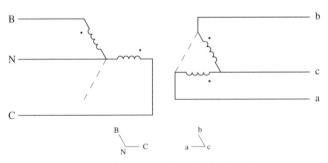

Figure 3.27 Open Wye–Open Delta Transformer

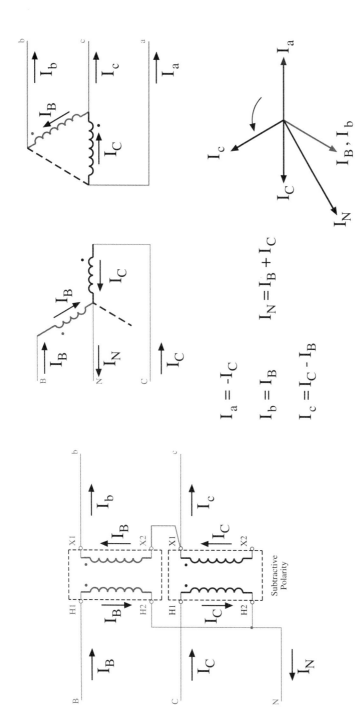

Figure 3.28 Open Wye–Open Delta Transformer Connections

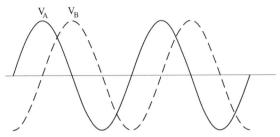

Figure 3.29 Two-Phase Voltage Sinusoids

3.7 TWO-PHASE TRANSFORMER CONNECTIONS

At times, a two-phase system must be derived from a three-phase source. This is required most often to supply two-phase motors. In a two-phase system, the two phase voltages are displaced by 90°.

It should be noted that in spite of the lack of perfect symmetry of the two-phase voltage waveforms, a perfectly constant power is produced by a two-phase machine. Figure 3.29 shows two equal voltage sinusoids displaced by 90°.

Next, a current sinusoid is introduced for each phase. To simplify the graphics, the current sinusoids are drawn in-phase with the voltage, as shown in Figure 3.30. Any power factor can be considered, but the case of unity is simplest.

Now, the power produced (or consumed) by each phase can be determined by multiplying the voltage and current sinusoids, point by point, for each phase, as shown in Figure 3.31. If a power factor other than unity is considered, the power factor ($\cos \theta$) must be multiplied along with the voltage and current. Note that the power waveform is a sinusoid that always remains positive and has a frequency twice that of the voltage or current.

Finally, the total power produced (or consumed) can be shown by adding P_A and P_B point by point. This results in a perfectly flat (constant) total power curve, as shown in Figure 3.32.

Obtaining a 90° phase shift from a system where the phases are displaced by 120° requires some clever transformer application. The most common method of

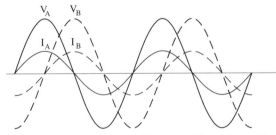

Figure 3.30 Two-Phase Voltages and Currents at Unity Power Factor ($\theta = 0$)

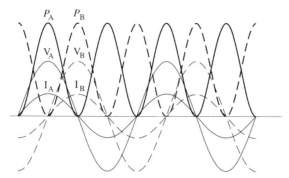

Figure 3.31 Two-Phase Power at Unity Power Factor

deriving a two-phase system from a three-phase source, or vice-versa, is to use a *T-* or *Scott connection*.

3.7.1 T-connection (Scott Connection)

The T-connection uses two single-phase transformers with taps called *Scott taps* at particular locations. The first transformer, or *main* transformer, has a 50% tap on the primary winding. The second transformer, the *teaser* transformer, has its primary winding tapped at $\sqrt{3}/2$ or 86.6%. The main transformer is connected line to line on the three-phase system, and the teaser transformer is connected from the 50% tap of the main transformer to the third line of the three-phase system through the 86.6% tap. The secondary windings are usually center tapped to allow for three-, four-, or five-wire secondaries. The wiring of the three-phase part of the T-connection is shown in Figure 3.33.

It is common to use identical transformers for both the main and teaser transformers. In Figure 3.33, H_4 indicates the nonpolarity end of the primary winding, H_3 is the 50% tap, H_2 is the 86.6% tap, and H_1 is the polarity end of the full winding.

A two-phase, three-wire circuit can be formed by connecting terminal X_1 of the teaser transformer to phase a (denoted by a_1) and terminal X_3 of the teaser

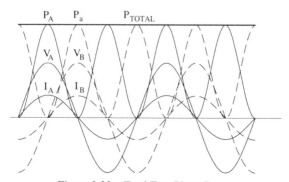

Figure 3.32 Total Two-Phase Power

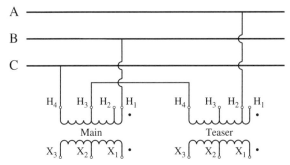

Figure 3.33   T-Connection Three-Phase Wiring

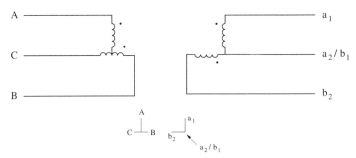

Figure 3.34   T-Connection (Three-Phase, Three-Wire to Two-Phase, Three-Wire)

transformer to phase b (denoted by b_1) and also to terminal X_1 of the main transformer. The T-connection used for a two-phase, three-wire circuit is shown in Figures 3.34 and 3.35.

A four-wire secondary can be formed by connecting terminals X_3 and X_1 of the teaser transformer across phase a and terminals X_3 and X_1 of the main transformer across phase b. The T-connection used for a two-phase, four-wire secondary is shown in Figures 3.36 and 3.37.

A five-wire secondary can be formed like the four-wire secondary with terminal X_2 of the teaser transformer grounded, making the neutral. The T-connection used for a two-phase, five-wire secondary is shown in Figures 3.38 and 3.39.

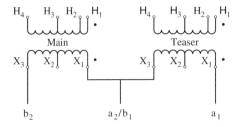

Figure 3.35   T-Connection (Two-Phase, Three-Wire) Two-Phase Connections

3.7 TWO-PHASE TRANSFORMER CONNECTIONS 91

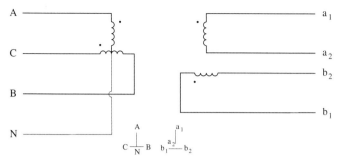

Figure 3.36 T-Connection (Three-Phase, Four-Wire to Two-Phase, Four-Wire)

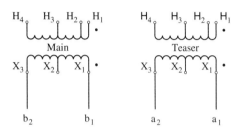

Figure 3.37 T-Connection (Two-Phase, Four-Wire) Two-Phase Connections

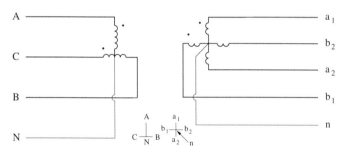

Figure 3.38 T-Connection (three-Phase, Four-Wire to Two-Phase, Five-Wire)

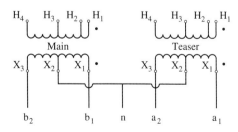

Figure 3.39 T-Connection (Two-Phase, Five-Wire) Two-Phase Connections

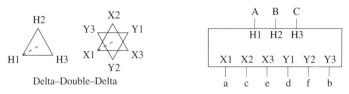

Figure 3.40 Delta–Double-Delta Connection

3.8 SIX-PHASE TRANSFORMER CONNECTIONS

Six-phase systems can be advantageous for power transmission and when rectification is required. A three-to-six-phase transformation can be made by building a transformer with delta-connected high-voltage windings and two delta-connected low-voltage windings, one with zero angular shift and the other with a 180° angular shift. This connection is called a delta–double-delta connection, and is shown in Figure 3.40.

Similarly, the two delta-connected delta windings can be replaced by two wye-connected windings, one with zero angular shift and the other with a 180° angular shift. This connection is called a delta–double-wye connection and is shown in Figure 3.41.

A wye-connected high-voltage winding can be used with two delta-connected low-voltage windings, one with zero angular shift and the other with a 180° angular shift. This connection is called a wye–double-delta connection and is shown in Figure 3.42.

A fourth option exist, which uses a wye-connected high-voltage winding with two wye-connected low-voltage windings, one with zero angular shift and the other

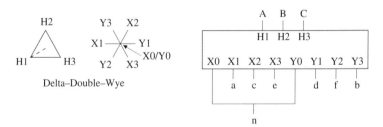

Figure 3.41 Delta–Double-Wye Connection

Figure 3.42 Wye–Double-Delta Connection

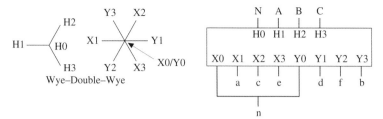

Figure 3.43 Wye–Double-Wye Connection

with a 180° angular shift. This connection is called a wye–double-wye connection and is shown in Figure 3.43.

3.9 TRANSFORMER PHASE SHIFTS

Delta–wye and wye–delta transformers always produce a 30° phase shift between the phasors in the high-voltage and low-voltage windings. The shift in a *standard* connection causes the low-voltage phasors to *lag* the high-voltage phasors. A *nonstandard* connection causes the low-voltage phasors to *lead* the high-voltage phasors.

Sometimes a phase shift other than 30° lagging or 30° leading is needed to match phasing when paralleling transformers. Consider the one-line diagram shown in Figure 3.44.

A positive sequence (a-b-c) phase sequencing is normally assumed. Note the orientation of the phasing symbols in Figure 3.44. The delta windings of T1 show phase a in the lower left corner of the delta. The dashed line in the delta denotes the position of the phase a-to-neutral phasor. In Figure 3.44, the phase a-to-neutral phasor for the high-voltage bus is oriented at 210° relative to the horizontal. The phase a-to-neutral phasor of the wye windings of T1 is oriented at 180°, indicating a 30° lag through the transformer. Although the horizontal leg of the wye phasors is not labeled as phase a, it is the only leg of the wye that can represent phase a because it is the only phasor that differs from the phase a-to-neutral phasor of the T1 delta windings by 30°.

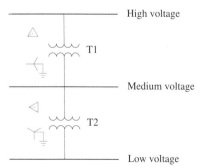

Figure 3.44 Transformer Phase Shifts on a Three-Bus System

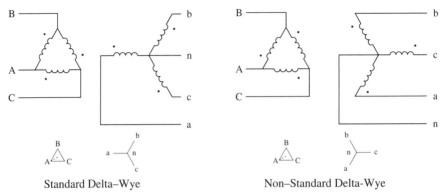

Standard Delta–Wye   Non–Standard Delta-Wye

Figure 3.45 Standard and Nonstandard Delta–Wye Connections

The orientation of the phase a-to-neutral phasor of the T2 delta must match the orientation of the phase a-to-neutral phasor of the T1 wye, because these two windings are connected to the same bus (Bus 2). The standard connection of T2 causes the phase a-to-neutral phasor of the T2 wye to lag the corresponding phasor of the T2 delta by 30°, orienting it at 150° with respect to the horizontal.

If a single transformer is required between buses 1 and 3, a 60° lag in phase is required. This is determined by comparing the phase a-to-neutral phasor of Bus 1 (210°) to the same phasor of Bus 3 (150°).

Phase shifts of any increment of 30° can be achieved by using a standard delta–wye, nonstandard delta–wye, standard wye–wye, or nonstandard wye–wye transformer. The difference between standard and nonstandard transformers lies in the way the windings are connected, as shown in Figure 3.45. The standard delta–wye shows the A-to-neutral phasor in the delta circuit at an orientation of 210°. The corresponding phasor in the wye circuit (a-to-neutral) is at a 180° orientation, producing a 30° lagging phase shift as one moves from the high-voltage circuit to the low-voltage circuit. The non-standard delta wye also shows the A-to-neutral phasor at 210°, but the a-neutral phasor is at 240°, producing a 30° leading phase shift as one moves from the high-voltage circuit to the low-voltage circuit.

Three-phase transformers must be configured as either standard or nonstandard when they are constructed because the connections that configure the windings are inside the tank and cannot be changed after the transformer is manufactured. Single-phase transformers, however, are connected externally, so they can be configured as either standard or nonstandard at the time of installation. While this can be a benefit, care must be taken when building three-phase transformer banks from single-phase transformers to assure that the winding polarities are connected correctly.

Table 3.1 shows the transformer connections necessary to produce a lagging phase shift of any multiple of 30°.

Some phase shifts require the *cross connection* of the low-voltage phases. This is where high-voltage phases A, B, and C are connected to bushings H1, H2, and H3, respectively, while the low-voltage phases are connected to noncorresponding bushings.

TABLE 3.1 Transformer Connections for Various Lagging Phase Shifts

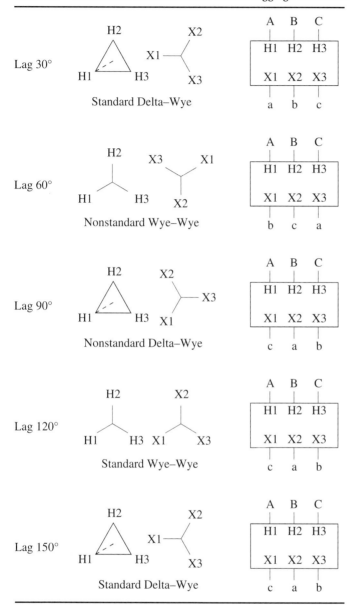

3.10 GROUNDING TRANSFORMERS

On some industrial distribution systems, particularly older installations, a system neutral is not available. Such an example is a three-wire delta source. When a system neutral is desired on a three-wire system, it can be obtained by using a *grounding transformer*.

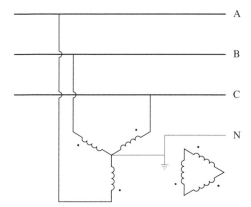

Figure 3.46 Wye–Delta Grounding Transformer

3.10.1 Wye–Delta

A wye–delta transformer can be used as a grounding transformer. The wye winding is connected to the source and its neutral is grounded, either solidly or through an impedance. The delta winding is necessary to provide a circulating path for zero-sequence current. It usually serves no load, but a small load could be connected to it. The voltage rating of the delta winding is arbitrary. The connection diagram of a wye–delta grounding transformer is shown in Figure 3.46.

3.10.2 Zig–Zag Connection

The preferred grounding transformer configuration is the *zig–zag connection* shown in Figure 3.47. A transformer with a zig–zag primary and no secondary offers high

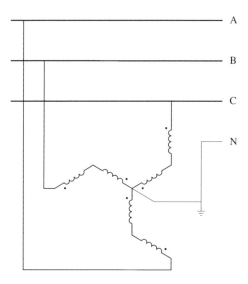

Figure 3.47 Zig–Zag Connection

impedance to positive- and negative-sequence currents and low impedance to zero-sequence current. Because of the subtractive winding polarities, the leakage flux is the only magnetic flux produced by zero-sequence currents, and the zero-sequence impedance is very low.

Positive- and negative-sequence currents produce a much higher magnetic flux intensity since the fluxes produced in the two sections of each branch of the zig–zag winding do not cancel as with the zero-sequence current. The larger flux gives rise to a larger impedance. This means that during normal conditions, only a small magnetizing current flows in the zig–zag windings. But when a ground fault occurs, a large ground current is allowed to flow. This ground current facilitates system protection by simplifying ground fault detection.

Another advantage of the zig–zag connection is the reduced kVA rating of the transformer. The kVA rating is defined as the product of the line-to-neutral voltage and the rated neutral current. Since the transformer is designed to carry rated current for just a short period of time, such as 30 seconds or 1 minute, a zig–zag grounding transformer has a significantly smaller kVA rating than a wye–delta grounding transformer sized for the same application.

3.11 FERRORESONANCE

Ferroresonance is a condition caused by a capacitive element, typically a length of insulated cable, in series with a lightly loaded transformer. This combination forms a resonant circuit. If the critical frequency of this resonant circuit coincides with the frequency of the third-harmonic magnetization current of the transformer (180 Hz for a 60 Hz system), ferroresonance will occur.

During ferroresonance, very high third harmonic voltages, possibly four to six times the normal system voltage will be present. These voltages can cause arresters to operate and can damage equipment. The effects of ferroresonance are exacerbated with three-phase transformers if the three phases supplying the transformer are not energized simultaneously. Consequently, gang-operated switching devices are recommended for three-phase transformers where ferroresonance may be a concern.

Switching single-phase transformers can also lead to ferroresonance conditions. Consider a typical 250 kVA single phase transformer. Its magnetizing reactance will be on the order of $j70$ kΩ. If the transformer is supplied by a 500-ft length of single-conductor AWG No. 2 cable having a capacitive reactance of approximately $-j70$ kΩ, a resonant circuit will be formed. Because of the nonlinear inductance of the iron core, the transformer and cable reactances do not have to be perfectly matched for resonant conditions to occur.

In special applications, series capacitors can be installed on a distribution circuit, often in conjunction with motor applications. When series capacitors are used, ferroresonance and a similar phenomenon called *subsynchronous resonance* (*SSR*) can occur. SSR will cause a motor to lock into a rotational speed less than the motor's rated speed. This can lead to excessive vibration and heating, and consequently, damage to the motor. With large synchronous machines, torsional forces due to SSR can be so severe as to break the rotor shaft. Paralleling a resistor across the series capacitor usually remedies the resonant condition. If this solution is not acceptable, an active filter may be required to shunt the subsynchronous current to ground.

Ferroresonance and other undesirable resonant conditions can be avoided with careful system design. If sufficient load exists on the transformer, ferroresonance will be quickly damped; therefore, applying load to the transformer prior to switching may avoid ferroresonance. Other practices that can be used to avoid resonant conditions include keeping the ratio of the cable reactance to the transformer magnetizing reactance high (much larger than 10). Grounded wye or grounded open-wye primary transformer connections also reduce the likelihood of ferroresonance, unless a secondary delta connection is used. Then, a grounding resistor in the neutral of the primary will reduce the magnitude of the resonant voltages.

SUMMARY

Proper voltage selection is critical for the economy, versatility, performance, and expandability of the industrial power system. Commonly used main distribution voltages include 13.8, 6.9, 4.16, and 2.4 kV. Utilization equipment frequently requires 480, 277, 240, 208, or 120 V.

Various three-phase transformer connections exist, each with advantages and disadvantages. *Delta–wye* and *wye–delta* transformers cause a 30° phase shift where the low-voltage side lags the high-voltage side in a *standard connection*. *Open delta* and *open wye* connections use only two single-phase transformers for a three-phase system. The *T-* or *Scott connection* transforms between three- and two-phase systems. Special transformer connections can also derive high phase-order systems, such as six-phase systems, from a three-phase system. *Grounding transformers*, connected either *wye-delta* or in a *zig-zag* configuration, can provide a grounded neutral on a three-wire system.

Ferroresonance occurs when the magnetizing susceptance of a transformer and the shunt capacitive susceptance of an insulated cable form a resonant circuit. It is most likely to occur when the transformer is lightly loaded or when the transformer is *single phased* by opening one high-voltage phase at a time instead of all three simultaneously.

Light load conditions aggravate ferroresonance, as load tends to be a damping factor. Ferroresonance can be predicted and design measures can be taken to avoid the condition. Most often, particular combinations of cable sizes and lengths and transformer kVA sizes are avoided.

FOR FURTHER READING

Dugan, R. C. Examples of ferroresonance in distribution systems. *Power Engineering Society General Meeting*, 2003, *IEEE*, vol. 2, pp.,1215 Vol. 2, July 13–17, 2003.
Farmer, R. G. and Agrawal, B. Subsynchronous resonance. *AccessScience*. McGraw-Hill Education, 2014.
IEEE Standard for Standard Terminal Markings and Connections for Distribution and Power Transformers, IEEE Standard C57.12.70, 2011.
Joffe, E. B., and Lock, K., *Grounds for Grounding: A Circuit to System Handbook*, Wiley-IEEE Press, 2010, ISBN 978-0-471-66008-8.

QUESTIONS

1. Discuss some issues that determine whether an industrial facility is supplied from the utility's transmission system or from a distribution feeder.
2. List some advantages and disadvantages of using the common distribution voltages 13.8, 6.9, 4.16, and 2.4 kV.
3. Why are system neutrals usually grounded? Give an example of a situation where the system neutral would not be grounded.
4. Construct a chart showing the common three-phase transformer connections. Discuss the advantages and disadvantages of each.
5. Discuss why generator step-up transformers are always configured as delta–wye.
6. When is it desirable to use a delta-connected low-voltage winding in a step-down transformer?
7. Why is an open delta–open delta transformer bank only capable of supplying 57.7% of the equivalent full delta rating instead of the intuitive 66.7%?
8. How does a two-phase system differ from an open delta or open wye system?
9. How is a six-phase system different from two three-phase systems?
10. Name some advantages and applications for high phase-order systems such as six-phase systems.
11. Why must two paralleled transformers have identical phasing? What happens if they do not?
12. In Figure 4.5, a normally open bus tie is proposed to connect 4.16 kV Bus 2 and 480 V Bus 3. What type of transformer connection should be used for this bus tie?
13. Describe the conditions necessary for ferroresonance to occur.
14. Give an example of how resonance can occur on a distribution system as
    a. a parallel resonant R-L-C circuit and
    b. a series resonant R-L-C circuit.
15. What design measures are taken to reduce hysteresis and eddy current losses in a transformer?

PROBLEMS

1. Draw a schematic diagram (similar to Figure 3.18) and express the corresponding phasor diagram for a *nonstandard* delta–wye transformer connection. Express the magnitudes and angles of the line currents supplying the delta-connected windings in terms of the magnitudes and angles of the line currents leaving the wye-connected windings.
2. The magnetizing reactance of a single-phase distribution transformer is $j\ 62$ kΩ. If a type of AWG 2/0 aluminum cable has a capacitive reactance of $-j\ 130$ Ω/ft, what length of cable will form a ferroresonant condition with the transformer?
3. A 4.2 Ω resistor is connected across the secondary terminals of a single-phase transformer rated 7200 V/240 V. What is the resistance as seen from the high-voltage circuit?

4. A delta–wye-connected transformer (standard connection) serves an unbalanced wye-connected load. The secondary line currents are 1.0 $\underline{/0°}$, 1.17 $\underline{/105°}$, and 0.86 $\underline{/245°}$ p.u., on phases a, b, and c, respectively. What are the primary line currents in per unit?

5. Expand Table 3.1 to include *leading* phase shifts of 30°, 60°, 90°, 120°, and 150°.

6. A delta–wye-connected transformer (nonstandard connection) serves an unbalanced delta-connected load. The primary line currents are 1.08 $\underline{/12°}$, 1.22 $\underline{/131°}$, and 0.92 $\underline{/212°}$ p.u. on phases a, b, and c, respectively. What are the secondary line currents in per unit?

7. What are the per-unit currents in the delta-connected load of Problem 6?

8. What is the inductance of a coil with a 2-ft diameter and 5-ft height consisting of 460 turns? *Be careful of units!*

9. Calculate the eddy current losses in a 1500-lb steel transformer core operating at 60 Hz with a maximum flux density of 1.5 teslas. The core laminations are 1.2 mm in thickness, and the steel has a resistivity of 47×10^{-8} Ω-m and a density of 7600 kg/m^3.

10. Repeat Problem 9 if the transformer is operated at 50 Hz.

11. An open delta–open delta transformer bank serves a balanced delta-connected load with an 86.6% lagging power factor. Find the power factor at which each of the two transformers operates.

12. Repeat Problem 11 for a balanced delta-connected load with an 86.6% *leading* power factor.

13. Show (graphically) that the real power delivered by a three-phase system is constant over time.

14. Show (graphically) how the real power delivered by a two-phase system changes as power factor varies.

15. The ratio of line-to-line voltage to line-to-neutral voltage is $\sqrt{3}$ for three-phase systems. What is that ratio for a six-phase system?

CHAPTER 4

FAULT CALCULATIONS

OBJECTIVES

- Recognize the data requirements for short-circuit calculations
- Be able to build first-cycle and contact-parting impedance networks
- Model a three-phase power system by developing the positive-, negative-, and zero-sequence networks
- Understand the behavior of zero-sequence current by knowing the zero-sequence impedance models for common power system equipment
- Apply symmetrical components to solve unbalanced three-phase power system problems
- Calculate fault current for three-phase, line-to-ground, double line-to-ground, and line-to-line faults on a three-phase system
- Calculate phase currents for one-line-open and two-lines-open faults on a three-phase system

4.1 OVERVIEW

Fault calculations are an important element of power engineering, as they provide information necessary to properly protect the power system, meaning both the equipment comprising the system and the personnel who work on or near it. Most fault conditions are asymmetrical, meaning a single-phase equivalent approach of analysis is not possible unless symmetry is somehow restored to the problem.

The restoration of symmetry to a seemingly asymmetric problem is achieved through the application of *symmetrical components*.

Although the method of symmetrical components has been known for almost a century, it is often shrouded in mystery. It is typically taught as an abstract mathematical exercise with very few if any ties to practicality. Even if the student masters the theory, application of the theory, particularly development of the zero-sequence network, often remains a challenge. In this chapter, we methodically develop the method of symmetrical components for a three-phase system. Particular attention is paid to *why* we do and what we do to solve the problems. Development of the sequence networks is done methodically and logically, leading to a complete understanding of

---

Industrial Power Distribution, Second Edition. Ralph E. Fehr, III.
© 2016 The Institute of Electrical and Electronics Engineers, Inc. Published 2016 by John Wiley & Sons, Inc.

how the method of symmetrical components works. It is crucial for understanding more complex topics involving symmetrical components, such as those explored in the power quality area.

4.2 TYPES OF FAULTS

A *short-circuit fault* occurs when a low-impedance path is formed between phases and/or a phase or phases and ground. The high current that flows through the low-impedance path can cause damage to any circuit elements in series with it, so the faulted portion of the system must be removed from service as quickly as possible to limit the extent of damage. Four different types of short-circuit faults can occur on a three-phase power system: three-phase, line-to-ground, line-to-line, and double line-to-ground. These fault types are shown in Figure 4.1.

The three-phase fault is the only fault type that produces balanced voltages and currents. Since it does, a three-phase fault can be analyzed easily using conventional circuit analysis methods applied to a single-phase equivalent of the three-phase system. The other fault types, however, result in unbalanced voltages and currents. Consequently, they must be analyzed using the method of *symmetrical components*. Because of the way we will develop symmetrical components for unbalanced phasors, we must designate phase A as the dissimilar phase for unbalanced faults. For example, in a line-to-ground fault, phase A is the faulted (dissimilar) phase and phases B and C remain unfaulted (similar). For line-to-line faults, phase A is the unfaulted (dissimilar) phase and phases B and C are faulted (similar).

Another class of unbalanced faults exists—open-circuit faults. If a three-phase circuit is protected by fuses, perhaps only one or two of the three fuses blow as a result of a short circuit or overload. Part of the system continues to operate, but in an unbalanced condition. This condition is sometimes called *single phasing*. Because of the asymmetry involved, symmetrical components are also used to analyze open-circuit faults.

In many—*but not all*—cases, the type of short-circuit fault that produces the highest current magnitude will be the three-phase fault. The line-to-ground fault often

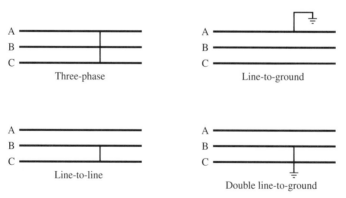

Figure 4.1 Types of Short-Circuit Faults

produces the lowest magnitude of the short-circuit faults. But grounding practices can drastically affect fault current magnitudes, so near generating stations containing impedance-grounded generators and near substations containing grounded delta–wye transformers, the line-to-ground fault is typically the highest magnitude short-circuit fault and the three-phase fault is the lowest. To be certain, all four short-circuit fault types should be analyzed when sizing protective devices to assure that the highest magnitude fault is being considered.

4.3 DATA PREPARATION

In order to perform any short-circuit calculation, certain data must be collected. A one-line diagram of the system that shows positive-, negative-, and zero-sequence impedances is necessary. This diagram should also note the positive-, negative-, and zero-sequence system impedance at the point of delivery and the X/R ratio of the utility source. Any normally open breakers or switches should be noted. Transformer data including kVA rating, impedance, voltage ratio including tap settings, winding configuration, and grounding information are required. Motor and generator data required includes subtransient and transient impedances for all synchronous machines, subtransient impedances for induction machines 50 hp and above, and rated speed of induction motors 250 hp and above. Cable and wire data includes type and size of conductor, length of circuit, type of raceway, spacing configuration, and distance to adjacent circuits in the vicinity.

The required transformer data can be found on the transformer nameplate with the exception of winding resistance. This value is typically not shown on the nameplate, but instead is noted indirectly in the transformer test report as watts of losses at various loadings. If a test report is not available, a typical X/R ratio can be obtained from Figure 4.2, which is based on IEEE standard C37.010.

Impedance data for generators can be found on the nameplate or in the generator test reports. Impedance data for motors, however, is sometimes difficult to obtain. IEEE standard 141, *IEEE Recommended Practice for Electric Power Distribution for Industrial Plants*, commonly referred to as the *IEEE Red Book*, suggests some typical reactance values for both induction and synchronous machines. These values

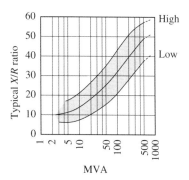

Figure 4.2 Typical X/R Ratios for Transformers

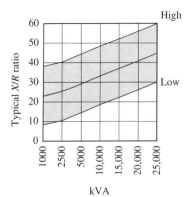

Figure 4.3 Typical X/R Ratios for Synchronous Machines

are in per unit on the machine's base and are summarized in Table D.1 in Appendix D. Note that transient reactances for induction machines are not listed, because the fault contribution from an induction machine is totally decayed after just a few cycles due to the quick collapse of the self-excited field. The typical range of X/R ratios for synchronous and induction machines can be found in Figures 4.3 and 4.4, respectively.

Calculating cable and wire impedance can be very complicated and tedious. A practical method of determining these impedances is to use tabular data compiled and presented in various publications, such as the *IEEE Red Book*. Resistance (R) is tabulated at a given frequency and given conductor temperature, typically 25°C. If the operating temperature of the conductor is significantly higher than 25°C, a correction can be made based on the temperature coefficient of resistivity (T). T is determined empirically by graphing a conductor's resistance as a function of temperature. Over the range of reasonable operating temperatures, resistance as a function of temperature should be essentially linear. If the inverse function is plotted (temperature as a function of resistance), the line can be extrapolated to determine the y-intercept. This value is T. Some values of T for commonly used conductors are shown in Table 4.1.

The resistance (R_{NEW}) at another temperature (t_{NEW}) can be calculated when the resistance (R_{OLD}) is known at a given temperature (t_{OLD}) using Eq. (4.1):

$$R_{NEW} = R_{OLD} \times \frac{T + t_{NEW}}{T + t_{OLD}} \qquad (4.1)$$

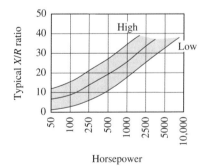

Figure 4.4 Typical X/R Ratios for Induction Machines

TABLE 4.1 Temperature Coefficient of Resistivity

Material	$T\ (°C)$
Annealed Copper (100% conductivity)	234.5
Hard-drawn Copper (97.3% conductivity)	241
Hard-drawn Aluminum (61% conductivity)	228

Inductive reactance of insulated cables is more complex, since it is a function of many variables, including the type of insulation, the spacing between conductors, and the type of raceway carrying the cables (magnetic or nonmagnetic). Because of all these variables, three charts are usually needed to determine reactance: one to determine the reactance at one foot conductor spacing (X_A), one to correct the reactance for the actual conductor spacing (X_B), and one to determine the *reactance factor* (*M*) which accounts for various constructions and installations. Using this method, the total three-phase impedance from line to neutral is

$$Z = R + j(X_A + X_B) \times M \tag{4.2}$$

Tables D.2, D.3, and D.4 in Appendix D showing X_A, X_B, and M were reprinted from IEEE Standard 242 (the IEEE Buff Book).

Several types of short-circuit calculations will be shown and explained throughout the remainder of this chapter. The one-line diagram shown in Figure 4.5 will be used to demonstrate the calculation methods.

A 10-MVA base will be used for all per-unit calculations in the following sections.

4.4 FIRST-CYCLE SYMMETRICAL CURRENT CALCULATIONS

The *first-cycle* or *momentary* short-circuit calculation determines the maximum magnitude of fault current that can ever be experienced at a particular location of a given power system. This maximum magnitude occurs just after inception of the fault. Immediately after inception, the fault current magnitude diminishes as a function of the *X/R* ratio of the system at the point of the fault. Small *X/R* ratios decay rather quickly as seen in Figure 4.6(a) while large *X/R* ratios decay more slowly as shown in Figure 4.6(b).

Current-carrying conductors produce magnetic fields, which exert mechanical forces on other current-carrying conductors. The first-cycle current represents the maximum mechanical forces to which the system components are subjected during the fault event. This information is required to design the bracing necessary to withstand the tremendous mechanical forces resulting from the enormous magnetic fields produced by the high current during the fault.

The high current drawn through the system from the inception of the fault until the fault is cleared produces a substantial voltage drop. If this voltage drop becomes excessive, operational problems will result, most notably contactor dropout. Running loads often rely on the system voltage to keep contactor coils held closed. If the voltage drops more than 10–15% below nominal, the holding coils can drop out,

106 CHAPTER 4 FAULT CALCULATIONS

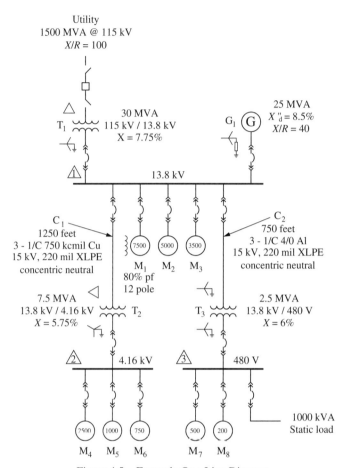

Figure 4.5 Example One-Line Diagram

thereby de-energizing the running load. Chapter 8 will examine undervoltage dropout and suggest means to mitigate it.

The first step in calculating first-cycle symmetrical current is to define the *first-cycle impedance diagram*. This is a version of the one-line diagram that shows the impedances present during the first cycle after the fault occurs. This includes *subtransient impedances* for all synchronous machines and for induction machines 50 hp and larger. Induction machines smaller than 50 hp have a negligible short-circuit contribution, so unless many small induction machines exist, they can be neglected.

The first-cycle reactance network for the one-line diagram in Figure 4.5 is shown in Figure 4.7. Sources and loads are connected to the reference bus. Sources are modeled as a reactance in series with an estimated pre-fault voltage of $1.0 \underline{/0°}$ p.u. In the absence of calculated pre-fault voltages, such as from a loadflow calculation, this estimated pre-fault voltage works well. The remainder of the system completes the network, with each numbered bus retaining its identity.

4.4 FIRST-CYCLE SYMMETRICAL CURRENT CALCULATIONS

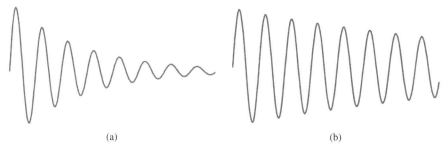

Figure 4.6 Fault Current Decay as a Function of X/R

The per-unit reactance calculations used to determine the values labeled in Figure 4.7 are shown in Eqs. (4.3) through (4.17). The system reactance is calculated by dividing the available fault MVA, which is a value supplied by the local utility, into the base MVA:

$$X_{\text{Utility}} = \frac{10}{1500} = 0.0067 \text{ p.u.} \tag{4.3}$$

The generator and transformer impedances are calculated by converting the given percent reactances to per unit on a 10-MVA base:

$$X_{G1} = 0.085 \left(\frac{10}{25}\right) = 0.0340 \text{ p.u.} \tag{4.4}$$

$$X_{T1} = 0.0775 \left(\frac{10}{30}\right) = 0.0258 \text{ p.u.} \tag{4.5}$$

$$X_{T2} = 0.0575 \left(\frac{10}{7.5}\right) = 0.0767 \text{ p.u.} \tag{4.6}$$

$$X_{T3} = 0.06 \left(\frac{10}{2.5}\right) = 0.2400 \text{ p.u.} \tag{4.7}$$

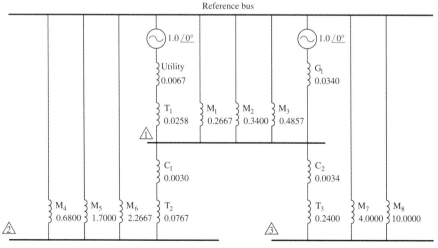

Figure 4.7 First-Cycle Reactance Network

The motor reactances are obtained from Table D.1 in Appendix D and converted to a 10-MVA base. An approximation of 1 kVA/hp is used as a conversion factor to account for efficiency and power factor. Typically, efficiency and power factor values are difficult to obtain for rotating machines, particularly induction motors operating below nameplate rating, as is often the case. This makes the 1 kVA = 1 hp approximation very useful.

$$X_{M1} = 0.20 \left(\frac{10}{7.5}\right) = 0.2667 \text{ p.u.} \tag{4.8}$$

$$X_{M2} = 0.17 \left(\frac{10}{5}\right) = 0.3400 \text{ p.u.} \tag{4.9}$$

$$X_{M3} = 0.17 \left(\frac{10}{3.5}\right) = 0.4857 \text{ p.u.} \tag{4.10}$$

$$X_{M4} = 0.17 \left(\frac{10}{2.5}\right) = 0.6800 \text{ p.u.} \tag{4.11}$$

$$X_{M5} = 0.17 \left(\frac{10}{1}\right) = 1.7000 \text{ p.u.} \tag{4.12}$$

$$X_{M6} = 0.17 \left(\frac{10}{0.75}\right) = 2.2667 \text{ p.u.} \tag{4.13}$$

$$X_{M7} = 0.20 \left(\frac{10}{0.5}\right) = 4.0000 \text{ p.u.} \tag{4.14}$$

$$X_{M8} = 0.20 \left(\frac{10}{0.2}\right) = 10.0000 \text{ p.u.} \tag{4.15}$$

The cable reactances are calculated from the values extracted from Table D.6 of Appendix D:

$$X_{C1} = 0.0459 \frac{\Omega}{\text{kft}} (1.25 \text{ kft}) \left(\frac{10}{13.8^2}\right) = 0.0030 \text{ p.u.} \tag{4.16}$$

$$X_{C2} = 0.0860 \frac{\Omega}{\text{kft}} (0.75 \text{ kft}) \left(\frac{10}{13.8^2}\right) = 0.0034 \text{ p.u.} \tag{4.17}$$

Next, a per-unit resistance network must be constructed. Often, resistance data is difficult to obtain, particularly for rotating machines. Typical X/R ratios for synchronous and induction machines are shown in Figures 4.3 and 4.4, respectively. This data can be used if specific resistance information is not available. The EMF sources have been omitted, since only a Thévenin equivalent resistance is needed from this network. The first-cycle resistance network is shown in Figure 4.8, and the calculations of the resistance values are shown in Eqs. (4.18) through (4.34):

$$R_{\text{Utility}} = \frac{X_{\text{Utility}}}{(X/R)_{\text{Utility}}} = \frac{0.0067}{100} = 0.000067 \text{ p.u.} \tag{4.18}$$

$$R_{G1} = \frac{X_{G1}}{(X/R)_{G1}} = \frac{0.0340}{50} = 0.000680 \text{ p.u.} \tag{4.19}$$

4.4 FIRST-CYCLE SYMMETRICAL CURRENT CALCULATIONS

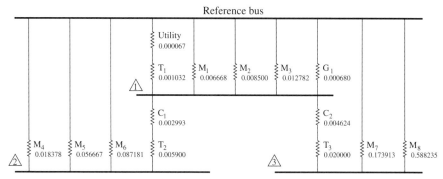

Figure 4.8 First-Cycle Resistance Network

The *X/R* ratios for the transformers are estimated from Figure 4.2. Slightly overestimating the *X/R* ratio adds conservatism to the calculations:

$$R_{T1} = \frac{X_{T1}}{(X/R)_{T1}} = \frac{0.0258}{25} = 0.001032 \text{ p.u.} \tag{4.20}$$

$$R_{T2} = \frac{X_{T2}}{(X/R)_{T2}} = \frac{0.0767}{15} = 0.005113 \text{ p.u.} \tag{4.21}$$

$$R_{T3} = \frac{X_{T3}}{(X/R)_{T3}} = \frac{0.2400}{12} = 0.020000 \text{ p.u.} \tag{4.22}$$

The *X/R* ratio for the synchronous motor is estimated from Figure 4.3:

$$R_{M1} = \frac{X_{M1}}{(X/R)_{M1}} = \frac{0.2667}{40} = 0.006668 \text{ p.u.} \tag{4.23}$$

The *X/R* ratios for the induction motors are estimated from Figure 4.4:

$$R_{M2} = \frac{X_{M2}}{(X/R)_{M2}} = \frac{0.3400}{40} = 0.008500 \text{ p.u.} \tag{4.24}$$

$$R_{M3} = \frac{X_{M3}}{(X/R)_{M3}} = \frac{0.4857}{38} = 0.012782 \text{ p.u.} \tag{4.25}$$

$$R_{M4} = \frac{X_{M4}}{(X/R)_{M4}} = \frac{0.6800}{37} = 0.018378 \text{ p.u.} \tag{4.26}$$

$$R_{M5} = \frac{X_{M5}}{(X/R)_{M5}} = \frac{1.7000}{30} = 0.056667 \text{ p.u.} \tag{4.27}$$

$$R_{M6} = \frac{X_{M6}}{(X/R)_{M6}} = \frac{2.2667}{26} = 0.087181 \text{ p.u.} \tag{4.28}$$

$$R_{M7} = \frac{X_{M7}}{(X/R)_{M7}} = \frac{4.0000}{23} = 0.173913 \text{ p.u.} \tag{4.29}$$

$$R_{M8} = \frac{X_{M8}}{(X/R)_{M8}} = \frac{10.0000}{17} = 0.588235 \text{ p.u.} \tag{4.30}$$

The resistances of the cables are obtained from Table D.6 in Appendix D, then converted to per unit:

$$R_{C1(\Omega)} = \left(0.0456 \frac{\Omega}{\text{kft}}\right)(1.25 \text{ kft}) = 0.057000 \, \Omega \tag{4.31}$$

$$R_{C1(\text{p.u.})} = (0.057000 \, \Omega) \frac{10 \text{ MVA}}{(13.8)^2} = 0.002993 \text{ p.u.} \tag{4.32}$$

$$R_{C2(\Omega)} = \left(0.1174 \frac{\Omega}{\text{kft}}\right)(0.75 \text{ kft}) = 0.088050 \, \Omega \tag{4.33}$$

$$R_{C2(\text{p.u.})} = (0.088050 \, \Omega) \frac{10 \text{ MVA}}{(13.8)^2} = 0.004624 \text{ p.u.} \tag{4.34}$$

It is important to construct and analyze two separate networks for resistance and reactance instead of combining them into a single network and reducing the complex impedances. When more than one source of fault current exists, each source has its own X/R ratio and thereby its own exponential time constant. Multiple exponential time constants cannot be mathematically combined into a single time constant to produce an effective X/R ratio for the entire system, but an effective X/R ratio describing the decay of the fault current can be approximated. If complex impedances represent the system and a Thévenin impedance is determined for the point of the fault, the X/R ratio of the Thévenin impedance can be considered. Unfortunately, this X/R ratio produces a fault current decay rate that is larger than the decay rate that actually occurs in the circuit. Using this X/R ratio to determine the interrupting current would result in an undercalculation of the current as the breaker contacts part, which could result in the failure of the device.

To avoid this situation, IEEE Standard C37.010 recommends separately calculating a Thévenin resistance and a Thévenin reactance for the system at the point of the fault, then dividing the resistance into the reactance to determine the effective X/R ratio for the system. Using this X/R ratio estimate to calculate interrupting current will always produce a value slightly more severe than the actual interrupting current, thereby yielding conservative results.

The following example of two fault current sources (the system and generator from Figure 4.9) shows the large variation in X/R ratio that results from using

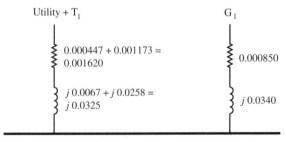

Figure 4.9 Impedances of Paralleled Sources

complex number calculations versus the recommended decoupled resistance/reactance method.

Using the recommended decoupled method which separately combines the resistances and the reactances,

$$R_{EQ(decoupled)} = \frac{(0.001240)(0.000850)}{0.001240 + 0.000850} = 0.000504 \text{ p.u.} \quad (4.35)$$

$$X_{EQ(decoupled)} = \frac{(0.0325)(0.0340)}{0.0325 + 0.0340} = 0.0166 \text{ p.u.} \quad (4.36)$$

The effective X/R ratio is

$$(X/R)_{effective(decoupled)} = \frac{X_{EQ(decoupled)}}{R_{EQ(decoupled)}} = \frac{0.0166}{0.000504} = 32.94 \quad (4.37)$$

If the effective X/R ratio is calculated by combining the complex impedances, a much different result is obtained.

$$Z_{SYSTEM} = 0.001240 + j0.0325 = 0.032524 \, \underline{/87.82°} \quad (4.38)$$

$$Z_{G1} = 0.000850 + j0.0340 = 0.034011 \, \underline{/88.57°} \quad (4.39)$$

$$Z_{EQ} = \frac{(0.032540 \, \underline{/87.15°})(0.034011 \, \underline{/88.57°})}{(0.032540 \, \underline{/87.15°}) + (0.034011 \, \underline{/88.57°})}$$

$$= \frac{0.001107 \, \underline{/175.72°}}{0.066546 \, \underline{/87.88°}} \quad (4.40)$$

$$= 0.016635 \, \underline{/87.84°}$$

Resolving Z_{EQ} into R_{EQ} and X_{EQ} yields

$$R_{EQ(complex)} = 0.016635 \cos 87.84° = 0.000627 \text{ p.u.} \quad (4.41)$$

$$X_{EQ(complex)} = 0.016635 \sin 87.84° = 0.016623 \text{ p.u.} \quad (4.42)$$

Comparing the results of Eqs. (4.35)–(4.41) and Eqs. (4.36)–(4.42), it is observed that while the reactance values are essentially identical, the resistance is over 12% larger when determined using complex impedances than when calculated separately. In actual calculations, the difference between the two calculation methods can greatly exceed 12%. This increase in resistance removes the conservatism from the X/R estimate, yielding a value too small to give accurate results. When the X/R ratio is calculated using the complex impedances, it is found to be

$$(X/R)_{effective(complex)} = \frac{X_{EQ(complex)}}{R_{EQ(complex)}} = \frac{0.016623}{0.000627} = 26.5 \quad (4.43)$$

Using an X/R ratio of 26.5 to determine the current magnitude at the time of contact parting results in an undercalculation of the interrupting current. The 32.94 value yields a slightly larger (conservative) current than actually is experienced.

The X/R ratio at the fault location is very important when selecting protective devices, as it accounts for the amount of DC offset (asymmetry) in the fault current when the circuit is interrupted. Applying Ohm's Law to this single reactance determines the current that flows during a three-phase fault. Unbalanced faults must be analyzed using the method of symmetrical components.

First-cycle fault current will be calculated for all three buses numbered in Figure 4.5. For Bus 1, the Thévenin equivalent resistance and reactance is determined. This can be done by paralleling the impedances of all sources of fault current: the utility source (Utility in series with T_1), the generator (G_1), the 13.8 kV motors (M_1, M_2, and M_3), the 4.16 kV switchgear (the parallel combination of M_4, M_5, and M_6 in series with T_2 and C_1), and the 480 V switchgear (the parallel combination of M_7 and M_8 in series with T_3 and C_2). Note that the 1000 kVA static load on the 480 V bus does not contribute fault current, since there is no stored energy to return to the circuit as with a rotating machine. The per-unit Thévenin resistance and reactance for Bus 1 are found to be

$$Z_{\text{Thévenin (Bus 1)}} = R_{\text{Thévenin(Bus 1)}} + jX_{\text{Thévenin(Bus 1)}} = 0.000464 + j0.0140 \text{ p.u.}$$
(4.44)

The Thévenin resistance and reactance for Bus 2 are determined by paralleling the 4.16 kV switchgear (the parallel combination of M_4, M_5, and M_6) with the rest of the system (M_7 paralleled with M_8, that result in series with T_3 and C_2, that result paralleled with G_1, Utility in series with T_1, M_1, M_2, and M_3, and that result in series with C_1 and T_2). The per-unit Thévenin resistance and reactance for Bus 2 are

$$Z_{\text{Thévenin(Bus 2)}} = R_{\text{Thévenin(Bus 2)}} + jX_{\text{Thévenin(Bus 2)}} = 0.004787 + j0.0742 \text{ p.u.}$$
(4.45)

The Thévenin resistance and reactance for Bus 3 are determined by paralleling the 480 V switchgear (the parallel combination of M_7 and M_8) with the rest of the system (the parallel combination of M_4, M_5, and M_6 in series with T_2 and C_1, that result paralleled with G_1, Utility in series with T_1, M_1, M_2, and M_3, and that result in series with C_2 and T_3). The per-unit Thévenin resistance and reactance for Bus 3 are

$$Z_{\text{Thévenin(Bus 3)}} = R_{\text{Thévenin(Bus 3)}} + jX_{\text{Thévenin(Bus 3)}} = 0.021665 + j0.2348 \text{ p.u.}$$
(4.46)

4.5 CONTACT-PARTING SYMMETRICAL CURRENT CALCULATIONS

Contact-parting current is that current present when the circuit breaker contacts separate to interrupt the current flow. Depending on the speed of the breaker, this is usually three to eight cycles after the commencement of the fault. Due to the resistance of the system, fault current at contact-parting time will be less than the current present immediately after the fault occurs, perhaps substantially less. This is because of the damping effect that the system resistance has on the fault current and is evident when looking at the effect of the system X/R ratio on the current amplitude, as shown in Figure 4.6.

The impedance values used to model synchronous machines for contact-parting calculations are different (larger) than those used in first-cycle current calculations. This increased impedance yields a smaller fault current value. In practical terms, *transient* impedances replace the subtransient impedances of synchronous machines used for first-cycle calculations. Induction machines are usually neglected when doing contact-parting calculations, because their self-excited fields are collapsed by the time the circuit breakers can open. This means induction machines, although fault contributors for first-cycle calculations, are insignificant contributors of short-circuit current after just a few cycles, so their impedances are modeled as infinite for contact-parting calculations.

A *contact-parting impedance diagram* is constructed. This diagram is similar to the first-cycle impedance diagram, with the subtransient impedances of synchronous machines replaced by transient impedances and the induction machines removed. The impedance diagram is reduced and evaluated to determine three-phase fault currents, as was the first-cycle impedance diagram. Unbalanced faults must be analyzed using the method of symmetrical components.

4.6 ANALYZING UNBALANCED SYSTEMS

In 1918, Charles Fortescue presented a paper to the American Institute of Electrical Engineers in Atlantic City describing how a system of unbalanced but related phasors can be represented by systems of balanced phasors. Using this principle, any unbalanced three-phase system can be represented by three balanced *sequence networks*. Incidentally, Fortescue's paper was selected by power engineering professionals as the number one high-impact paper in power engineering of the twentieth century (see the References at the end of this chapter).

The theory of symmetrical components and the synthesis of sequence networks for three-phase power systems are instrumental for solving most unbalanced problems such as asymmetrical short-circuit and open-circuit faults. Symmetrical components and sequence networks are also vital for understanding the unbalanced operating conditions of an otherwise balanced system, and the behavior and influence of harmonic voltages and currents.

Unfortunately, the theory of symmetrical components is often learned as a set of abstract algebraic equations, into which known values are substituted and, hopefully, out of which the correct answer emerges. Sequence networks are often synthesized using a *building block approach*, where documented sequence impedance models of various power system elements are connected together much like building blocks to form the sequence network. This method often leads to errors, since topological errors in connecting the *blocks* are easy to make, and topological errors in the sequence network will often produce inaccurate results. But even if the networks are properly constructed, the engineer often lacks the insight and level of understanding to thoroughly comprehend the behavior of the system.

The novel approach for understanding symmetrical components and synthesizing sequence networks presented in this chapter enlightens the engineer to the reasons behind the behaviors observed on an unbalanced three-phase power system. It is this enlightenment that allows the engineer to fully understand the behavior of the three-phase system under unbalanced conditions. As an additional benefit to applying these approaches, commonly made errors in unbalanced system calculations will be significantly decreased if not totally eliminated.

Many power system calculations involve analysis of a balanced three-phase system. When this is the case, only one phase needs to be analyzed. The symmetry of the problem allows the behavior of the other two phases to be determined based on the calculated behavior of the first phase. This single-phase equivalent approach greatly simplifies the calculation process.

But when the conditions to be analyzed result in an unbalanced system of voltage and current phasors, the single-phase equivalent approach cannot be directly applied. Such an example is determining the system response to an unbalanced short-circuit fault, such as a line-to-ground fault. The option of analyzing the unbalanced system as a three-phase problem is not an appealing one, since the resulting mathematics would be cumbersome and very tedious to solve. Using a single-phase approach would be possible if the unbalanced phasors could be resolved into balanced components. Charles Fortescue's theory of symmetrical components shows us that resolving an unbalanced set of voltage or current phasors into a set of balanced components is always possible. Before developing the symmetrical components of an unbalanced set of three-phase phasors, let us look at a more straightforward example of resolving a vector into components.

4.7 PHYSICAL EXAMPLE OF VECTOR COMPONENTS

A basic problem in statics involves calculating the reaction at the attachment point of a cantilevered beam when subjected to a vertical loading, as shown in Figure 4.10.

The downward force **F** tries to produce a clockwise rotation of the beam about the attachment point with the wall. The rotational force is called a *moment*. The wall will, hopefully, produce a moment equal in magnitude and opposite in direction to the moment produced by force **F**. The net moment will then be zero, and the beam will remain stationary.

The moment produced by force **F** is the product of the force magnitude and the perpendicular distance from the force to the attachment point, or $\mathbf{M} = \mathbf{F} \times \mathbf{d}$, or in scalar terms, $|\mathbf{M}| = |\mathbf{F}| |\mathbf{d}|$.

A slight modification to this problem leads to a complication. Figure 4.11 shows another force **F** applied to the same beam, but this time, **F** is not directed downward, but at an angle θ with respect to the vertical.

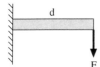

Figure 4.10 Vertical Loading on Cantilevered Beam

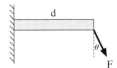

Figure 4.11 Loading Applied to Cantilevered Beam at Angle θ

Now, the resulting moment at the attachment point is not simply the product of |**F**| and |**d**|, since the vector from the attachment point to the point of application of the force, which represents **d**, and the force **F** are not orthogonal or perpendicular. In order to calculate the moment, we can resolve the force **F** into two components: one perpendicular to the **d** vector (the component we need to calculate the moment) and one parallel to the **d** vector (which produces no moment). The resolution of **F** into these components is shown in Figure 4.12.

Now, the moment can be calculated as |**F**$_V$| |**d**|. Note that **F** could have been resolved into components other than **F**$_V$ and **F**$_H$, but **F**$_V$ and **F**$_H$ met the criteria required by the moment equation (**F**$_V$ is perpendicular to **d**, thus determining the moment, and **F**$_H$ is parallel to **d**, which results in no moment).

Trigonometry can be used to express **F**$_V$ and **F**$_H$ in terms of **F** and θ:

$$\mathbf{F_V} = \mathbf{F}\cos\theta \qquad (4.47)$$

$$\mathbf{F_H} = \mathbf{F}\sin\theta \qquad (4.48)$$

Equations (4.47) and (4.48) allow transformation from the original parameters **F** and θ into a component environment to facilitate calculation of the moment. A constraint which must be enforced is that the vector sum of **F**$_V$ and **F**$_H$ equals the original force **F**:

$$\mathbf{F} = \mathbf{F_V} + \mathbf{F_H} \qquad (4.49)$$

The angle θ can also be determined if the components **F**$_V$ and **F**$_H$ are known:

$$\theta = \tan^{-1}\left(\frac{|\mathbf{F_H}|}{|\mathbf{F_V}|}\right) \qquad (4.50)$$

Equations (4.49) and (4.50) allow transformation from the component environment back to the original parameters of **F** and θ.

So, beginning in the **F** and θ environment, we can convert to the component environment of **F**$_V$ and **F**$_H$ to facilitate calculation of the moment. This same concept can be applied for developing a system of components to facilitate the calculation of unbalanced voltages and currents on a normally balanced three-phase system. Not all three-phase systems are balanced by design. If the location of the derived neutral point does not coincide with the balanced neutral point, as shown in Figure 4.13, the

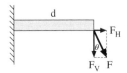

Figure 4.12 Force **F** Resolved into Appropriate Components

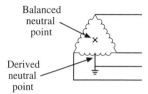

Figure 4.13 Inherently Unbalanced System

system is inherently unbalanced. Note that if the three-phase system is unbalanced by design, such as a four-wire delta system, this method of applying symmetrical components cannot be used.

4.8 APPLICATION OF SYMMETRICAL COMPONENTS TO A THREE-PHASE POWER SYSTEM

Now we will apply the same methodology as we did to resolve force **F** on the beam into suitable components to a normally balanced three-phase system which is operating in an unbalanced mode. We can consider either voltage or current phasors, since the same methodology applies to both. Since current (the flow of charges) is more easily envisioned than voltage (the difference of two electric potentials), we will use current for this example. Consider the unbalanced set of current phasors shown in Figure 4.14.

The phasors are rotating *counterclockwise* at the radian fundamental frequency of the system. For a 60-Hz system, this would be $\omega = 2\pi(60) \approx 377$ rad/s. All phasors rotate counterclockwise—*always*. This is because angles are measured in the counterclockwise direction by convention and is a fact that must be remembered at all times to fully comprehend the concept of symmetrical components.

Since the current phasors are *unbalanced*, meaning that each may have a different magnitude and different angular separation from the other two phasors, we cannot analyze the system taking a single-phase equivalent approach. But by resolving the unbalanced phasors into a suitable set of components, we could then perform a single-phase equivalent analysis of the system, greatly simplifying the analysis process.

Recalling the previous beam example, the necessary requirements for the components devised to represent the force **F** were (1) one component must be perpendicular to the distance vector **d** and (2) the other component must be parallel to the distance vector **d**. The constraint was that the vector sum of the two components

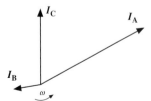

Figure 4.14 Unbalanced Current Phasors

4.8 APPLICATION OF SYMMETRICAL COMPONENTS TO A THREE-PHASE POWER SYSTEM

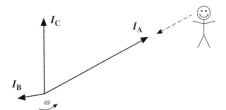

Figure 4.15 Phase Sequencing – A–B–C, or Positive Sequence

equals the original force **F**. Since the three-phase system of current phasors has more degrees of freedom than the beam example, it turns out that we need not two but three sets of components to represent the unbalanced phasors. The necessary requirements for the components of the unbalanced currents are (1) the magnitudes of each of the phasors of a given set of components are equal and (2) the angular separation between any two phasors in a given set of components is equal. These requirements placed on the components give us the name *symmetrical components*.

Before determining the symmetrical components of the unbalanced phasors, we need to understand the concept of *phase sequencing*. Quite often, phase sequencing is referred to as *phase rotation*, but this terminology is very misleading and is technically incorrect. In fact, all phasors *rotate* in the counterclockwise direction—*always*. But while the direction of *rotation* never changes, the *sequencing* of the phasors may change.

Referring to Figure 4.15, the observer is looking directly at the phase A phasor. The next phasor to come around and point at the observer is the one from phase B. Finally, the phase C phasor will rotate by the observer. This defines an A–B–C, or *positive sequence*, set of components.

In contrast, Figure 4.16 shows an example of an A–C–B, or *negative sequence*, set of components.

Notice that in both Figures 4.15 and 4.16, the phasors are *rotating* in the same (counterclockwise) direction. It is the rotational order, or *sequencing*, of the phasors that differs.

Now we can define the three sets of balanced components to represent the unbalanced set of current phasors shown in Figure 4.14. The first set will be a balanced set of phasors having the same phase sequencing as the unbalanced currents (A–B–C). We will call this set the *positive-sequence components* and denote the positive-sequence values with the subscript 1, as shown in Figure 4.17.

The second set of components will be a balanced set of phasors having the opposite sequencing as the original, or A–C–B phase sequencing. We will call this set the *negative-sequence components* and will denote these phasors with the subscript 2, as shown in Figure 4.18.

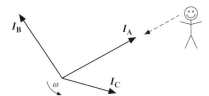

Figure 4.16 Phase Sequencing – A–C–B, or Negative Sequence

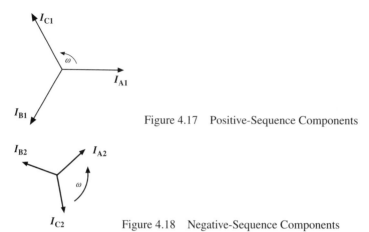

Figure 4.17 Positive-Sequence Components

Figure 4.18 Negative-Sequence Components

Although the magnitudes of the positive-sequence and negative-sequence phasors differ, each set contains three phasors of equal magnitude. Also each phasor in both the positive-sequence and negative-sequence set of components is separated from the other two phasors by equal angular displacements (120°). This fulfills the requirement that each set of components be balanced or symmetric.

The third set of components must also be balanced. We will choose a set of components with equal magnitudes, and *no* angular displacement between the phases. Note that an angular separation of zero also fulfills the definition of *balanced*, since the same angular displacement exists between any two of the three phasors. This set of components is the *zero sequence*, and we use the subscript 0 to denote them, as shown in Figure 4.19.

Studying the sequence components shown in Figures 4.8 through 4.10, it is apparent that we can exploit the symmetry of the systems to simplify the nomenclature. We can define an operator a such that multiplying any phasor by a simply rotates the original phasor by 120°. Thinking in terms of polar coordinates, it becomes obvious that the a operator must have a magnitude of 1, or multiplying a phasor by a would rescale the phasor. To achieve the 120° rotation, the angle of the a operator must be 120°, since angles are additive when multiplying numbers in polar form. Therefore,

$$a \equiv 1\,\underline{/120°} \tag{4.51}$$

The a operator can also be expressed in rectangular form as

$$a = \frac{1}{2} + j\frac{\sqrt{3}}{2}, \tag{4.52}$$

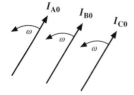

Figure 4.19 Zero-Sequence Components

4.8 APPLICATION OF SYMMETRICAL COMPONENTS TO A THREE-PHASE POWER SYSTEM

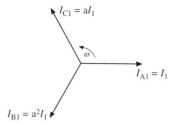

Figure 4.20 Positive-Sequence Components Expressed in Terms of Phase A Quantities

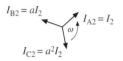

Figure 4.21 Negative-Sequence Components Expressed in Terms of Phase A Quantities

where $j \equiv \sqrt{-1}$. Just as a can be thought of as a 120° rotator, j can be viewed as a 90° rotator. This is an important insight, since it makes the problem less algebraic and more visual and intuitive.

The following powers of the j and a operators are helpful in visualizing how multiplying a phasor by a power of one of these operators manipulates the original phasor. It should be remembered that squaring an operator repeats its rotation twice, cubing an operator repeats its rotation three times, so on.

$$j^2 = 1\,\underline{/180°} = -1 \tag{4.53}$$

$$j^3 = 1\,\underline{/270°} = -j \tag{4.54}$$

$$j^4 = 1\,\underline{/360°} = 1 \tag{4.55}$$

$$a^2 = 1\,\underline{/240°} = \frac{-1}{2} - j\frac{\sqrt{3}}{2} \tag{4.56}$$

$$a^3 = 1\,\underline{/360°} = 1 \tag{4.57}$$

Using the a operator, we can eliminate the double subscript notation used in Figures 4.8 through 4.10 by expressing each phasor in terms of the phase A phasor. This process brings us to a single-phase equivalent of the original system—the goal we were attempting to attain. It is important to remember that the phasors are being redefined in terms of the phase A phasor—this fact will become very important when modeling faults.

Relabeling Figures 4.8 through 4.10 in terms of the phase A quantities by dropping the A subscript and also incorporating the a operator has been done in Figures 4.20 through 4.22.

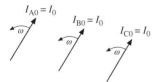

Figure 4.22 Zero-Sequence Components Expressed in Terms of Phase A Quantities

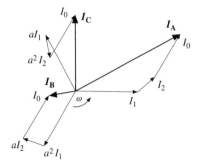

Figure 4.23 Sum of Sequence Components

Of course, the symmetrical components must satisfy the constraint that their vector sum equals the original set of unbalanced phasors. This is shown graphically in Figure 4.23.

Now that a suitable set of components to represent the unbalanced currents has been developed; a set of transformation equations similar to Eqs. (4.47) through (4.50) must be established.

Expressing the unbalanced currents as the sum of their components provides three of the necessary six transformation equations:

$$I_A = I_0 + I_1 + I_2 \tag{4.58}$$
$$I_B = I_0 + a^2 I_1 + a I_2 \tag{4.59}$$
$$I_C = I_0 + a I_1 + a^2 I_2 \tag{4.60}$$

The remaining three equations are obtained by solving Eqs. (4.58) through (4.60) for I_0, I_1, and I_2. This is cumbersome using algebra, but the matrix methods offered by linear algebra greatly simplify the process of obtaining solutions for I_0, I_1, and I_2.

The first step is to rewrite Eqs. (4.58) through (4.60) as a single matrix equation:

$$\begin{bmatrix} I_A \\ I_B \\ I_C \end{bmatrix} = \begin{bmatrix} 1 & 1 & 1 \\ 1 & a^2 & a \\ 1 & a & a^2 \end{bmatrix} \cdot \begin{bmatrix} I_0 \\ I_1 \\ I_2 \end{bmatrix} \tag{4.61}$$

Multiplying both sides of Eq. (4.61) by the inverse of the square coefficient matrix gives us the solution for I_0, I_1, and I_2:

$$\begin{bmatrix} 1 & 1 & 1 \\ 1 & a^2 & a \\ 1 & a & a^2 \end{bmatrix}^{-1} \cdot \begin{bmatrix} I_A \\ I_B \\ I_C \end{bmatrix} = \begin{bmatrix} 1 & 1 & 1 \\ 1 & a^2 & a \\ 1 & a & a^2 \end{bmatrix}^{-1} \cdot \begin{bmatrix} 1 & 1 & 1 \\ 1 & a^2 & a \\ 1 & a & a^2 \end{bmatrix} \cdot \begin{bmatrix} I_0 \\ I_1 \\ I_2 \end{bmatrix} \tag{4.62}$$

Simplifying and bringing the I_0, I_1, and I_2 vector to the left side of the equal sign,

$$\begin{bmatrix} I_0 \\ I_1 \\ I_2 \end{bmatrix} = \begin{bmatrix} 1 & 1 & 1 \\ 1 & a^2 & a \\ 1 & a & a^2 \end{bmatrix}^{-1} \cdot \begin{bmatrix} I_A \\ I_B \\ I_C \end{bmatrix} \tag{4.63}$$

4.9 ELECTRICAL CHARACTERISTICS OF THE SEQUENCE CURRENTS

Inverting the square coefficient matrix gives

$$\begin{bmatrix} I_0 \\ I_1 \\ I_2 \end{bmatrix} = \frac{1}{3} \begin{bmatrix} 1 & 1 & 1 \\ 1 & a & a^2 \\ 1 & a^2 & a \end{bmatrix} \cdot \begin{bmatrix} I_A \\ I_B \\ I_C \end{bmatrix} \quad (4.64)$$

Writing Eq. (4.64) as three algebraic equations gives us the last three necessary transformation equations:

$$I_0 = \frac{1}{3}(I_A + I_B + I_C) \quad (4.65)$$

$$I_1 = \frac{1}{3}(I_A + aI_B + a^2 I_C) \quad (4.66)$$

$$I_2 = \frac{1}{3}(I_A + a^2 I_B + aI_C) \quad (4.67)$$

Using Eqs. (4.58) through (4.60) and Eqs. (4.65) through (4.67) as transformation equations, we can apply the theory of symmetrical components to unbalanced conditions on otherwise balanced three-phase power systems. But before we do, we must understand the electrical characteristics of the sequence currents so we can comprehend how they behave in the three-phase system. This knowledge is essential for the proper synthesis of the sequence networks.

4.9 ELECTRICAL CHARACTERISTICS OF THE SEQUENCE CURRENTS

Figure 4.24 shows a wye-connected source supplying an unspecified load. The load is required so a closed path exists for current to flow.

If the top wire is carrying a current I from the source to the load, we know the only way that current can flow is if there is a return path back to the source. When there is an angular displacement between the line currents, the middle and bottom wires serve as the return path for the source current flowing in the top wire. Writing a node equation using the cloud representing the load as the node, we see that $I = x + y$. This relationship can be verified graphically by drawing the three line currents in the time domain. At a specific time $t = T$, the instantaneous values of the currents in the three wires sum to zero, thus honoring Kirchhoff's current law. This relationship can be seen in Figure 4.25.

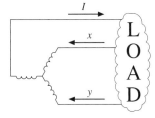

Figure 4.24 Wye-Connected Source Serving Unspecified Load

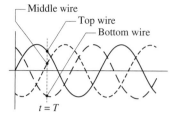

Figure 4.25   Line Currents in Time Domain

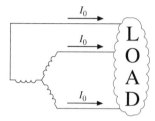

Figure 4.26   Zero-Sequence Current

But the ability to return the current from the source to the load in the top wire (I) on the bottom two wires requires a phase difference, in this case 120°, between the line currents. In the case of the positive- and negative-sequence currents, there is a 120° angular displacement between the top, middle, and bottom wire currents. So for the positive and negative sequence, *a current supplied by one-phase conductor is returned to the source by the other two.* This relation is always true in three-phase circuits.

The zero-sequence current, however, behaves differently. There is no angular displacement between the phases, so whatever instantaneous current flows in the top wire also must flow in the middle and bottom wires.

Figure 4.26 shows a total of 3 I_0 delivered from the source to the load. The only way this current can flow is if it can return to the source. *The zero-sequence current is supplied to the load on the phase conductors, but it cannot return to the source on the phase conductors.*

A fourth conductor must be present to serve as the return path. This fourth conductor can be the neutral. The neutral returns the zero-sequence current supplied by each phase conductor, or 3 I_0. *If a fourth conductor (return path) does not exist, zero-sequence current will not flow.* This is always the case with three-phase circuits.

Figure 4.27 shows the complete circuit path of the zero-sequence current, being supplied on the phase conductors and being returned on the neutral. Many engineers

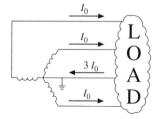

Figure 4.27   Zero-Sequence Current with Return Path

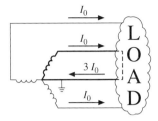

Figure 4.28 Single-Phase Equivalent for Zero-Sequence Current

say that a "ground connection" is required for zero-sequence current to flow. This is really not the case, because a connection to earth is not a requirement for zero-sequence current to flow. It would flow just fine in an ungrounded (isolated) neutral, but it is very unusual to not ground the neutral for safety reasons.

Sometimes the terms *neutral* and *ground* are [incorrectly] used interchangeably. We will make the assumption that the neutral conductor is always grounded but will refer to the return conductor for zero-sequence current as the *neutral* conductor. Keep in mind that with a multiple-point grounded neutral, as is common on utility systems, the neutral conductor is electrically in parallel with the earth, so the zero-sequence current is returned by both the neutral conductor and the earth. In fact, other parallel paths typically exist in the zero-sequence return path, including conduit, structural steel, fences, pipelines—any conductive path between the load and the source.

The circuit depicted in Figure 4.27 becomes interesting when trying to apply a single-phase equivalent approach. The "single-phase" circuit is highlighted in Figure 4.28.

Note that since the neutral conductor returns not only the zero-sequence current from the phase we are considering as our single-phase equivalent (phase A) but also the zero-sequence current from the other two phases. This causes a problem when trying to analyze the single-phase equivalent, since the current supplied (I_0) and the current returned ($3\,I_0$) are different. This problem must be fixed and can be quite simply with a little algebra. For the single-phase equivalent to be valid, the correct voltage drop must be calculated for the neutral return path. If the series reactance in the return path is X_N, the voltage drop for the neutral return path is found using Ohm's Law:

$$V = (3I_0) \times X_N \tag{4.68}$$

Forcing the current in the neutral return path of the single-phase equivalent circuit to equal the current supplied by the single phase (I_0), the coefficient 3 must be removed from the current I_0. Simply discarding this coefficient would change the calculated voltage drop for the neutral return path, thus invalidating the single-phase equivalent circuit. But the calculated voltage drop remains correct if the coefficient is simply grouped with the other term (X_N). This is shown in Eq. (4.69):

$$V = I_0 \times (3X_N) \tag{4.69}$$

This subtle algebraic change has a significant physical interpretation. Any impedance in the neutral return path is subjected to three times the zero-sequence current as is flowing in each of the phase conductors; therefore, to provide the proper voltage drop, any impedance in the neutral portion of the circuit must be *tripled* when modeling the circuit as sequence networks. And since zero-sequence current is the only current component that can flow in the neutral, this condition applies only to the zero-sequence network.

4.10 SEQUENCE NETWORKS

Now that the need for modeling unbalanced currents as symmetrical components is understood, the concept of *sequence networks* must be introduced. When a current I flows through an impedance Z, the current should be interpreted as the sum of three balanced components. For phase A,

$$I_A = I_0 + I_1 + I_2 \tag{4.70}$$

Each component of current can experience a different effective value of impedance. This rather abstract concept must be accepted, although the underlying reasons are not easily understood. Although far from a perfect analogy, one might consider a current containing several harmonic components. Each harmonic component experiences a different resistance value when flowing through a wire. This is due to the fact that AC resistance is a function of frequency. The sequence currents I_0, I_1, and I_2 are all at the system fundamental frequency, so the analogy is not perfect, but like the harmonic currents, the symmetrical components can each experience a different impedance value in a given portion of a system. Ohm's law can therefore be stated for each sequence component:

$$V_0 = I_0 \times Z_0 \tag{4.71}$$
$$V_1 = I_1 \times Z_1 \tag{4.72}$$
$$V_2 = I_2 \times Z_2 \tag{4.73}$$

Since each component of current can experience a different impedance, three different impedance networks must be developed for the system to be analyzed. Since most studies of unbalanced systems involve short-circuit fault calculations, it is common to neglect the resistive portion of the impedance, since its effect on the short-circuit current magnitude is very small. For that reason, we will proceed to develop a positive-, a negative-, and a zero-sequence reactance network. Consider the one-line diagram shown in Figure 4.29.

The positive-sequence reactance network is developed directly from the one-line diagram of the system. First, a Positive-Sequence Reference Bus is drawn. By convention, this bus is drawn at the top of the diagram. Although merely a convention, being consistent with this practice will facilitate both the proper network topology and the correct interconnection of the networks when the fault calculation is done. After the Positive-Sequence Reference Bus is drawn, all *sources* and *loads* capable of storing energy (fault current contributors) on the one-line diagram are connected to it. This typically means utility system interconnections, generators, and motors. Capacitor banks are also a source of fault current, so their reactance must also be connected to the reference bus.

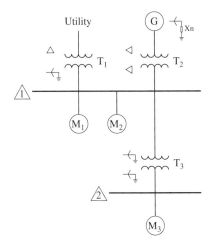

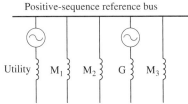

Figure 4.29 Example One-Line Diagram

Figure 4.30 First Stage of Positive-Sequence Network Development

The source impedances are modeled in series with an EMF source representing the pre-fault voltage at that point in the system. Since this voltage is generally not known unless a power flow calculation is performed, it is often assumed as 1.0 p.u. and is assigned the reference angle of zero degrees. The first stage of the positive-sequence network construction is shown in Figure 4.30.

Next, the other components from the one-line diagram are modeled as reactances. Transformers T1 and T2 are drawn, and the location of Bus 1 is established. This can be seen in Figure 4.31.

Finally, Transformer T3 and the location of Bus 2 are established, completing the positive-sequence network as shown in Figure 4.32.

Note that each numeric reactance value is the positive-sequence reactance for that component.

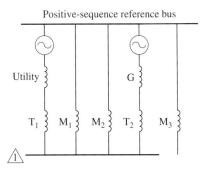

Figure 4.31 Next Stage of Positive-Sequence Network Development

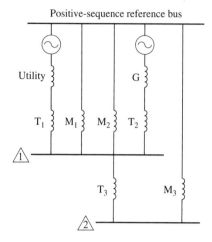

Figure 4.32 Positive-Sequence Network

The negative-sequence network can be developed directly from the positive-sequence network. The following three steps, when applied to the positive-sequence network, will yield the negative-sequence network:

1. Remove (short circuit) the EMF sources
2. Relabel the reference bus as the "Negative-Sequence Reference Bus"
3. Change the numeric values of the reactances from the positive-sequence values to the negative-sequence values

In the actual system, only positive-sequence voltages are generated. Therefore, all voltage sources will appear in the positive-sequence network only. Note that the reactances behind the generated voltages still appear in the negative-sequence network as negative-sequence reactances.

Most power system components are *bilateral*, meaning that their behavior when current flows through them in one direction is the same as when the direction of current flow reverses. Rotating machines are an exception. Due to the direction of the rotor rotation, the behavior of the magnetic flux across the air gap changes with the direction of the current flow. This means that, in general, rotating machines will have different negative-sequence reactance values than positive-sequence values. The numeric difference increases as the machine speed decreases and as the pole faces become more salient, so for cylindrical rotor high-speed machines, $X_1 \approx X_2$, but X_1 and X_2 can become substantially different as the number of poles increases and the poles become more salient.

The negative-sequence network for the one-line diagram shown in Figure 4.29 is shown in Figure 4.33.

The zero-sequence network can be developed from the negative-sequence network, but the modification steps are a bit different:

1. Relabel the reference bus as the "Zero-Sequence Reference Bus"
2. Change the numeric values of the reactances from the negative-sequence values to the zero-sequence values

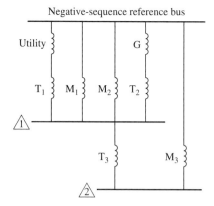

Figure 4.33 Negative-Sequence Network

3. Add *three times* the grounding impedance to the numeric reactance value of any machine that is grounded through an impedance
4. Adjust the topology of the network to force proper zero-sequence current behavior

The fourth step is a very important one and will be addressed in detail, but first, a few comments about steps 2 and 3. The negative- and zero-sequence reactance values are substantially different for most components. A notable exception is the two-winding power transformer, where all three sequence reactances equal the leakage reactance of the transformer. Transmission and distribution lines have zero-sequence reactances that are higher than their positive- and negative-sequence reactances. This is due to the impedance of the earth current return path and is explained by Carson's equations (see the References at the end of this chapter). Rotating machines, on the other hand, have zero-sequence reactances that are much lower than their positive- and negative-sequence reactances due to the large magnitude of the in-phase zero-sequence flux across the air gap.

Step 3 states the need to increase the numeric reactance value of any machine that is grounded through an impedance by three times the value of the grounding impedance. The grounding impedance must be included in the zero-sequence network and is not included in the other two sequence networks, because only zero-sequence current can flow to ground. The other two sequence currents can flow only in the phase conductors. The reason for tripling the grounding impedance is shown by Eq. (4.69). Using impedance grounding can substantially reduce the magnitudes of ground faults. In generator circuits, a grounding resistor coupled through a single-phase distribution transformer is used to limit the zero-sequence current, typically to 5 A. It should be noted that grounding impedance has absolutely no effect of the magnitudes of three-phase or line-to-line faults. The use of isolated-phase bus in the generator circuit eliminates the possibility of an ungrounded fault by surrounded each phase conductor with a grounded aluminum shield.

The fourth step is to adjust the topology of the zero-sequence network to force the current flowing in that network to behave like zero-sequence current. Section 4.9 explains that zero-sequence can only flow in the parts of a circuit that have a

128 CHAPTER 4 FAULT CALCULATIONS

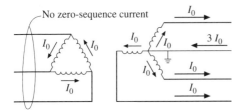

Figure 4.34 Delta–Wye Transformer Zero-Sequence Current Behavior

fourth conductor to serve as a return path. This means that delta and ungrounded wye portions of the system will not allow zero-sequence current to flow. The network topology must be altered to reflect this fact.

These alterations are best understood by example. But before attempting an example, it will be helpful to analyze the zero-sequence current behavior of the delta–wye transformer.

Begin analyzing the delta–wye transformer shown in Figure 4.34 with the wye circuit. In order for zero-sequence current to flow to the load on the phase conductors, the total zero-sequence current furnished ($3\,I_0$) must return on the neutral.

The per-unit zero-sequence current flowing in each of the wye-connected transformer windings must also flow in the corresponding delta-connected windings. Writing the node equation at each corner of the delta makes it apparent that no zero-sequence current can flow out of the delta onto the lines. The zero-sequence current flows in the wye-connected windings and circulates in the delta-connected windings.

Note that if the circulating current in the delta is in the form of a third harmonic current, the resulting temperature rise due to the higher frequency (and skin effect) may be problematic. Third harmonic currents behave like zero-sequence currents, as will be shown in Chapter 12.

In the zero-sequence reactance diagram, the zero-sequence current must be blocked from exiting the delta onto the lines of the three-wire circuit. This is accomplished by introducing an open circuit on the delta side of the transformer. But that open circuit would also prevent zero-sequence current from flowing through the transformer impedance (windings), and we can see in Figure 4.34 that this is incorrect. So a short circuit back to the reference bus allows the zero-sequence current to flow from the wye circuit, through the transformer reactance, and to the reference bus, while the open circuit prevents the zero-sequence current from flowing out of the transformer to the delta circuit. The short circuit to the reference bus simulates the zero-sequence current circulation in the delta-connected windings. This is what Table 4.2 shows as the required topology alteration for a delta connection (Open Circuit AND Short

TABLE 4.2 Zero-Sequence Network Alteration Rules

Connection	Alteration
Grounded Wye	None
Wye	Open Circuit
Delta	Open Circuit AND Short Circuit to Reference Bus

4.10 SEQUENCE NETWORKS 129

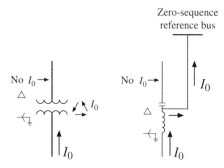

Figure 4.35 Delta–Wye Transformer Zero-Sequence Circuit Model

Circuit to Reference Bus). Figure 4.35 shows the zero-sequence circuit model for a delta–grounded wye transformer.

Using the alteration rules summarized in Table 4.2, the topology of each machine (transformer, generator, and motor) reactance can be altered to allow proper zero-sequence behavior. Note that the connection types of the motors shown in Figure 4.29 are not specified. This is not a problem, because motors are virtually always wired as a three-wire connection (either delta or wye—*not* grounded wye). According to Table 4.2, both the delta and wye connections involve an open circuit, and this open circuit effectively removes the motor from the zero-sequence network, as motor impedances are loads and are always connected directly to the reference bus.

Begin developing the zero-sequence network from the negative-sequence network by relabeling the reference bus and changing the numeric reactance values from the negative-sequence values to the zero-sequence values. Next, transfer the phasing symbols (deltas, wyes, and grounded wyes) from the one-line diagram to the zero-sequence reactance diagram, as shown in Figure 4.36.

At this point, any impedance grounded devices, such as the generator, must have their zero-sequence reactance increased by three times the grounding impedance. Figure 4.36 shows the impedance of the generator increased from its original value of G to $G + 3 X_n$.

Finally, the topology alteration rules of Table 4.2 can be applied. Any grounded wye circuit is left unaltered, since zero-sequence current can flow in a four-wire

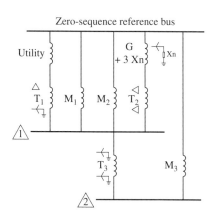

Figure 4.36 First Step of Zero-Sequence Reactance Network Development

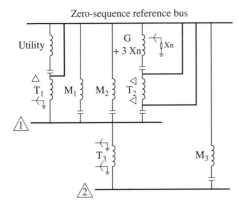

Figure 4.37 Zero-Sequence Network

circuit. Ungrounded wye devices are open-circuited, because the lack of return path will prevent zero-sequence current from flowing in a three-wire circuit. And delta-connected components are altered to include both an open circuit (to prevent zero-sequence current from flowing on the three-wire circuit) and a short circuit to the reference bus (to simulate the circulating path provided by the delta-connected windings). The resulting zero-sequence network is shown in Figure 4.37.

Fault current magnitudes will be calculated at Bus 1 of the system pictured in Figure 4.5 for all four types of short-circuit faults. To facilitate the fault calculations, the Thévenin reactance of the positive-, negative-, and zero-sequence networks now will be determined for Bus 1.

Referring to Figure 4.7, the Thévenin reactance of the positive-sequence network for a fault at Bus 1 is found by short-circuiting all EMF sources, then determining the equivalent reactance from the fault location (Bus 1) to the reference bus. Multiple parallel paths exist from Bus 1 to the reference bus. One path is the series combination of the utility reactance and the reactance of transformer T_1. Another is the generator reactance G_1. The three motors connected to Bus 1 (M_1, M_2, and M_3) each provide a path from Bus 1 to the reference bus. Two more parallel paths exist. The first is through Bus 2: C_1, T_2, and the parallel combination of M_4, M_5, and M_6, all combined in series. The second is through Bus 3: C_2, T_3, and the parallel combination of M_7 and M_8, all combined in series. Symbolically, the Thévenin reactance at Bus 1 can be shown as

$$X_{1\text{Thevenin (Bus 1)}} = (\text{Utility} + T_1) \| G_1 \| M_1 \| M_2 \| M_3 \| [C_1 + T_2 \\ + (M_4 \| M_5 \| M_6)] \| [C_2 + T_3 + (M_7 \| M_8)] \quad (4.74)$$

Numerically,

$$X_{1(\text{Bus 1})} = (0.0067 + 0.0258) \| 0.0340 \| 0.2667 \| 0.3400 \| \\ 0.4857 \| [0.0030 + 0.0767 + (0.6800 \| 1.7000 \| 2.6667)] \| \quad (4.75) \\ [0.0034 + 0.2400 + (4.0000 \| 10.0000)]$$

Combining the paralleled motor reactances,

$$X_{1(\text{Bus 1})} = (0.0067 + 0.0258) \| (0.0277) \| 0.4857 \| \\ [0.0030 + 0.0767 + (0.4109)] \| [0.0034 + 0.2400 + (2.8571)] \quad (4.76)$$

Combining the series reactances,

$$X_{1(\text{Bus 1})} = 0.0325 \| 0.0277 \| 0.4857 \| 0.4906 \| 3.1005 \tag{4.77}$$

Finally, combining these five reactances in parallel,

$$X_{1(\text{Bus 1})} = j\,0.0140 \text{ p.u.} \tag{4.78}$$

To determine the negative-sequence reactance diagram for the system shown in Figure 4.5, start with the positive-sequence network (Figure 4.7), short-circuit the EMF sources, relabel the reference bus as the "Negative-Sequence Reference Bus," then be sure that all reactance values are numerically equal to the negative-sequence reactances. Only rotating machines can have X_2 values that differ from their X_1 values, so in our system, we will change the reactance values of the generator and the synchronous motor.

Negative-sequence machine impedances are highly dependent on the design of the machine and must be provided, usually in a machine test report. Using a negative-sequence reactance of 9% on its own bases, the generator will have a per-unit negative-sequence reactance of

$$X_{2(G1)} = 0.09 \left(\frac{10}{25}\right) = 0.0360 \text{ p.u.} \tag{4.79}$$

We will use a negative-sequence reactance of 22.5% on its own bases for the synchronous motor; its per-unit negative-sequence reactance becomes

$$X_{2(M1)} = 0.225 \left(\frac{10}{7.5}\right) = 0.3000 \text{ p.u.} \tag{4.80}$$

The negative-sequence reactance network for the system shown in Figure 4.5 is shown in Figure 4.38.

The Thévenin reactance of the negative-sequence network at Bus 1 is determined using the same method as was used to reduce the positive-sequence network to its Thévenin equivalent. Remember that some impedance values may differ between

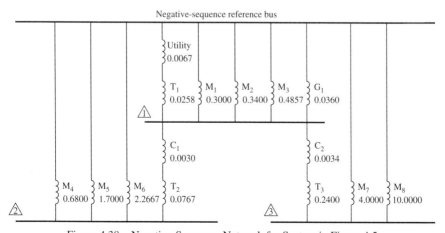

Figure 4.38 Negative-Sequence Network for System in Figure 4.5

the positive- and negative-sequence networks, but in practice the values usually do not differ by much. When the negative-sequence network shown in Figure 4.38 is simplified to a Thévenin reactance at Bus 1, that reactance is found to be

$$X_{2(\text{Bus 1})} = j\,0.0144 \text{ p.u.} \tag{4.81}$$

To determine the zero-sequence reactance diagram for the system shown in Figure 4.5, start with the negative-sequence network (Figure 4.38), relabel the reference bus as the "Zero-Sequence Reference Bus," and change the reactance values to reflect the zero-sequence reactances.

Next, the numeric values of the reactances must be changed to reflect the zero-sequence reactance values. Two-winding power transformers have X_0 values equal to their X_1 and X_2 values, so no changes are needed.

The utility source will have a different X_0 value than its X_1 and X_2 values. The value of the utility's X_0 is predominantly a function of distance from the delta–grounded wye substation transformer supplying the feeder to which the industrial system is connected. Since this parameter can vary widely, the utility X_0 must be obtained from the utility company. However, if a delta–wye transformer is used to supply the industrial system, which is the case in our example and is a common utility practice, the utility X_0 becomes immaterial, as we will see later in this example. So for our example, the utility's X_0 is not needed.

The zero-sequence reactances of rotating machines will be substantially lower than their X_1 and X_2 values. The X_0 value of a rotating machine is not nameplate data, so its value must either be obtained from test report data or estimated. A reasonable approximation of X_0 for a rotating machine is about 1/3 of its positive-sequence reactance. But, unless the rotating machine is grounded (which is usually done for generators only), the X_0 value becomes immaterial, as we will see later in this example. So for our example, we will use an X_0 value of 0.0140 for generator G_1 and will disregard the X_0 values for the ungrounded motors.

Cables have zero-sequence reactances larger than their positive-sequence values. The X_0 values for cables should be obtained from manufacturer data, as they can vary widely. In absence of accurate data, a reasonable estimate of 2.5 to 4.0 times the X_1 value may be used. For our example, we will use X_0 values of 0.0105 and 0.0120 for C_1 and C_2, respectively. The zero-sequence network up to this point can be seen in Figure 4.39.

Next, take the grounding impedance for any impedance-grounded machine, triple it, and add that value to the zero-sequence reactance of that machine. In this example, we will use an X_n value of 0.0120 for generator G_1. Last, transfer the phasing symbols from the one-line diagram to the zero-sequence reactance diagram and modify the topology of the network per Table 4.2. The completed zero-sequence network can be seen in Figure 4.40.

Note that motors are connected either in ungrounded wye or in delta—usually not in grounded wye. This is because under normal operation, the motor represents a balanced three-phase load, so no current would flow on the neutral. The neutral conductor would only come into play during ground faults, in which case its presence would increase fault current magnitude. Since the neutral conductor would serve no benefit in a motor circuit and would actually worsen the effects of a ground

4.10 SEQUENCE NETWORKS   133

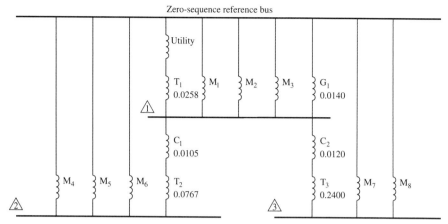

Figure 4.39 Partially Developed Zero-Sequence Network for System in Figure 4.5

fault, three-wire circuits are virtually always used for motor applications. Three-wire circuits always involve an open-circuit in the zero-sequence network, which isolates the motor from the rest of the system. For this reason, motors are not considered as sources of zero-sequence current.

The delta–wye transformer supplying this system from the utility also introduces an open circuit in the zero-sequence network thus isolating the utility system from the industrial system in the zero-sequence network. This makes the utility's X_0 immaterial when analyzing the industrial system.

The Thévenin equivalent for a fault at Bus 1 of Figure 4.40 is found by combining the reactance of transformer T_1 in parallel with the impedance of the generator, including three times its grounding reactance, as every other reactance in the system is isolated by an open circuit in the zero-sequence network. This result can be seen in Figure 4.41.

Note that Bus 2 is removed from the network because of the open circuit introduced by the delta winding of transformer T_2. Similarly, the delta winding

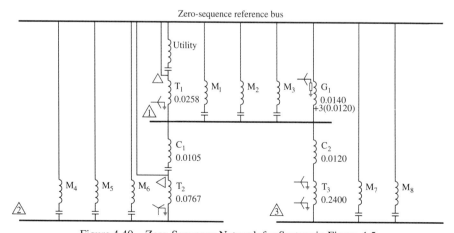

Figure 4.40 Zero-Sequence Network for System in Figure 4.5

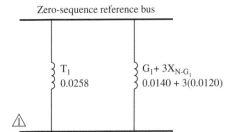

Figure 4.41 Zero-Sequence Reactance Network for Fault at Bus 1 of Figure 4.5

of transformer T_1 removes the system reactance from the network. Bus 3 is also removed from the network, because there is an infinite impedance between Bus 3 and the zero-sequence reference bus. Simplifying the network shown in Figure 4.41 yields a Thévenin reactance of

$$X_{0(\text{Bus 1})} = j\,0.0170 \text{ p.u.} \tag{4.82}$$

4.11 SHORT-CIRCUIT FAULTS

A short-circuit fault occurs when one phase is electrically connected, either solidly or through impedance, to another phase or phases and/or ground. The *three-phase fault* is the only symmetrical short-circuit fault on a three-phase system. The other three types of short-circuit faults possible on a three-phase system are asymmetric and must be analyzed as unbalanced faults using the method of symmetrical components.

In practice, fault calculations are usually done with the assistance of computer software developed specifically for that purpose. Such software is typically user-friendly, requiring minimal time to become proficient in its use. It should be noted that the same data preparation steps need to be taken whether performing the fault calculations manually or by computer. And the need to verify the validity of computer-based calculations cannot be emphasized enough. A quick manual calculation is an easy and effective means of validating the output of a computer-based calculation. When under time constraints, it may be tempting to skip the manual verification step, but the ramifications of this shortcut can be severe. Good engineering practice requires a means of checking the validity of a computer calculation, and a quick hand calculation is an excellent means of doing so.

4.11.1 Three-Phase Fault

A three-phase short-circuit fault involves the shorting of all three phases, as shown in Figure 4.42.

Since the three-phase fault is symmetric, it is analyzed by considering the positive-sequence network only. The Z_F impedance represents the impedance of

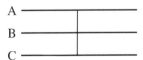

Figure 4.42 Three-Phase Fault

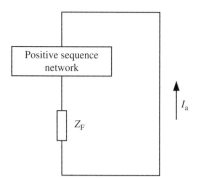

Figure 4.43   Three-Phase Fault Circuit Model

the fault, such as arcing impedance. Typically, this value is assumed to be zero to give a worst-case fault current magnitude, but incorporating a nonzero value into the calculation gives a more realistic value of the fault current that is likely to occur during the fault, which is useful for coordinating protective devices. Typical arcing impedance for a medium-voltage system usually falls in the range of 25–40 Ω. The three-phase short-circuit fault model is shown in Figure 4.43.

To calculate the fault current magnitude for a three-phase fault at Bus 1 of the system shown in Figure 4.5, the three-phase fault circuit model is constructed, as shown in Figure 4.44. Fault impedance is neglected to calculate the maximum fault current.

The phase a fault current is

$$I_A = \frac{1\,\underline{/0°}}{0.0140\,\underline{/90°}} = 71.429\,\underline{/-90°}\ \text{p.u.} \tag{4.83}$$

The base current at Bus 1 is

$$I_{\text{base (Bus 1)}} = \frac{10,000\,\text{kVA}}{(13.8\,\text{kV})\sqrt{3}} = 418\,\text{A} \tag{4.84}$$

Therefore, the phase a fault current at Bus 1 for a three-phase fault becomes

$$I_A = (71.429\,\underline{/-90°}\ \text{p.u.})(418\,\text{A}) = 29.86\,\underline{/-90°}\ \text{kA} \tag{4.85}$$

Assuming positive (a–b–c) sequencing, by symmetry

$$I_B = 29.86\,\underline{/150°}\ \text{kA} \tag{4.86}$$

and

$$I_C = 29.86\,\underline{/30°}\ \text{kA} \tag{4.87}$$

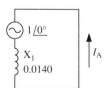

Figure 4.44   Three-Phase Fault at Bus 1 of Figure 4.5

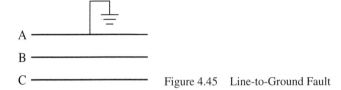

Figure 4.45 Line-to-Ground Fault

4.11.2 Line-to-Ground Fault

A line-to-ground fault, shown in Figure 4.45, requires placement of the fault on phase a, since that is the phase that was used for the single-phase equivalent when we developed our symmetrical component relationships.

When the fault occurs, no fault current will flow on the unfaulted phases, so

$$I_b = I_c = 0 \tag{4.88}$$

Using Eqs. (4.13) and (4.14), currents I_b and I_c can be set equal to each other:

$$I_0 + a^2 I_1 + a I_2 = I_0 + a I_1 + a^2 I_2 \tag{4.89}$$

Subtracting the I_0 term on each side of the equation and collecting similar terms,

$$a^2 I_1 - a I_1 = a^2 I_2 - a I_2 \tag{4.90}$$

Factoring gives

$$I_1(a^2 - a) = I_2(a^2 - a) \tag{4.91}$$

Finally, dividing out the common term produces

$$I_1 = I_2 \tag{4.92}$$

But we also know that I_b and I_c must equal zero, so setting one of them equal to zero gives

$$I_0 + a^2 I_1 + a I_2 = 0 \tag{4.93}$$

Substituting Eq. (4.92) results in

$$I_0 + a^2 I_1 + a I_1 = 0 \tag{4.94}$$

Equation (4.94) shows that the positive-sequence current phasor rotated by 240°, added to the positive-sequence current phasor rotated by 120°, must equal zero after being added with I_0. This can only be true if I_0 is equal to the positive-sequence current, because then Eq. (4.94) would become

$$I_1 + a^2 I_1 + a I_1 = 0 \tag{4.95}$$

Any phasor added to itself rotated by 240° and again to itself rotated by 120° forms an equilateral triangle, indicating a sum of zero. So

$$I_0 = I_1 \tag{4.96}$$

Between Eqs. (4.92) and (4.96), we can see that

$$I_0 = I_1 = I_2 \tag{4.97}$$

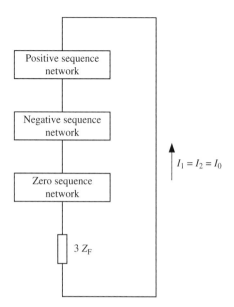

Figure 4.46 Line-to-Ground Fault Circuit Model

For equal currents to flow through each sequence network, the three sequence networks must be *in series*.

Therefore, a line-to-ground fault is modeled by connecting the positive-, negative-, and zero-sequence networks in series. The phase a current for a line-to-ground fault equals three times the sequence current that flows in the series-connected circuit.

$$I_{a(L-G\ fault)} = 3I_1 \qquad (4.98)$$

The line-to-ground fault circuit model is shown in Figure 4.46.

The fault impedance is tripled since it lies in the path of the zero-sequence current (see Eq. 4.69).

To calculate the fault current magnitude for a line-to-ground fault at Bus 1 of the system shown in Figure 4.5, the line-to-ground fault circuit model is constructed, as shown in Figure 4.47.

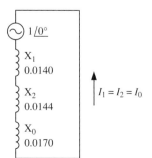

Figure 4.47 Line-to-Ground Fault at Bus 1 of Figure 4.5

138 CHAPTER 4 FAULT CALCULATIONS

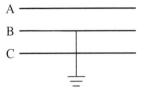

Figure 4.48 Double Line-to-Ground Fault

Next, the three series reactances are combined to a single reactance of 0.0454 p.u. The phase a positive-sequence current is calculated by dividing this reactance into the prefault voltage of 1 $\underline{/0°}$ p.u.

$$I_1 = \frac{1 \underline{/0°}}{0.0454 \underline{/90°}} = 22.0264 \underline{/-90°} \text{ p.u.} \qquad (4.99)$$

The total phase a current for a line-to-ground fault is three times the positive-sequence current as stated in Eq. (4.84).

$$I_a = 3(22.0264 \underline{/-90°}) = 66.0793 \underline{/-90°} \text{ p.u.} \qquad (4.100)$$

Since phases b and c are unfaulted, no fault current flows in them and ($I_b = I_c = 0$). All fault current flows in phase a and returns via the neutral.

Using the base current for Bus 1 as calculated in Eq. (4.84), the total symmetrical line-to-ground first-cycle fault current magnitude at Bus 1 is

$$I_{L-G \text{ (Bus 1)}} = (66.0793 \text{ p.u.})(418 \text{ A}) = 27.62 \text{ kA} \qquad (4.101)$$

4.11.3 Double Line-to-Ground Fault

A double line-to-ground fault, shown in Figure 4.48, requires placement of the fault on phases b and c, since phase a must be the dissimilar (in this case, the unfaulted) phase, as it was the phase that was used to develop our symmetrical component relationships.

When the fault occurs, the faulted phases will have a voltage equal to zero, so

$$V_b = V_c = 0 \qquad (4.102)$$

Using the voltage-equivalent forms of Eqs. (4.59) and (4.60), voltages V_b and V_c can be set equal to each other:

$$V_0 + a^2 V_1 + a V_2 = V_0 + a V_1 + a^2 V_2 \qquad (4.103)$$

Subtracting the V_0 term on each side of the equation and collecting similar terms,

$$a^2 V_1 - a V_1 = a^2 V_2 - a V_2 \qquad (4.104)$$

Factoring gives

$$V_1(a^2 - a) = V_2(a^2 - a) \qquad (4.105)$$

Finally, dividing out the common term produces

$$V_1 = V_2 \qquad (4.106)$$

4.11 SHORT-CIRCUIT FAULTS

But we also know that V_b and V_c must equal zero, so setting one of them equal to zero gives

$$V_0 + a^2 V_1 + a V_2 = 0 \qquad (4.107)$$

Substituting Eq. (4.41) results in

$$V_0 + a^2 V_1 + a V_1 = 0 \qquad (4.108)$$

Equation (4.108) shows that the positive-sequence voltage phasor, rotated by 240°, added to the positive-sequence voltage phasor, rotated by 120°, must equal zero after being added with V_0. This can only be true if V_0 is equal to the positive-sequence voltage, because then Eq. (4.108) would become

$$V_1 + a^2 V_1 + a V_1 = 0 \qquad (4.109)$$

Again, any phasor added to itself rotated by 240° and again to itself rotated by 120° forms an equilateral triangle, indicating a sum of zero. So

$$V_0 = V_1 \qquad (4.110)$$

Between Eqs. (4.106) and (4.110), we can see that

$$V_0 = V_1 = V_2 \qquad (4.111)$$

For equal voltages to appear across each sequence network, the three sequence networks must be *in parallel*.

Therefore, a double line-to-ground fault is modeled by connecting the positive-, negative-, and zero-sequence networks in parallel. The current flowing through the positive-sequence network is the positive-sequence current. Similarly, the current flowing through the negative-sequence network is the negative-sequence current, and the zero-sequence current flows through the zero-sequence network.

The double line-to-ground fault circuit model is shown in Figure 4.49.

To calculate the fault current magnitude for a double line-to-ground fault at Bus 1 of the system shown in Figure 4.5, the double line-to-ground fault circuit model is constructed, as shown in Figure 4.50.

Basic circuit analysis techniques are used to determine the sequence currents. The X_0 and X_2 reactances can be combined in parallel, then added in series with X_1

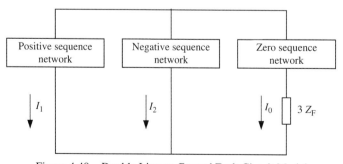

Figure 4.49 Double Line-to-Ground Fault Circuit Model

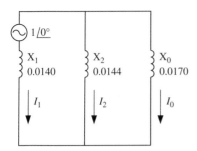

Figure 4.50 Double Line-to-Ground Fault at Bus 1 of Figure 4.5

to determine current I_1 by dividing that resultant reactance into the prefault voltage of $1 \underline{/0°}$.

$$I_1 = \frac{V}{\frac{X_0 X_2}{X_0 + X_2} + X_1} = \frac{1 \underline{/0°}}{0.0218 \underline{/90°}} = 45.8796 \underline{/-90°} \text{ p.u.} \quad (4.112)$$

The negative-sequence current then can be determined by applying the current divider equation to I_1, which divides between the negative- and zero-sequence branches. Note that I_1 will always be 180° out of phase from I_2 and I_0 due to the defined directions of the currents in Figure 4.49.

$$I_2 = I_1 \frac{X_0}{X_0 + X_2} = (45.8796 \underline{/90°})(0.5414)$$
$$= 24.8393 \underline{/90°} \text{ p.u.} \quad (4.113)$$

Finally, the zero-sequence current is found by writing the node equation at the bottom node of Figure 4.50.

$$I_0 = -(I_1 + I_2) = -(45.8796 \underline{/-90°} + 24.8393 \underline{/90°}) = 21.0403 \underline{/90°} \text{ p.u.} \quad (4.114)$$

Once the sequence currents have been determined, the phase currents can be calculated by applying Eq. (4.61).

$$\begin{bmatrix} I_a \\ I_b \\ I_c \end{bmatrix} = \begin{bmatrix} 1 & 1 & 1 \\ 1 & a^2 & a \\ 1 & a & a^2 \end{bmatrix} \cdot \begin{bmatrix} 21.0403 \underline{/90°} \\ 45.8796 \underline{/-90°} \\ 24.8393 \underline{/90°} \end{bmatrix} \quad (4.115)$$

Rewriting as algebraic equations,

$$I_a = I_0 + I_1 + I_2 = 21.0403 \underline{/90°} + 45.8796 \underline{/-90°} + 24.8393 \underline{/90°} = 0 \quad (4.116)$$

$$I_b = I_0 + a^2 I_1 + a I_2$$
$$= 21.0403 \underline{/90°} + 45.8796 \underline{/150°} + 24.8393 \underline{/210°} \quad (4.117)$$
$$= 68.8980 \underline{/152.74°} \text{ p.u.}$$

4.11 SHORT-CIRCUIT FAULTS

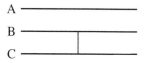

Figure 4.51 Line-to-Line Fault

$$I_c = I_0 + aI_1 + a^2I_2$$
$$= 21.0403 \underline{/90°} + 45.8796 \underline{/30°} + 24.8393 \underline{/330°} \quad (4.118)$$
$$= 68.8980 \underline{/27.26°} \text{ p.u.}$$

Note that the angles of current I_b and I_c are supplementary angles (152.74° + 27.26° = 180°). This will always be the case with double line-to-ground faults.

Applying the base current at Bus 1 of 418 A per Eq. (4.84) to the results of Eq. (4.116) through (4.118) gives

$$I_a = (0)(418\,\text{A}) = 0\,\text{kA} \quad (4.119)$$

$$I_b = (68.8980\,\text{p.u.})(418\,\text{A}) = 28.80\,\text{kA} \quad (4.120)$$

$$I_c = (68.8980\,\text{p.u.})(418\,\text{A}) = 28.80\,\text{kA} \quad (4.121)$$

Equation (4.119) logically yields a zero result, since phase a is unfaulted. All fault current flows in phases b and c.

4.11.4 Line-to-Line Fault

A line-to-line fault, as shown in Figure 4.51, behaves similarly to a double line-to-ground, except for the fact that zero-sequence current cannot flow since there is no return path for it.

So a line-to-line fault is modeled by connecting the positive- and negative-sequence reactances in parallel. The zero-sequence network is open circuited and does not appear in the circuit model. Any fault impedance Z_F is added in series between the networks. Because of the parallel connection of the positive- and negative-sequence networks, the positive- and negative-sequence currents are equal but opposite in sign. The line-to-line fault circuit model is shown in Figure 4.52.

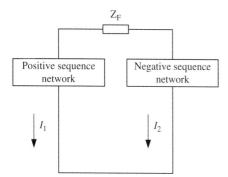

Figure 4.52 Line-to-Line Fault Circuit Model

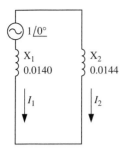

Figure 4.53 Line-to-Line Fault at Bus 1 of Figure 4.5

The fault current magnitude for a line-to-line fault at Bus 1 of the system shown in Figure 4.5 can be determined by constructing the circuit model shown in Figure 4.53.

Adding the two reactances and dividing that sum into the positive-sequence voltage V determines the positive-sequence current I_1.

$$I_1 = \frac{V}{X_1 + X_2} = \frac{1\ \underline{/0°}}{0.0284\ \underline{/90°}} = 35.2113\ \underline{/-90°}\ \text{p.u.} \qquad (4.122)$$

The negative-sequence current is simply 180° out of phase with I_1.

$$I_2 = -I_1 = 35.2113\ \underline{/90°}\ \text{p.u.} \qquad (4.123)$$

The phase currents can be found by using Eqs. (4.58) through (4.60). Since there is no zero-sequence current, $I_0 = 0$.

$$I_a = I_1 + I_2 = 35.2113\ \underline{/-90°} + 35.2113\ \underline{/90°} = 0 \qquad (4.124)$$

$$\begin{aligned} I_b &= a^2 I_1 + a I_2 = 35.2113\ \underline{/150°} + 35.2113\ \underline{/210°} \\ &= 60.9878\ \underline{/180°}\ \text{p.u.} \end{aligned} \qquad (4.125)$$

$$\begin{aligned} I_c &= a I_1 + a^2 I_2 = 35.0877\ \underline{/30°} + 35.0877\ \underline{/330°} \\ &= 60.9878\ \underline{/0°}\ \text{p.u.} \end{aligned} \qquad (4.126)$$

Applying the base current at Bus 1 of 418 A per Eq. (4.84) to the results of Eq. (4.124) through (4.126) gives

$$I_a = (0)(418\ \text{A}) = 0\ \text{kA} \qquad (4.127)$$

$$I_b = (60.9878\ \text{p.u.})(418\ \text{A}) = 25.49\ \text{kA} \qquad (4.128)$$

$$I_c = (60.9878\ \text{p.u.})(418\ \text{A}) = 25.49\ \text{kA} \qquad (4.129)$$

These results are also consistent with logic. Since phase a is unfaulted, all fault current flows in phases b and c, and since all the fault current supplied by phase b must return on phase c, their equal magnitudes and opposite polarities are a necessity.

4.12 OPEN-CIRCUIT FAULTS

An open-circuit fault occurs when one or two phases lose continuity and become open circuited. This can happen when fuses blow or when conductors burn or break. Opening all three phases of a three-phase circuit simply de-energizes the circuit, but opening one or two phases results in an unbalanced condition that must be analyzed using symmetrical components.

The positive-, negative-, and zero-sequence networks are constructed and the open-circuit point is identified in each of the three networks. The networks are then connected in a manner defined by the type of open-circuit fault to be analyzed, and circuit analysis techniques are used to calculate the sequence currents. Phase currents are then calculated by using Eqs. (4.58) through (4.60).

It should be noted that when analyzing an open-circuit fault, the Thevenin reduction of each sequence network is *not* taken from the fault location back to the reference bus, as with a short-circuit fault. Instead, the Thevenin equivalent is taken from one side of the open-circuit fault location, through the entire network, to the other side of the fault location. The reference bus plays no role in open-circuit fault analysis. The determination of the Thevenin equivalents will become apparent by studying the example calculations that follow.

4.12.1 One-Line-Open Fault

A one-line-open fault, as shown in Figure 4.54, requires placement of the open on phase a, since phase a must be the dissimilar (in this case, the faulted) phase since it was the phase that was used to develop our symmetrical component relationships.

When the fault occurs, the unfaulted phases will have a voltage across the fault location equal to zero, so

$$V_b = V_c = 0 \tag{4.130}$$

Using the voltage-equivalent forms of Eq. (4.59) and (4.60), voltages V_b and V_c can be set equal to each other.

$$V_0 + a^2 V_1 + a V_2 = V_0 + a V_1 + a^2 V_2 \tag{4.131}$$

Subtracting the V_0 term on each side of the equation and collecting similar terms,

$$a^2 V_1 - a V_1 = a^2 V_2 - a V_2 \tag{4.132}$$

Factoring gives

$$V_1(a^2 - a) = V_2(a^2 - a) \tag{4.133}$$

Finally, dividing out the common term produces

$$V_1 = V_2 \tag{4.134}$$

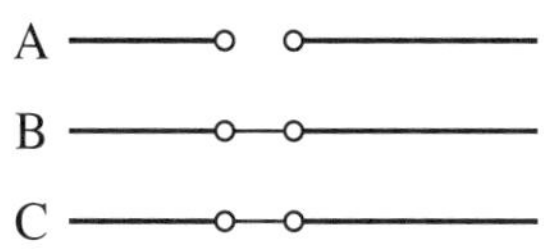

Figure 4.54 One-Line-Open Fault

But we also know that V_b and V_c must equal zero, so setting one of them equal to zero gives

$$V_0 + a^2 V_1 + a V_2 = 0 \qquad (4.135)$$

Substituting Eq. (4.134) results in

$$V_0 + a^2 V_1 + a V_1 = 0 \qquad (4.136)$$

Equation (4.136) shows that the positive-sequence voltage phasor, rotated by 240°, added to the positive-sequence voltage phasor, rotated by 120°, must equal zero after being added with V_0. This can only be true if V_0 is equal to the positive-sequence voltage, because then Eq. (4.136) would become

$$V_1 + a^2 V_1 + a V_1 = 0 \qquad (4.137)$$

Any phasor added to itself rotated by 240° and again to itself rotated by 120° forms an equilateral triangle, indicating a sum of zero. So

$$V_0 = V_1 \qquad (4.138)$$

Between Eqs. (4.134) and (4.138), we can see that

$$V_0 = V_1 = V_2 \qquad (4.139)$$

For equal voltages to appear across each sequence network, the three sequence networks must be *in parallel*.

Therefore, a one-line-open fault is modeled by connecting the positive-, negative-, and zero-sequence networks in parallel. The current flowing through the positive-sequence network is the positive-sequence current. Similarly, the current flowing through the negative-sequence network is the negative-sequence current, and the zero-sequence current flows through the zero-sequence network.

The one-line-open fault circuit model is shown in Figure 4.55.

To calculate the line currents during a one-line-open fault between the utility source and Transformer T_1 of Figure 4.5, the one-line-open fault circuit model is constructed, as shown in Figure 4.56.

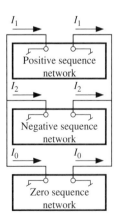

Figure 4.55   One-Line-Open Fault Circuit Model

4.12 OPEN-CIRCUIT FAULTS **145**

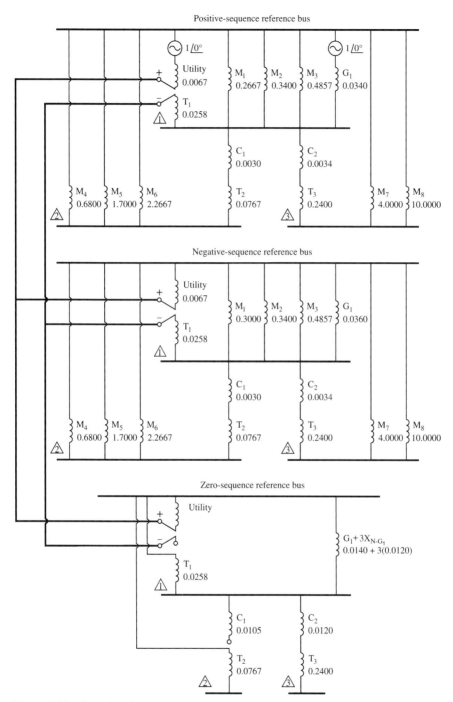

Figure 4.56 One-Line-Open Fault Between Utility Source and Transformer T_1 of Figure 4.5

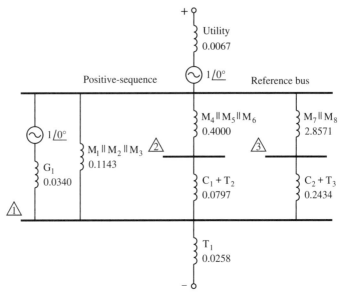

Figure 4.57 Redrawn Positive-Sequence Network

The positive-sequence network can be redrawn with the two points between which the network is open circuited at the top and bottom of the diagram and marked with a "+" and "−", as shown in Figure 4.57.

This redrawn circuit easily can be reduced to a single impedance of 0.0572 p.u. The negative-sequence network can be redrawn and reduced similarly to yield

$$X_2 = 0.0582 \text{ p.u.} \tag{4.140}$$

Because of the delta winding in transformer T_1, the zero-sequence network is open-circuited, so

$$X_0 = \text{infinite} \tag{4.141}$$

These three sequence impedances now can be cascaded in parallel as shown in Figure 4.13, resulting in the circuit shown in Figure 4.58.

Solving for currents I_1 and I_2,

$$I_1 = \frac{1 \angle 0°}{0.1154 \angle 90°} = 8.6655 \angle{-90°} \text{ p.u.} \tag{4.142}$$

$$I_2 = -I_1 = 8.6655 \angle{-90°} \text{ p.u.} \tag{4.143}$$

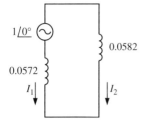

Figure 4.58 Simplified Circuit Model for One-Line-Open Fault

Because of the open circuit in the zero-sequence network,
$$I_0 = 0 \tag{4.144}$$
Now Eqs. (4.58) through (4.60) can be applied to calculate the line currents.
$$\begin{aligned} I_a &= I_0 + I_1 + I_2 \\ &= 0 + 8.6655 \:\underline{/-90°} + 8.6655 \:\underline{/90°} \\ &= 0 \end{aligned} \tag{4.145}$$

$$\begin{aligned} I_b &= I_0 + a^2 I_1 + a I_2 \\ &= 0 + 8.6655 \:\underline{/150°} + 8.6655 \:\underline{/210°} \\ &= 15.0091 \:\underline{/180°} \text{ p.u.} \end{aligned} \tag{4.146}$$

$$\begin{aligned} I_c &= I_0 + a I_1 + a^2 I_2 \\ &= 0 + 8.6655 \:\underline{/30°} + 8.6655 \:\underline{/330°} \\ &= 15.0091 \:\underline{/0°} \text{ p.u.} \end{aligned} \tag{4.147}$$

The base current at the high-voltage terminals of transformer T_1 is
$$I_{\text{base}(115\,\text{kV})} = \frac{10,000\,\text{kVA}}{115\,\text{kV}\sqrt{3}} = 50.2044\,\text{A} \tag{4.148}$$
So the line currents on the high-voltage side of transformer T_1 are
$$I_a = 0 \tag{4.149}$$
$$I_b = (15.0091\,\text{p.u.})(50.2044\,\text{A}) = 754\,\text{A} \tag{4.150}$$
$$I_c = (15.0091\,\text{p.u.})(50.2044\,\text{A}) = 754\,\text{A} \tag{4.151}$$

4.12.2 Two-Lines-Open Fault

A two-lines-open fault requires that phase a be the dissimilar (non-faulted) phase as shown in Figure 4.59, since that is the phase that was used to develop our symmetrical component relationships.

When the fault occurs, no current will flow on the opened phases, so
$$I_b = I_c = 0 \tag{4.152}$$
Using Eq. (4.59) and (4.60), currents I_b and I_c can be set equal to each other.
$$I_0 + a^2 I_1 + a I_2 = I_0 + a I_1 + a^2 I_2 \tag{4.153}$$
Subtracting the I_0 term on each side of the equation and collecting similar terms,
$$a^2 I_1 - a I_1 = a^2 I_2 - a I_2 \tag{4.154}$$
Factoring gives
$$I_1(a^2 - a) = I_2(a^2 - a) \tag{4.155}$$

A ──o o──────
B ──o o──
C ──o o────── Figure 4.59   Two-Lines-Open Fault

Finally, dividing out the common term produces

$$I_1 = I_2 \tag{4.156}$$

But we also know that I_b and I_c must equal zero, so setting one of them equal to zero gives

$$I_0 + a^2 I_1 + a I_2 = 0 \tag{4.157}$$

Substituting Eq. (4.156) results in

$$I_0 + a^2 I_1 + a I_1 = 0 \tag{4.158}$$

As was seen in previous examples, for Eq. (4.158) to hold,

$$I_0 = I_1 \tag{4.159}$$

Referring to Eq. (4.156), this means that

$$I_0 = I_1 = I_2 \tag{4.160}$$

For equal currents to flow through each sequence network, the three sequence networks must be *in series*.

Therefore, a line-to-ground fault is modeled by connecting the positive-, negative-, and zero-sequence networks in series. The phase a current for a line-to-ground fault equals three times the sequence current that flows in the series-connected circuit.

$$I_{a(2LO\ fault)} = 3I_1 \tag{4.161}$$

The line-to-ground fault circuit model is shown in Figure 4.60.

To calculate the line currents during a two-lines-open fault between the utility source and Transformer T_1 of Figure 4.5, the one-line-open fault circuit model is constructed, as shown in Figure 4.61.

When the three sequence networks are cascaded in series, the open circuit in the zero-sequence network prevents current from flowing in all of the sequence networks.

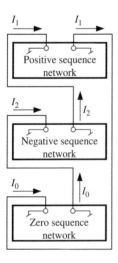

Figure 4.60   Two-Lines-Open Fault Circuit Model

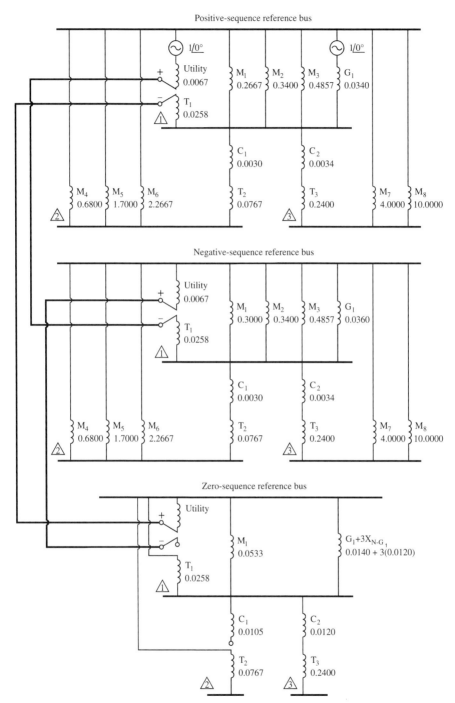

Figure 4.61 Two-Lines-Open Fault Between Utility Source and Transformer T_1 of Figure 4.5

This means that all sequence and line currents are zero. So, when phases b and c are opened on the high-voltage side of transformer T_1 and phase a remains energized, no current flows through transformer T_1.

SUMMARY

Fault current calculations can be done for *first-cycle* current magnitude or *interrupting* current magnitude. Interrupting current is the smaller of the two values, since fault current decays as a function of the *X/R* ratio of the circuit at the point of the fault. Of the various types of faults that can occur on a three-phase system, only the three-phase fault is balanced. While the three-phase fault *usually* has the highest current magnitude of all short-circuit faults, this is not always the case. Current magnitudes for *all* fault types must be checked before sizing protective devices.

A substantial amount of data is required to perform a short-circuit study, and not all the data comes from the same sources. Equipment nameplates are an excellent starting point for data collection, but test reports and perhaps standards, such as IEEE standard 141, may be necessary to gather or estimate all required data.

Unbalanced faults can be analyzed by the method of *symmetrical components*, which allows a single-phase equivalent approach to be taken. Sequence networks are developed for the system being studied. The sequence networks are interconnected in specific ways to simulate the various fault types.

Understanding the behavior of the sequence currents, in particular that of the zero-sequence current, is crucial to understanding the behavior of power systems both during unbalanced faults and in the presence of harmonics.

FOR FURTHER READING

Carson, J. R. Wave propagation in overhead wires with ground return. Bell System Technical Journal, Vol. 5, pp. 539–554, October 1926.

Clarke, E. *Circuit Analysis of A-C Power Systems*, Vol. 1, John Wiley & Sons, 1943.

Clarke, E. *Circuit Analysis of A-C Power Systems*, Vol. 2, John Wiley & Sons, 1950.

Electrical Transmission and Distribution Reference Book, Westinghouse Electric Corporation, Central Station Engineers, East Pittsburgh, Pennsylvania, 1950.

Fortescue, C. Method of symmetrical co-ordinates applied to the solution of polyphase networks. Transactions of the American Institute of Electrical Engineers, Vol. 37, pp. 1027–1140, 1918.

IEEE Application Guide for AC High-Voltage Circuit Breakers Rated on a Symmetrical Current Basis, IEEE Standard C37.010, 1999.

IEEE Recommended Practice for Calculating Short-Circuit Currents in Industrial and Commercial Power Systems (Violet Book), IEEE Standard 551, 2006.

IEEE Recommended Practice for Electric Power Distribution for Industrial Plants (Red Book), IEEE Standard 141, 1993.

IEEE Recommended Practice for Industrial and Commercial Power Systems Analysis (Brown Book), IEEE Standard 399, 1997.

IEEE Recommended Practice for Protection and Coordination of Industrial and Commercial Power Systems (Buff Book), IEEE Standard 242, 2001.

Joffe, E. B., and Lock, K. *Grounds for Grounding: A Circuit to System Handbook*, Wiley-IEEE Press, 2010. ISBN: 978-0-471-66008-8.

Venkata, S. S., Heydt, G. T., and Balijepalli, N. "High impact papers in power engineering (1900-1999)." In: Proceedings 2000 *North American Power Symposium*, vol. 1, October 2000.

QUESTIONS

1. What are some of the reasons that a fault must be removed quickly from a power system?
2. Describe some methods of arc extinction used by various interrupting devices, and describe the results if arc extinction is not successful.
3. What causes an interrupting device to fail when its interrupting rating is exceeded?
4. It is typical to add new loads, and consequently increase the SCA, at an industrial plant. Also, the supplying utility may make changes to their system that increases the SCA at the point of delivery to the plant. How can these situations be managed so that the interrupting ratings of protective devices in the plant are not exceeded?
5. Why is the approximation 1 kVA = 1 hp often used to estimate motor load if 1 hp is only 0.746 kW?
6. When estimating X/R ratios from graphs, is it preferable to err on the high or low side? Why?
7. Which is larger: first-cycle fault current or contact-parting fault current? Why?
8. Is a current-limiting fuse effective if it blows $1/2$ cycle after the occurrence of a fault? Why or why not?
9. Which component(s) of current can flow in the phase conductors of a three-phase system?
10. Which component(s) of current can flow in the neutral conductor of a three-phase system?
11. Suggest a practical method of measuring zero-sequence current in a three-phase system.
12. If a delta-(grounded) wye transformer is protected by fuses on the high-voltage (delta) side, and two fuses are found to be blown, what event most likely led to the blowing of the two fuses?
13. In the scenario of Problem 12, if only one high-side fuse was blown, suggest an event that could have led to that outcome.
14. Should cable impedance be considered when calculating fault current? Why or why not?
15. Why is neglecting utility system impedance when developing sequence networks for fault calculations a reasonable assumption?
16. Impedance grounding reduces fault current magnitude for which kind(s) of fault(s)?
17. For fault types whose magnitudes are not reduced by impedance grounding, what *strategy* is used to guard against these fault types?
18. How could one determine a typical value for fault impedance on a given system?

19. Negative-sequence reactances are typically the same as positive-sequence reactances for a given power system element. What type(s) of devices can have different values for their positive- and negative-sequence reactances?

20. Assuming that a generator is connected to the system through a delta–grounded wye generator step-up transformer, Is the generator's zero-sequence reactance necessary to create a model for analyzing fault conditions outside of the power plant? Explain why or why not.

PROBLEMS

1. The resistance of a hard-drawn copper conductor is 2.35 Ω at 40°C.
   a. What is the resistance of the conductor at 80°C?
   b. What is the percent increase in resistance over this 40°C temperature increase?

2. To what temperature must a hard-drawn aluminum conductor operating at 30°C rise such that its resistance increases by 10%?

3. Given the following line currents, calculate I_0, I_1, and I_2.
   $I_a = 427 \underline{/0°}$ A
   $I_b = 401 \underline{/232°}$ A
   $I_c = 457 \underline{/129°}$ A

4. Given the following sequence currents, calculate I_a, I_b, and I_c.
   $I_0 = 0.09 \underline{/37°}$ p.u.
   $I_1 = 1.12 \underline{/20°}$ p.u.
   $I_2 = 0.17 \underline{/326°}$ p.u.

5. Given the following line-to-neutral voltages, calculate V_0, V_1, and V_2
   $V_{a-n} = 8.23 \underline{/0°}$ kV
   $V_{b-n} = 7.39 \underline{/255°}$ kV
   $V_{c-n} = 2.95 \underline{/94°}$ kV

6. Given the following sequence voltages, calculate V_{a-n}, V_{b-n}, and V_{c-n}.
   $V_0 = 0.36 \underline{/123°}$ p.u.
   $V_1 = 0.99 \underline{/349°}$ p.u.
   $V_2 = 0.78 \underline{/222°}$ p.u.

7. Find the line-to-line voltages for Problem 6.

8. What is the largest magnitude of current that can flow in the neutral conductor of a four-wire system in terms of the currents flowing in the phases?

9. Please refer to Figure 4.5. A second 4.16 kV switchgear bus (Bus 4) is fed from Bus 1 through 500 ft of 3-1/c 500 kcmil copper cable. The 13.8 kV/4.16 kV, 5 MVA transformer at the load end of the cable has a 5.5% impedence. The 4.16 kV switchgear supplies three 1000-hp induction motors.
   a. Sketch a one-line diagram showing the existing system plus the new 4.16 kV switchgear bus.

b. Draw the first-cycle reactance network for the existing system plus the new 4.16 kV switchgear bus.

   c. Draw the first-cycle resistance network for the existing system plus the new 4.16 kV switchgear bus.

   d. Determine the Thévenin reactance and resistance at Bus 4.

   e. What is the approximated X/R ratio at Bus 4?

10. Please refer to Problem 9.

   a. What is the magnitude of a three-phase fault applied at Bus 4 in RMS amperes?

   b. What is the magnitude of a line-to-ground fault applied at Bus 4 in RMS amperes?

   c. What is the magnitude of a double line-to-ground fault applied at Bus 4 in RMS amperes?

   d. What is the magnitude of a line-to-line fault applied at Bus 4 in RMS amperes?

11. Please refer to Figure 4.5. If a second 4.16 kV bus identical to Bus 2 is added to Bus 1, what is the effect on the SCA at Bus 1?

12. Please refer to Figure 4.5. If transformer T_1 is replaced with a 45-MVA unit having a reactance of 5.25%, what is the effect on the SCA at Bus 1?

13. Two identical delta–grounded wye transformers are connected in parallel to an infinite bus. The transformers are rated 20 MVA, 115 kV/13.8 kV with $X = 6\%$.

   a. Find the fault current in kiloampere when a line-to-ground fault occurs at the transformer secondary bushings.

   b. Repeat Part a assuming that the neutral-to-ground connection of one of the transformers becomes open-circuited.

   c. What is the percent change in fault current magnitude from Part a to Part b?

   d. What changes in transformer protection could result due to the change in fault current magnitude?

14. An impedance-grounded wye generator ($X_1 = X_2 = 0.12, X_0 = 0.04, X_n = 0.03$) is connected to an infinite bus through a delta–grounded wye generator step-up (GSU) transformer with $X = 5\%$.

   a. What is the per-unit fault current magnitude at the generator terminals when a line-to-ground fault occurs there?

   b. What would be the answer to Part a if the generator was solidly grounded?

   c. What is the percent change in fault current magnitude from Part a to Part b?

15. a. If the GSU in Problem 14 was replaced by a grounded wye–grounded wye transformer with $X = 5\%$, what would be the per-unit fault current magnitude at the generator terminals when a line-to-ground fault occurs there?

   b. What is the difference in the zero-sequence network when the grounded wye–grounded wye transformer replaces the delta–grounded wye transformer?

16. If the ground connection at the X_0 bushing of the GSU in Problem 14 becomes disconnected, what would be the per-unit fault current magnitude at the generator terminals when a line-to-ground fault occurs there?

154 CHAPTER 4 FAULT CALCULATIONS

17. Please refer to Figure 4.5. A one-line-open fault occurs at the high-voltage terminals of transformer T3. Find the currents flowing in phases b and c after phase a opens.

18. Please refer to Figure 4.5. A two-lines-open fault occurs at the high-voltage terminals of transformer T_3. Find the current flowing in phase a after phases b and c open.

19. Please refer to Figure 4.5. A one-line-open fault occurs at the high-voltage terminals of transformer T_2. Find the currents flowing in phases b and c after phase a opens.

20. Please refer to Figure 4.5. A two-lines-open fault occurs at the high-voltage terminals of transformer T_2. Find the current flowing in phase a after phases b and c open.

CHAPTER 5

PROTECTIVE DEVICE SELECTION AND COORDINATION

OBJECTIVES

- Know how to size power circuit breakers, fused low-voltage circuit breakers, molded-case circuit breakers, medium-voltage fuses, current-limiting fuses, and low-voltage fuses per ANSI/IEEE C37 standards
- Understand the basic concepts of fuse coordination
- Be able to properly coordinate a fuse with another fuse, accounting for the effects of predamage, tolerances, ambient temperature, and preloading

5.1 OVERVIEW

Regardless of how reliably a power system is designed and how well it is maintained, malfunctions will occur. The most likely element of the system to fail is the *insulation*. This could mean a breakdown of the plastic- or rubber-based insulation of low- or medium-voltage power cable, the enamel- or epoxy-based insulation of a rotating machine, the cellulose insulation of a liquid immersed transformer, the air surrounding bare conductors, or devices such as cable terminations, equipment bushings, or surge arresters.

Many different insulation systems are temperature dependent. At a sufficiently high temperature, they will undergo physical changes and lose their dielectric integrity. Since the heat produced by flowing current is directly proportional to the square of the current, ampacity ratings on particular devices must be carefully observed and maintained. Any loading in excess of that rating will decrease the life of the insulation. Depending on the type of the insulation, this decrease could be significant, even for small to moderate overloads.

When insulation fails, the results can be catastrophic. Tremendous amounts of energy are released very quickly, and explosion and fire are not uncommon. To minimize the risk of catastrophic failure, *protective devices* must be installed in the power system. These devices must be able to *detect* any type of problem that is likely to occur, then *safely de-energize* the effected part of the circuit. Ideally, the

---

Industrial Power Distribution, Second Edition. Ralph E. Fehr, III.
© 2016 The Institute of Electrical and Electronics Engineers, Inc. Published 2016 by John Wiley & Sons, Inc.

problem should be detected and removed from the system as quickly as possible. Also, only the failed component(s) should be removed from service. The protection engineer strives to achieve these ideals, but often, compromises must be made. One compromise that should *never* be made is to decrease the level of safety designed into the system. Protecting equipment and, more importantly, personnel needs to be the engineer's top priority.

Insulation failure leads to a significant reduction in impedance between an energized conductor and a conductor at a different potential, either another operating voltage or ground. This reduction in impedance gives rise to a very high current flow. The high current causes a very large voltage drop-perhaps to the point of voltage collapse. But it is the high current that can cause serious damage if not interrupted quickly.

The most common devices to interrupt fault current are *fuses* and *circuit breakers*. These devices interrupt fault current by creating an open circuit in the fault current path. In the case of the fuse, the open circuit is created by a fusible element in the fuse assembly melting due to the heat created by the fault current. As the current path starts to break, arcing will occur across the open circuit. While the arcing persists, current still flows. The arc poses a larger impedance than the conducting material it is replacing, but not large enough to stop the current flow. Although the arcing will stop when the current waveform crosses zero, the fuse must provide some means of arc extinction to prevent the arc from reestablishing, or the arcing will continue until enough energy is released to cause the fuse to fail catastrophically. Since energy equals the product of current squared and time, high current faults must be interrupted very quickly to keep the fault energy within the ratings of the fuse.

Different technologies exist for arc extinction in fuses. In some types of fuses, the arc is drawn in a medium that is designed to dissipate heat, thereby extinguishing the arc. This medium is often sand or boric acid. Other fuse designs use stored energy in a spring to rapidly pull the fuse element apart as it starts to melt.

An important concept to remember when working with fuses is the fact that fuses are not precise devices. Although much sophisticated engineering goes into the design of the fuse, each time a fuse is subjected to currents that are higher than usual but not high enough to blow the fuse, the fusible element degrades. This degradation is known as *predamage*. Predamage causes a fuse to blow at a current level *lower than* for which it was designed to operate. Fortunately, this is erring on the conservative side, but it can cause operating difficulties by blowing unnecessarily (nuisance operations) and surprising unwary engineers.

The *circuit breaker* mechanically opens contacts to create an open circuit. Alternating current circuit breakers rely on zero-crossings of the current waveform to interrupt the arc. When the current waveform crosses zero, which it does 120 times every second on a 60 Hz power system, there is no current flow between the opening contacts, and therefore no arc. The arc will restrike as the voltage magnitude increases unless something is done to prevent it. Many different methods of arc extinction are used in circuit breakers; in fact, breakers are classified by the arc extinction method used: oil, air blast, vacuum, sulfur hexafluoride (SF_6) gas, etc.

At distribution voltages, air blast technology has been successfully used for years. Although this technology performed well, air blast circuit breakers did require a substantial amount of maintenance to maintain reliable operation. Sulfur hexafluoride gas, once only seen at very high transmission voltages, had started to emerge in the medium-voltage circuit breaker market but has been largely surpassed by vacuum technology. Today, vacuum circuit breakers have become the industry standard technology for voltages up to 35 kV, largely due to their relatively low maintenance requirement compared to the older technologies.

Fault current is sinusoidal at the power system frequency but is not necessarily symmetric about the horizontal axis. This is due to a direct current (DC) offset, which occurs if the fault begins at any time other than a zero current crossing. This DC component decays based on the ratio of inductive reactance to resistance (X to R ratio, or X/R) in the system at the point of the fault. The higher the X/R, the longer the DC component persists. Depending on the angular displacement of the current waveform when the fault occurs, the DC component can be substantial, capable of doubling the magnitude of the symmetrical current at the instant of the fault. This becomes critical when calculating first-cycle fault currents and interrupting current requirements for very fast breakers. Both the *symmetric* fault current and an *asymmetry factor* must be determined. The asymmetry factor is *absolutely essential* when sizing protective devices. Many times, a device may be capable of interrupting the symmetrical fault current but not the increased current magnitude caused by the asymmetry. An example of an asymmetric short-circuit waveform where the fault occurs at $t = 0$ is shown in Figure 5.1.

Changes were implemented in the electric power industry in 1964 defining how short-circuit calculations are performed. Prior to that time, asymmetric fault current calculations were performed, which determined the absolute current maximum, sometimes called *total current*. After 1964, American National Standards Institute (ANSI) standard C37.010 became cross-listed by IEEE and required the calculation of symmetric fault current and a multiplying factor to account for the asymmetry caused by the DC offset. This is the approach that will be used in this text. Please note that fault calculations done prior to the acceptance of ANSI/IEEE C37.010-1964, or fault calculations done in foreign countries, may differ from this approach.

Two different fault current values are of interest to the engineer. The *first-cycle symmetrical current* is the maximum symmetrical current the system experiences. It is used, along with a multiplier to account for the asymmetry due to the DC offset, to determine the short-circuit withstand requirements of the system components.

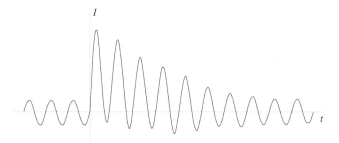

Figure 5.1 Asymmetric Short-Circuit Current

The *contact parting symmetrical current* is used in conjunction with an asymmetry multiplier to determine the actual current that a circuit breaker must interrupt. This value is less than the first-cycle value because the fault current decays between the time the fault occurs and the time that the breaker contacts begin to part.

The key factors that determine how much decay occurs are the X/R ratio at the point of the fault, which determines the rate of decay, and the time it takes for the contacts to open. The fastest breakers available have a contact parting time of approximately 1.5 cycles. It takes another cycle for the protective relays to detect the fault and issues a trip signal to the breaker, so a total of approximately 2.5 cycles elapses from the occurrence of the fault until the breaker contacts begin to open. This brief period is enough time for essentially all of the fault contribution from induction motors to decay. Because of the externally provided excitation, synchronous machines tend to contribute fault current substantially longer than induction machines. Also realize that this 2.5 cycle value applies only to the fastest circuit breakers. Some breakers have contact parting times in excess of five cycles. The longer it takes for circuit breakers to operate, the less fault current they must interrupt, but the more total energy they must dissipate during the fault.

5.2 POWER CIRCUIT BREAKER SELECTION

Medium-voltage power circuit breakers have several ratings that must be observed when specifying a breaker. A *continuous current* rating is the maximum sustained *root-mean-square (RMS)* load current the breaker can handle without exceeding temperature limits. When circuit breakers are installed in nonair-conditioned environments, the maximum continuous current the breaker is allowed to handle may have to be less than that prescribed by the continuous current rating. This derating can sometimes be avoided by providing cooling fans for the breakers.

The *interrupting rating* of a breaker is the maximum RMS current that can be safely interrupted by the device. It is crucial that this value is never exceeded, because doing so could result in catastrophic failure of the device during a fault.

If the load current greatly exceeds the continuous current rating of the largest available breaker, two breakers can be paralleled to serve the load, provided that the interrupting rating of either breaker is not exceeded. When a fault occurs, the two paralleled breakers will not open at the same precise instant. One breaker will open first, forcing the second breaker to carry all the load current for a fraction of a cycle then interrupt the entire current flow. The practice of paralleling breakers should be used only when there is no acceptable alternative.

An *interrupting time* rating is the maximum time, expressed in 60 Hz cycles, between the energization of the trip coil and the time the current flow is interrupted. After the main contacts begin to separate, arcing occurs and continues to occur until extinguished by the breaker. This arcing time is included in the interrupting time. Older circuit breakers had interrupting times on the order of eight cycles. Modern breakers can be purchased with two cycle interrupting times. The faster the breaker operates, the higher the current that must be interrupted. This is because fault current magnitude decays as a function of the X/R ratio at the point of the fault. On the other hand, the faster a breaker operates, the less total energy must be dissipated in the breaker during the fault.

Contact parting time is the interrupting time plus any additional time delay, including intentional time delay, introduced by the protective relays. This is the time that determines how much current the breaker must interrupt.

When selecting medium-voltage power circuit breakers, the sources of fault current must be considered carefully. Sources far from the fault will not decay like sources close to the fault. This is because if a generator is more than one transformation away, or if the impedance from the fault to the generator is more than 1.5 times the generator's subtransient reactance, the fault will appear as load current, so the generator simply furnishes fault current as if it was load current. If the generator is electrically close to the fault, however, the effective generator impedance changes by transitioning from a low subtransient impedance to a higher transient impedance before returning to an even higher synchronous impedance. This increase in impedance after the fault occurs leads to decay in the fault current supplied.

Another way to view the behavior of a remote generator is to consider the equivalent *X/R* ratio at the fault location. When a generator is located close to the fault location, the effective *X/R* ratio is relatively low, governed mainly by the generator's small subtransient reactance. But as the distance from the generator to the fault increases, much more reactance contributes to a larger *X/R* ratio. This additional reactance is a function of transformers, cables, and overhead lines. The larger *X/R* ratio greatly slows the decay of the fault current.

The *closing and latching* capability of a circuit breaker defines the breaker's ability to withstand the extreme magnetic forces that occur during the first cycle of a short circuit. Closing and latching ratings can be expressed in *RMS amperes* or *peak (or crest) amperes*.

Closing and latching capabilities are determined by Eqs. (5.1) and (5.2):

$$\text{Close and Latch (RMS)} = 1.6 \text{ Maximum Symmetrical RMS Fault Current} \quad (5.1)$$

$$\text{Close and Latch (peak or crest)} = 2.7 \times \text{Maximum Symmetrical RMS Fault Current} \quad (5.2)$$

The first-cycle symmetrical fault current at Bus 1 of the example shown in Figure 4.5 can be calculated by dividing the Thévenin reactance at Bus 1 from Eq. (4.44) into an assumed pre-fault voltage of 1 p.u.:

$$I_{\text{SC SYM (Bus 1)}} = \frac{1 \angle 0°}{0.0140 \angle 90°} = 71.429 \angle -90° \text{ p.u.} \quad (5.3)$$

The base current at Bus 1 is

$$I_{\text{Base (Bus 1)}} = \frac{10,000}{\sqrt{3}\,(13.8)} = 418 \text{ A}. \quad (5.4)$$

Therefore, the first-cycle symmetrical fault current at Bus 1 in amperes is

$$I_{\text{SC-SYM(Bus1)}} = (71.429)(418) = 29{,}857 \text{ A} = 29.9 \text{ kA}. \tag{5.5}$$

Applying Eqs. (5.1) and (5.2), the closing and latching capability of a circuit breaker installed at Bus 1 is determined to be

$$\text{Close and Latch (RMS)} = 1.6 \times 29.9 = 47.8 \text{ kA}. \tag{5.6}$$

$$\text{Close and Latch (peak or crest)} = 2.7 \times 29.9 = 80.6 \text{ kA}. \tag{5.7}$$

5.3 FUSED LOW-VOLTAGE CIRCUIT BREAKER SELECTION

If the interrupting rating of a particular circuit breaker is inadequate but the continuous current rating is sufficient, one option is to use an integrally fused low-voltage circuit breaker. The integrated current-limiting fuse may prove to be less costly than selecting a larger circuit breaker frame size to achieve an adequate interrupting rating. A fused low-voltage circuit breaker is shown in Figure 5.2.

Fused low-voltage power circuit breakers are rated based on first-cycle symmetrical current per the IEEE C37 standards. Although the device rating is based on symmetrical current, a certain amount of asymmetry is allowed for in the short-circuit current waveform. The asymmetry allowed in the symmetrical current rating of fused low-voltage circuit breakers corresponds to an X/R ratio of 4.9. If the X/R ratio at the point of the fault exceeds 4.9, a multiplying factor specified by IEEE standard

Figure 5.2 Fused Low-Voltage Circuit Breaker. (*Photo courtesy of GE Energy Management Industrial Solutions.*)

5.3 FUSED LOW-VOLTAGE CIRCUIT BREAKER SELECTION

C37.13 must be used to adjust the calculated short-circuit current. The calculation method for this multiplying factor is shown in Eq. (5.8):

$$\mathrm{MF_{LV\ fused\ bkr}} = \frac{\sqrt{2\ e^{-2\pi/X/R} + 1}}{1.25} \quad \text{for } X/R > 4.9 \tag{5.8}$$

The first-cycle symmetrical fault current at Bus 3 of the example shown in Figure 4.5 can be calculated by dividing the Thévenin reactance at Bus 3 from Eq. (4.46) into an assumed pre-fault voltage of 1 p.u.:

$$I_{\mathrm{SC\ SYM\ (Bus\ 3)}} = \frac{1\ \underline{/0°}}{0.2348\ \underline{/90°}} = 4.2589\ \underline{/-90°}\ \text{p.u.} \tag{5.9}$$

The base current at Bus 3 is

$$I_{\mathrm{Base\ (Bus\ 3)}} = \frac{10,000}{\sqrt{3}\ (0.48)} = 12.028\ \mathrm{kA}. \tag{5.10}$$

Therefore, the first-cycle symmetrical fault current at Bus 3 in kiloamperes is

$$I_{\mathrm{SC-SYM(Bus3)}} = (4.2589)(12.028) = 51.226\ \mathrm{kA}. \tag{5.11}$$

Next, the X/R ratio at Bus 3 must be calculated:

$$\left(\frac{X}{R}\right)_{\mathrm{Bus\ 3\ (firstcycle)}} = \frac{0.2348}{0.021665} = 10.8 \tag{5.12}$$

Since the X/R ratio calculated in Eq. (5.12) is greater than the threshold stipulated in Eq. (5.8), the short-circuit current waveform will decay more slowly than accounted for by the circuit breaker design. Therefore, the multiplying factor defined in Eq. (5.8) must be applied.

$$\mathrm{MF_{LV\ fused\ bkr}} = \frac{\sqrt{2e^{-2\pi/10.8} + 1}}{1.25} = 1.164 \tag{5.13}$$

Adjusting the first-cycle symmetrical fault current for the additional asymmetry caused by the high X/R ratio yields

$$I_{\mathrm{SC-SYM(Bus3)-Adj}} = (51.226)(1.164) = 59.627\ \mathrm{kA}. \tag{5.14}$$

Any fused low-voltage circuit breaker installed at Bus 3 must have a symmetrical rating of at least 59.7 kA.

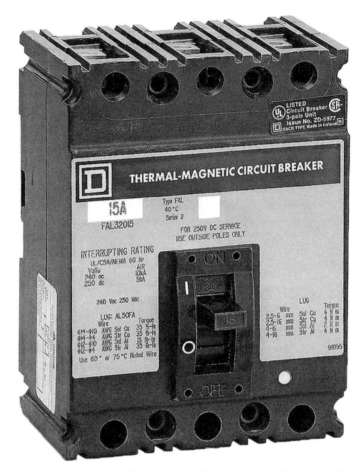

Figure 5.3 Molded-Case Circuit Breaker *Photo courtesy of Schneider Electric*

5.4 MOLDED-CASE CIRCUIT BREAKER SELECTION

Molded-case circuit breakers, as shown in Figure 5.3, are commonly used to protect low-voltage circuits.

Molded-case circuit breakers are rated based on first-cycle symmetrical current per the IEEE C37 standards. A degree of asymmetry corresponding to an X/R ratio of 6.6 is built into the rating of molded-case circuit breakers. But if the X/R ratio at the point of the fault is 6.6 or more, the calculated short-circuit current must be adjusted by a multiplying factor to account for the slower rate of fault current decay due to the higher X/R ratio. This multiplying factor is specified by IEEE standard C37.13 and is expressed by Eq. (5.15):

$$\text{MF}_{\text{molded-case bkr}} = \frac{\sqrt{2}\ (e^{-\pi/X/R} + 1)}{2.29} \text{ for } X/R > 6.6 \qquad (5.15)$$

Per Eq. (5.11), the first-cycle symmetrical fault current at Bus 3 is 51.226 kA and according to Eq. (4.46), the X/R ratio at Bus 3 is 10.8. Since the X/R ratio exceeds the threshold stated in Eq. (5.15), the multiplying factor expressed in Eq. (5.15) must be applied.

$$\text{MF}_{\text{molded case bkr}} = \frac{\sqrt{2}(e^{-\pi/10.8} + 1)}{2.29} = 1.079 \qquad (5.16)$$

Adjusting the first-cycle symmetrical fault current for the additional asymmetry caused by the high X/R ratio yields

$$I_{\text{SC-SYM(Bus3)-Adj}} = (51.226)(1.079) = 55.273 \text{ kA} \qquad (5.17)$$

Any molded-case circuit breaker installed at Bus 3 must have a symmetrical rating of at least 55.3 kA.

5.5 MEDIUM-VOLTAGE FUSE SELECTION

Various types of medium-voltage fuses exist for different applications. A *power fuse*, as shown in Figure 5.4, is an expulsion-type device.

Figure 5.4 Power Fuses *Photo courtesy of S&C Electric Company*

Modern power fuses are filled with powdered boric acid. When the fuse link melts, the heat of the arc forms deionized steam which is expelled from the interrupting chamber. This expulsion of gases extinguishes the arc but produces a loud explosion-like sound in the process. Because of the noise and the expulsion of hot gases and possibly flame during operation, expulsion-type power fuses are limited to outdoor applications.

Similar in operation to the power fuse is the fused cutout, shown in Figure 5.5. The fused cutout contains a fusible link inside a fiberglass tube. After the fuse operates, only the fusible link needs to be replaced. A typical fusible link is shown in Figure 5.6.

Power fuses are tested in a circuit having an X/R ratio of 15. If the X/R ratio at the point of the fault exceeds 15, a multiplying factor must be used to adjust the calculated short-circuit current for the slower rate of decay than accounted for by

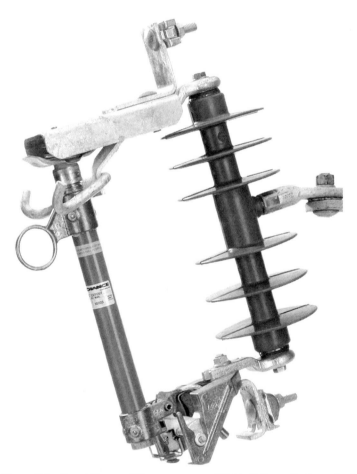

Figure 5.5   Fused Cutout. (*Photo courtesy of Hubbell Power Systems, Inc.*)

5.5 MEDIUM-VOLTAGE FUSE SELECTION

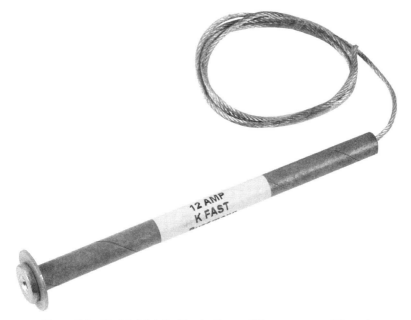

Figure 5.6 Fusible Link for Use in Cutout. (*Photo courtesy of Eaton.*)

the manufacturer due to the high *X/R* ratio. This multiplying factor is calculated per Eq. (5.18):

$$\text{MF}_{\text{medium-voltage fuse}} = \frac{\sqrt{2e^{-2\pi/X/R} + 1}}{1.52} \quad \text{for } X/R > 15 \quad (5.18)$$

From Eq. (4.85), the first-cycle symmetrical fault current at Bus 1 is 29.86 kA. Next, the *X/R* ratio at Bus 1 must be calculated:

$$\left(\frac{X}{R}\right)_{\text{Bus1(firstcycle)}} = \frac{0.0140}{0.000464} = 30.2 \quad (5.19)$$

Since the *X/R* ratio calculated in Eq. (5.19) is greater than the threshold stipulated in Eq. (5.18), a slower rate of decay in the short-circuit current waveform is present than accounted for by the fuse design. Therefore, the multiplying factor defined in Eq. (5.18) must be applied.

$$MF_{\text{medium-voltage fuse}} = \frac{\sqrt{2e^{-2\pi/30.2} + 1}}{1.52} = 1.066 \quad (5.20)$$

Adjusting the first-cycle symmetrical fault current for the slower decay rate caused by the high X/R ratio yields

$$I_{\text{SC-SYM(Bus1)-Adj}} = (29.857)(1.066) = 31.828 \text{ kA} \tag{5.21}$$

Any medium-voltage fuse installed at Bus 1 must have a symmetrical rating of at least 31.9 kA.

5.6 CURRENT-LIMITING FUSE SELECTION

Distribution current-limiting fuses are typically filled with sand to absorb the thermal energy produced by the arc that forms as the fusible element melts. Current-limiting fuses do not protect against overloads and low-magnitude faults. Instead, they operate very quickly, in less than 1/4 cycle, to interrupt high-magnitude faults before the fault current reaches its peak. This is necessary to protect downstream devices against very high fault currents. A current-limiting fuse is shown in Figure 5.7.

Current-limiting fuses are tested in a circuit having an X/R ratio of 10. If the X/R ratio at the point of the fault exceeds 10, a multiplying factor must be used to adjust the calculated short-circuit current. This multiplying factor is calculated per Eq. (5.22):

$$\text{MF}_{\text{current-limiting fuse}} = \frac{\sqrt{2e^{-2\pi/X/R} + 1}}{1.44} \text{ for } X/R > 10 \tag{5.22}$$

The first-cycle symmetrical fault current at Bus 2 of the example shown in Figure 4.5 can be calculated by dividing the Thévenin reactance at Bus 2 from Eq. (4.45) into an assumed pre-fault voltage of 1 p.u.

$$I_{\text{SC-SYM(Bus 2)}} = \frac{1\ \underline{/0°}}{0.0742\ \underline{/90°}} = 13.477\ \underline{/-90°} \text{ p.u.} \tag{5.23}$$

The base current at Bus 2 is

$$I_{\text{Base (Bus 2)}} = \frac{10,000}{\sqrt{3}\ (4.16)} = 1388 \text{ A} \tag{5.24}$$

Therefore, the first-cycle symmetrical fault current at Bus 2 in kiloamperes is

$$I_{\text{SC-SYM(Bus2)}} = (13.477)(1388) = 18.706 \text{ kA}. \tag{5.25}$$

Figure 5.7 Current-Limiting Fuse. (*Photo courtesy of Eaton.*)

Next, the *X/R* ratio at Bus 2 must be calculated:

$$\left(\frac{X}{R}\right)_{\text{Bus 2 (firstcycle)}} = \frac{0.0742}{0.004787} = 15.5 \qquad (5.26)$$

Since the *X/R* ratio calculated in Eq. (5.26) is greater than the threshold stipulated in Eq. (5.22), a slower decay rate will be present in the short-circuit current waveform than accounted for by the fuse rating. Therefore, the multiplying factor defined in Eq. (5.22) must be applied:

$$\text{MF}_{\text{current-limiting fuse}} = \frac{\sqrt{2e^{-2\pi/15.5} + 1}}{1.44} = 1.061 \qquad (5.27)$$

Adjusting the first-cycle symmetrical fault current for the slower decay rate caused by the high X/R ratio yields

$$I_{\text{SC-SYM(Bus2)-Adj}} = (18.706)(1.061) = 19.847 \text{ kA}. \tag{5.28}$$

Any current-limiting fuse installed at Bus 2 must have a symmetrical rating of at least 19.9 kA.

5.7 LOW-VOLTAGE FUSE SELECTION

Low-voltage fuses, as the one shown is Figure 5.8, are tested in a circuit having an X/R ratio of 4.9.

If the X/R ratio at the point of the fault exceeds 4.9, a multiplying factor must be used to adjust the calculated short-circuit current for the slower decay rate caused by the high X/R ratio. This multiplying factor is calculated per Eq. (5.29):

$$\text{MF}_{\text{LV fuse}} = \frac{\sqrt{2e^{-2\pi/X/R} + 1}}{1.25} \quad \text{for } X/R > 4.9 \tag{5.29}$$

Figure 5.8 Low-Voltage Fuse. (*Photo courtesy of Eaton.*)

Per Eq. (5.11), the first-cycle symmetrical fault current at Bus 3 is 51.226 kA, and the X/R ratio at Bus 3 is 10.8 according to Eq. (4.46). Since the X/R ratio exceeds the threshold stated in Eq. (5.29), the multiplying factor expressed in Eq. (5.29) must be applied:

$$MF_{LV\ fuse} = \frac{\sqrt{2e^{-2\pi/10.8} + 1}}{1.25} = 1.164 \qquad (5.30)$$

Adjusting the first-cycle symmetrical fault current for the slower decay rate caused by the high X/R ratio yields

$$I_{SC-SYM(Bus3)-Adj} = (51.226)(1.164) = 59.627\ \text{kA}. \qquad (5.31)$$

Any low-voltage fuse installed at Bus 3 must have a symmetrical rating of at least 59.7 kA.

5.8 OVERCURRENT DEVICE COORDINATION

A sound protection philosophy requires all short-circuit faults to be cleared quickly with minimal disruption to the system. To assure that all faults are cleared, it is desirable to have more than one protective device sense a given fault. Then, if the desired device is unable to clear the fault, a backup device is ready to do the clearing. To assure that only one device operates to clear the fault while other devices are aware of the fault, timing is used to determine which device clears the fault. The application of this timing scheme is referred to as *coordination*.

The operating time for a protective device is inversely proportional to the magnitude of overcurrent. If the overcurrent is slight, such as with an overload, a relatively long time is required for the protective device to operate. But when very large overcurrents exist, such as with a short-circuit fault, the device operation is very fast. Device operating time as a function of current is described by a *time–current characteristic*, or TCC, as shown in Figure 5.9.

Note that both the current and time axes are logarithmic. The nonlinear timescale must be interpreted carefully, as a given vertical distance near the bottom of the graph represents a very small time increment. The same vertical distance further up the y-axis represents much more time.

Of the four curves shown in Figure 5.9, Curve D is the fastest device, because for any value of current between 300 and 10,000 A, the curve for device D lies below the other curves, indicating less operating time.

In Figure 5.10, Curve A is clearly faster than Curve B for magnitudes of current below 700 A. Curve B is clearly the faster curve for magnitudes of current in

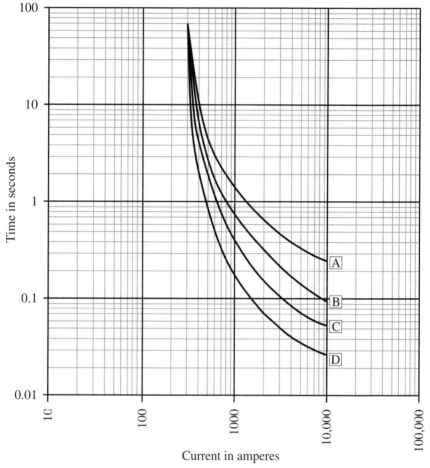

Figure 5.9 Time–Current Characteristic

excess of 900 A, but between 700 and 900 A, the faster device cannot be determined for certain since the curves are closer together than the manufacturing tolerances of the fuse. This region of uncertain coordination should be avoided because, even though a fault with a magnitude in the uncertain region will be cleared, we cannot be sure which device will do the clearing. It is quite possible that the nonpreferred device will operate, resulting in a larger area of service interruption than is necessary.

Uncertain coordination can be avoided by selecting two TCCs that are more parallel to each other. Doing so will either eliminate the crossing of the TCCs or will force the crossing to occur to the right of the maximum fault current magnitude—a part of the domain in which we would never operate.

5.8 OVERCURRENT DEVICE COORDINATION   171

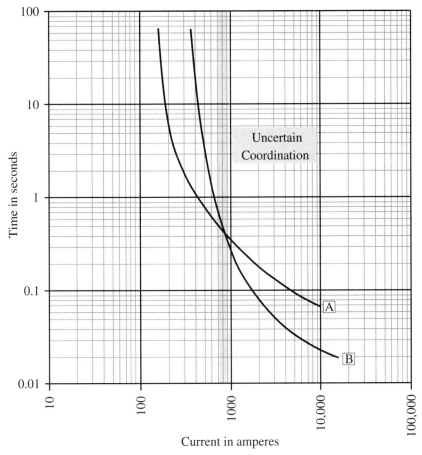

Figure 5.10   Uncertain Coordination

Fuses are overcurrent devices that are commonly used in both industrial and utility applications. When the current flowing through a fuse exceeds a particular value for a specified amount of time, the fuse element begins to melt. By a given amount of time for a given current, the fuse element will be melted and the fault will be cleared. These operating characteristics for a fuse suggest a TCC made up of two parts. The point at which the melting process may begin must be specified. This part of the TCC is called the *minimum melt* characteristic. Additionally, the point at which clearing can be assured must also be specified. This is known as the *total clear* characteristic. A typical fuse TCC is shown in Figure 5.11.

Many different types of fuse links have been developed over the years, each having a slightly different shape to its TCC. The different TCC shapes are helpful when coordinating with other protective devices. Commonly used fuse links include the T-link and K-link tin elements, various silver elements that melt at considerably

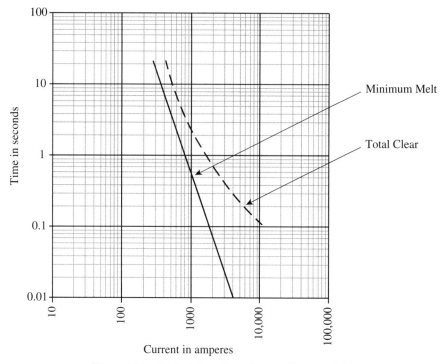

Figure 5.11 Typical Fuse Time-Current Characteristic

higher temperatures than tin elements, and dual-element fuses containing both fast and time-delayed elements.

When coordinating one fuse with another, it is important to assure that the downstream fuse is able to clear the fault before the upstream fuse element begins to melt. If not, a condition known as *predamage* may occur to the upstream fuse. Predamage permanently alters the TCC of a fuse by reducing its operating time, making its precise operating time unpredictable.

Figure 5.12 shows two fuse TCCs. If the fault current exceeds 6 kA, the upstream fuse may be damaged before the downstream fuse is able to clear the fault, but for current magnitudes less than 6 kA, the coordination is acceptable.

Predamage is one of the four factors that affect the coordination of two fuses in series. *Tolerances* must also be considered to assure proper coordination. Although fuses are engineered to conform to specific TCCs, a certain amount of tolerance is incorporated into the curves. The *ambient temperature* of the fuse also affects its operating time, since the hotter the fuse, the smaller the temperature rise required to reach the element's melting point. The amount of current flowing through the fuse prior to the fault, or *preloading*, also affects the fuse's operating time by affecting the temperature of the fuse link itself. The effects of predamage, tolerances, ambient temperature, and preloading on fuse operating time can be difficult to consider since

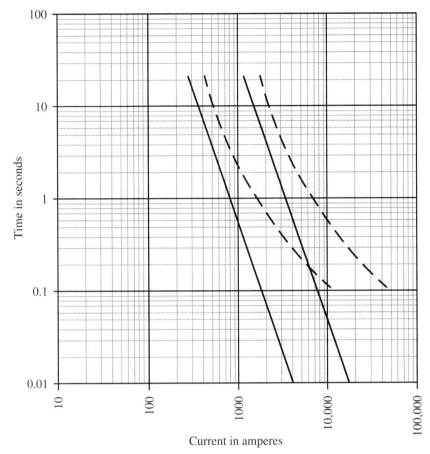

Figure 5.12   Two Fuse TCCs

their effects are difficult to quantify. This difficulty led engineers to develop a rule of thumb called *the 75% method* to estimate the effect of the four factors mentioned above on the coordination of fuses in series. The 75% method requires the total clearing time of the downstream fuse to be no more than 75% of the minimum melting time of the upstream link. This margin assures adequate time to clear the fault with the downstream fuse without damaging the upstream fuse.

Figure 5.13 shows an upstream and downstream fuse. At a calculated fault current of 5 kA, the minimum melting time of the upstream fuse is 0.3 seconds. The total clearing time of the downstream fuse at that same fault current magnitude is 0.22 seconds. Applying the 75% method, $0.22 < 0.3 \times 75\% = 0.225$. This means that these fuses are properly coordinated for fault currents less than or equal to 5 kA. Exceeding 5 kA of fault current puts one into a region of uncertain coordination.

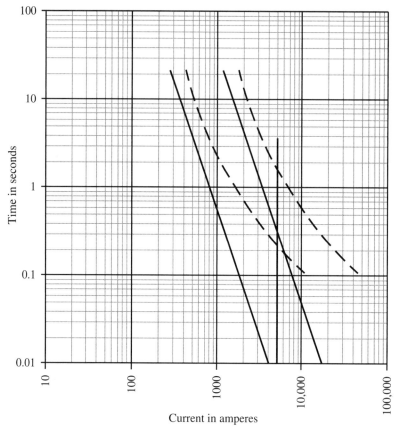

Figure 5.13 The 75% Method

5.9 SUMMARY

The X/R ratio at the point of a short-circuit fault determines the degree of asymmetry and the rate of decay of the short-circuit current waveform. Protective devices are designed for a specific X/R ratio so a realistic interrupting current magnitude can be determined and used as a design criterion for the device. If the actual X/R of the system exceeds the design X/R ratio for the device, a multiplying factor must be calculated to increase the required interrupting rating. If the actual X/R ratio of the system is less than the design X/R ratio for the device, no adjustment to the calculated fault current magnitude is made. Each class of protective device has its own design X/R ratio and formula for calculating the multiplying factor, as summarized in Table 5.1.

When a short-circuit fault occurs, the main objective is to clear the fault quickly. To assure that all faults are cleared, the protection is designed so that multiple protective devices can sense any given short-circuit fault. By designing certain protective devices to be faster than others, the particular protective device to clear a

TABLE 5.1 Protective Device Design X/R Ratios and Multiplying Factors

Device	Design X/R	Multiplying Factor When Exceeding Design X/R
Fused low-voltage circuit breaker	4.9	$MF_{\text{LV fused bkr}} = \dfrac{\sqrt{2e^{-2\pi/X_R} + 1}}{1.25}$ for $X/R > 4.9$
Molded-case circuit breaker	6.6	$MF_{\text{molded-case bkr}} = \dfrac{\sqrt{2}\,(e^{-\pi/X_R} + 1)}{2.29}$ for $X/R > 6.6$
Medium voltage fuse	15	$MF_{\text{medium-voltage fuse}} = \dfrac{\sqrt{2e^{-2\pi/X_R} + 1}}{1.52}$ for $X/R > 15$
Current-limiting fuse	10	$MF_{\text{current-limiting fuse}} = \dfrac{\sqrt{2e^{-2\pi/X_R} + 1}}{1.44}$ for $X/R > 10$
Low voltage fuse	4.9	$MF_{\text{LV fuse}} = \dfrac{\sqrt{2e^{-2\pi/X_R} + 1}}{1.25}$ for $X/R > 4.9$

given fault can be specified. In the event that the preferred device is unable to clear the fault, a backup device will do the clearing in a slightly longer period of time.

Time-current characteristics, or *TCCs*, are log–log plots of time versus current. These plots are used to determine the operating time of a protective device for a given current magnitude. TCCs are also used to verify proper coordination between multiple protective devices.

Factors such as *predamage*, *tolerances*, *ambient temperature*, and *preloading* can affect the operating time of a fuse. The 75% method can be used to account for these factors by assuring the total clearing time of the downstream fuse is no more than 75% of the minimum melting time of the upstream fuse.

FOR FURTHER READING

IEEE Application Guide for AC High-Voltage Circuit Breakers Rated on a Symmetrical Current Basis, IEEE Standard C37.010, 1999.

IEEE Standard for Low-Voltage AC Power Circuit Breakers Used in Enclosures, IEEE Standard C37.13, 1990.

QUESTIONS

1. Why are different types of protective devices designed for different X/R ratios?
2. If the X/R ratio of your system exceeds the design X/R ratio of the protective device you wish to implement, what adjustments must be made when specifying the protective device's interrupting rating?

3. If the *X/R* ratio of your system is less than the design *X/R* ratio of the protective device you wish to implement, what adjustments must be made when specifying the protective device's interrupting rating?

4. Would you expect higher *X/R* ratios on a lower voltage system or on a higher voltage system? Why?

5. How have *X/R* ratios at a given voltage level evolved (changed) over the years? Do you expect this trend to continue?

6. Can a multiplying factor used to adjust the interrupting rating of a protective device be less than one? Why or why not?

7. Why should more than one protective device be able to sense a given short-circuit fault?

8. If multiple protective devices can sense a short-circuit fault, what procedure is employed to assure that only one device operates?

9. List four parameters that affect the interrupting time of a fuse and explain the significance of each.

10. How can the parameters from Question 9 be quantitatively considered when coordinating fuses?

PROBLEMS

1. The maximum first-cycle symmetrical RMS fault current on a particular bus is 17,500 A.
   a. What is the minimum close and latch RMS rating of a power circuit breaker installed at that bus?
   b. Is a circuit breaker with a close and latch peak (crest) rating of 50 kA adequate for that bus?

2. A 480 V switchgear bus has a maximum first-cycle symmetrical fault current of 15,000 A and an *X/R* ratio of 12. What is the minimum symmetrical rating of a fused low-voltage circuit breaker installed in that switchgear?

3. A 480 V bus has a maximum first-cycle symmetrical fault current magnitude of 18,500 A. The *X/R* ratio is 14. Is a molded-case circuit breaker rated at 20 kA adequate for installation on this bus?

4. A 2.4 kV bus has an *X/R* ratio of 25. What is the maximum first-cycle symmetrical fault current that can be safely interrupted on that bus by a fuse rated at 25 kA?

5. A 4.16 kV bus has 24,000 A of first-cycle symmetrical fault current available. How high can the *X/R* ratio be so that a current-limiting fuse rated at 25 kA can be installed on this bus?

6. A 480 V bus has a maximum first-cycle symmetrical fault current of 24,000 A and an unusually high *X/R* ratio of 22. What is the minimum interrupting rating for a fuse installed on this bus?

7. A 10-T fuse link is used to protect a transformer. A second fuse, which will see a total load of no more than 30 A, must be applied upstream from the 10-T fuse. A third fuse, which

will see a total load of no more than 60 A, must be applied upstream from the second fuse. The third fuse must be a standard-speed fuse link.

Maximum fault current levels at the 10-T fuse location, the second fuse location, and the third fuse location are 700, 1000, and 2000 A, respectively. Disregard the effects of predamage, tolerances, ambient temperature, and preloading.

Recommend a fuse size and type for the second and third fuses. Construct a TCC showing all three fuses to verify proper coordination.

8. A 30-K fuse link is used to protect a transformer. A second fuse, which will see a total load of no more than 85 A, must be applied upstream from the 30-K fuse. A third fuse, which will see a total load of no more than 125 A, must be applied upstream from the second fuse. The third fuse must be a standard-speed fuse link.

Maximum fault current levels at the 30-K fuse location, the second fuse location, and the third fuse location are 1000, 2000, and 4000 A, respectively. Disregard the effects of predamage, tolerances, ambient temperature, and preloading.

Recommend a fuse size and type for the second and third fuses. Construct a TCC showing all three fuses to verify proper coordination.

9. Repeat Problem 7 applying the 75% method.
10. Repeat Problem 8 applying the 75% method.

CHAPTER 6

RACEWAY DESIGN

OBJECTIVES

- Understand the significance of *pulling tension* and *sidewall pressure* and how to calculate them
- Recognize essential design criteria when designing raceway systems
- Design conduit and tray systems, verifying that cable parameters will not be exceeded during installation

6.1 OVERVIEW

Power, control, and instrumentation cables are installed in *raceways* such as *conduit* or *cable tray* to protect the cables from physical damage. *Conduit* is metallic or nonmetallic tubing specifically designed to protect electrical cables. Metallic conduit can be *magnetic* (galvanized steel) or *nonmagnetic* (aluminum or stainless steel). Since magnetic conduit allows induced eddy currents to flow, the ability to dissipate heat is different for magnetic conduit than for nonmagnetic conduit. Also, the inductive reactance of the circuit in the conduit is influenced by the magnetic properties of the conduit. Nonmetallic conduit, made of such materials as fiber-based composites or polyvinyl chloride, is often encased in concrete to form a *duct bank*.

Cable tray is a prefabricated structure often resembling a ladder. Aluminum is the most common material used to fabricate cable trays, but steel (raw, galvanized, or stainless) and fiber-reinforced plastic trays can also be purchased. Some manufacturers also offer vinyl-coated steel as a material option. When installed horizontally (with the rails parallel to the floor or ceiling), cables are laid between the rails and are supported by the rungs. When cable tray is installed vertically (with the rails running from the floor to the ceiling), the cables must be fastened to the rungs to distribute the cable weight uniformly. Power cables installed in horizontal sections of tray must also be secured to the tray to keep them in place when subjected to the mechanical forces caused by a short circuit. In harsh industrial environments, lids are sometimes installed on cable trays to keep contaminants away from the cables. If lids are used, the natural cooling effect of the cables is reduced greatly. Consequently, the cables in a covered tray must be derated.

Industrial Power Distribution, Second Edition. Ralph E. Fehr, III.
© 2016 The Institute of Electrical and Electronics Engineers, Inc. Published 2016 by John Wiley & Sons, Inc.

180 CHAPTER 6 RACEWAY DESIGN

Figure 6.1 Power Cables in Ladder-type Cable Tray. (*Photo Courtesy of Tampa Electric Company.*)

Ladder-type cable tray, as shown in Figure 6.1, is very strong and is suitable for support spans of up to 30 ft.

A full range of accessories to join tray sections and support them from walls or ceilings is available from many manufacturers. These accessories include couplings to form three- and four-way intersections, as shown in Figure 6.2.

Ladder-type cable tray is the most commonly used type, although other types are available. *Solid-bottom* cable tray is nonventilated and gives delicate cables added protection and continuous support. Solid-bottom tray often is used for communications and instrumentation cables and usually is designed for support spans of 5 to 12 feet.

Trough, *channel*, and *wire mesh* cable tray are also available. Their support spans are typically short (4–10 ft) and their uses are specialized. *Single rail* cable tray typically is mounted on walls and is used where installation ease and freedom to move cables is critical. Instead of pulling cables through the cable tray system, single rail tray enables the cables to be laid into the tray from the side.

The National Electrical Code (NEC) states which kinds of cables can be installed in the same raceway, and which cables must be separated. In general, low- and medium-voltage cables must be run in their own conduits. Power and control cables can be mixed if installed in tray and if a metallic partition separates the two cable types.

Since metallic raceways can carry ground currents, grounding and bonding of the raceways are critical. Aluminum and steel cable tray must have bonding straps bridging the connections between sections and metallic conduit sections either must be threaded or have bonding straps between them.

Figure 6.2 Three-way Intersection in Ladder Raceway. (*Photo courtesy of Tampa Electric Company*.)

Typically, the primary goal of raceway design is to minimize the length of the raceway. While this sounds simple, it seldom is in practice. Inevitably, there will be obstructions such as structural steel, piping, and pieces of machinery between the beginning point and the ending point of the proposed raceway. When obstructions are encountered, the raceway is usually routed around the obstacle. This not only increases the length of the raceway, but also introduces additional turns in it. Bends are usually designed at either 45° or 90° to keep a neat appearance. Bends in raceways, particularly conduit systems, can greatly complicate the installation of the cables. With cable tray, bends require special pieces that add expense and complicate installation. Bends in conduit, on the other hand, increase the tension required to pull the cables through the conduit. It is a good design practice to minimize the number of bends per pull. The NEC allows up to four 90° bends, or equivalent, per pull. The limitation of four 90° bends does not assure that design parameters of the cable such as *maximum pulling tension* or *sidewall pressure* will not be exceeded. These quantities must be calculated to assure that the maximum values are not exceeded.

6.2 CONDUIT AND DUCT SYSTEMS

When designing conduit and duct systems, a conduit size must be selected. Conduit shall be sized properly so cables can be pulled without damaging the cable jacket or insulation and so adequate cooling exists when the cables are energized. The NEC limits conduit fill to 40%, meaning the sum of the cross-sectional areas of all cables in a conduit cannot exceed 40% of the cross-sectional area defined by the conduit inside diameter. In addition to the percent fill, a quantity called the *jam ratio* must also be determined. The jam ratio predicts the likelihood that the cables will jam by lining up along the conduit's inside diameter. If the cables jam in the conduit during pulling, the potential for cable damage is high. The jam ratio for three single conductor cables of the same diameter is defined as the ratio of the conduit inside diameter to the single conductor cable outside diameter, as shown in Eq. (6.1):

$$j = \frac{ID_{conduit}}{OD_{cable}} \quad (6.1)$$

Jamming cannot occur if the jam ratio is greater than 3.0 and normally does not occur when the jam ratio is less than 2.8. Jam ratios between 2.8 and 3.0 should be avoided, since this is a "danger zone" where jamming is likely to occur. Consider three equal-sized cables in a conduit sized to the NEC 40% fill limit. The jam ratio can be calculated as follows:

$$A_{cables} = 3\pi \left(\frac{OD_{cable}}{2}\right)^2 = \frac{3\pi}{4}(OD_{cable})^2 \quad (6.2)$$

$$A_{conduit} = \pi \left(\frac{ID_{conduit}}{2}\right)^2 = \frac{\pi}{4}(ID_{conduit})^2 \quad (6.3)$$

$$\frac{3\pi}{4}(OD_{cable})^2 = 0.40 \left(\frac{\pi}{4}\right)(ID_{conduit})^2 \quad (6.4)$$

$$(OD_{cable})^2 = \frac{0.40}{3}(ID_{conduit})^2 = 0.133(ID_{conduit})^2 \quad (6.5)$$

$$OD_{cable} = \sqrt{0.133\,(ID_{conduit})^2} = 0.365\,ID_{conduit} \quad (6.6)$$

$$j = \frac{ID_{conduit}}{0.365\,ID_{conduit}} = 2.740 \quad (6.7)$$

This jam ratio is slightly less than the 2.8 threshold. To avoid jam ratios in the "danger zone," conduit fills between 1/3.0 (or 33.3%) and 1/2.8 (or 35.7%) should be avoided.

A minimum bending radius must be determined according to the cables being pulled. *Shielded* cables contain a metallic layer just beneath the jacket to distribute evenly the electric field gradient throughout the insulation. This shield is sometimes provided in the form of a thin copper tape, or *tape shield*, which is grounded at the cable terminations. Another shielding option is a *concentric neutral*, in which strands of copper conductor are placed just beneath the jacket. These strands are used for the neutral in a four-wire system.

Typically, in three-phase circuits, each cable has enough copper strands in its concentric neutral to make up one-third of the system neutral. Tape-shielded cables require a separate neutral conductor in addition to the phase conductors. Shielding is required with cables above 5 kV, is optional with the 5kV class, and is seldom used with cables below 5 kV. Figure 6.3 shows the construction of an unshielded cable, while Figures 6.4 and 6.5 show the construction of tape-shielded and concentric neutral cables, respectively.

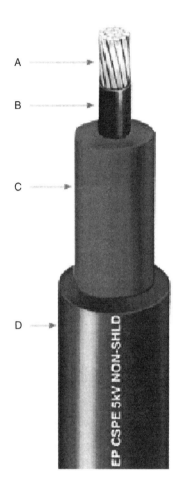

Figure 6.3 Unshielded Cable. (*Photo courtesy of The Okonite Company.*) *Notes:* A, Conductor; B, Semiconductor Layer; C, Ethylene Propylene Rubber (EPR) Insulation; D, Jacket

Unless otherwise specified by the cable manufacturer, a minimum bending radius of five times the cable outside diameter can be assumed for unshielded cables. Shielded cables are assumed to have a minimum bending radius of 12 times the cable outside diameter. Of course, using information from the manufacturer is preferable to these rules of thumb.

If more than the four 90° bends allowed by the NEC are required in the raceway, *pull boxes* should be installed as shown in Figure 6.6.

A pull box is a piece of hardware with a removable cover that is inserted in the conduit raceway. Cable is pulled into the pull box and out through the removable cover. Then the cable is pulled back into the conduit to the next pull box or to the end of the raceway. This effectively allows a long pull with many bends to become two or more shorter pulls with fewer bends.

The pull box shown in Figure 6.6 shows a 90° change of direction in the conduit. Pull boxes also can be built for straight-through pulls, or for U-pulls where a 180° change in direction is needed. Sizing of pull boxes is governed by the NEC.

Figure 6.4 Copper Tape-Shielded Cable. (*Photo courtesy of The Okonite Company.*) *Notes:* A, Conductor; B, Inner Semiconductor Layer; C, Ethylene Propylene Rubber (EPR) Insulation; D, Outer Semiconductor Layer; E, Tape Shield; F, Jacket

The width of a pull box for a straight-through pull is determined by the size of the largest conduit and by the space required by the locknuts and bushings, which secure the conduit to the box. The length cannot be less than eight times the diameter of the largest conduit.

When an angle or U-pull is required, the minimum length requirement used for straight pulls also determines the minimum width. Additionally, the inside length of the pull box cannot be less than six times the largest conduit plus the sum of the diameters of all additional conduits entering the box. In the case of an angle pull, the diagonal distance between the centers of the conduits where they enter and exit the box must be at least six times the conduit diameter. The wall of the pull box with the most conduit penetrations must be used to size the box. Examples of pull box sizing are shown in Figure 6.7.

Pulling cables through a conduit raceway requires skill and care. A variety of basket grips and pulling eyes are available to attach the pulling rope to the cable(s). Proper pulling grip selection depends on the required pulling tension and is crucial

Figure 6.5　Concentric Neutral-Shielded Cable. (*Photo courtesy of The Okonite Company.*) *Notes:* A, Conductor; B, Inner Semiconductor Layer; C, Ethylene Propylene Rubber (EPR) Insulation; D, Outer Semiconductor Layer; E, Concentric Neutral; F, Jacket

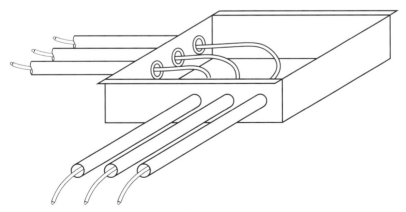

Figure 6.6　Pull Box in Conduit Raceway

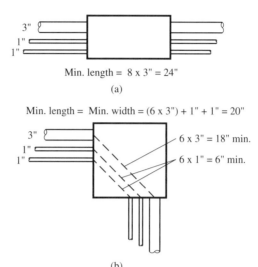

Figure 6.7 Pull Box Sizing

to prevent cable and equipment damage and injury to installation personnel. Two particular quantities must be calculated before the cable is pulled into conduit and carefully monitored during the pulling process. These quantities are *pulling tension* and *sidewall pressure*. *Pulling tension* is the tensile force that must be applied to the cable to overcome friction as the cable is pulled through the raceway. *Sidewall pressure* is the crushing force applied to the cable by the conduit in the radial direction as the cable is pulled around a bend. Exceeding the maximum allowable pulling tension is seldom a problem for large power cables since that maximum value is usually quite large (in the thousands of pounds). Control and instrument cables, however, tend to have much lower maximum allowable pulling tensions. It is not uncommon to exceed the maximum allowable pulling tension for these types of cables in long raceways. Sidewall pressure is usually the controlling factor when designing conduit raceways for large power cables.

When calculations show that the maximum allowable pulling tension for a cable or group of cables will be exceeded, several measures can be taken to reduce the required pulling tension. One is to reduce the coefficient of friction between the conduit and the cable. Various types of lubricants exist for this purpose, each reducing the coefficient of friction to a specific value. The coefficient of friction can range from more than 0.5 for dry cable to less than 0.2 for well-lubricated cable. Suitable lubricants include wire soaps, waxes, and synthetic polymer compounds. Many installers prefer the polymer compounds because the soaps and waxes leave a residue on the cable which can be messy and, in some cases, combustible. If the calculated tensions are unacceptably high, it may be necessary to select a different lubricant with a lower coefficient of friction to reduce the required pulling tension to an acceptable value.

If the coefficient of friction cannot be reduced sufficiently, reducing the length or changing the direction of the pull may be successful. Installing pull boxes in the

Cradled   Triangular   Figure 6.8   Cradled and Triangular Configurations

raceway can reduce the length of the pull. Cable is pulled from pull box to pull box, so if a pull box is installed midway in a 300-ft run of conduit, the installation becomes two pulls of 150 ft each. If the conduit is encased in an underground duct bank, a *manhole* is needed to serve as a pull point. Unless the geometry of the raceway is symmetric, different tensions will be required to pull cable from end A to end B and from end B to end A. Pulling calculations should be done for pulling in both directions to determine which direction of pull will require the lesser tension.

When more than one cable is pulled in a conduit, a *weight correction factor* (w_c) is needed to account for the additional friction forces that exist between the cables. The most common case is to install three cables of the same size in a conduit. The weight correction factor depends on whether the cables are arranged in a triangular or cradled configuration. An illustration of these two configurations is shown in Figure 6.8.

In the triangular configuration, the top cable may experience very little tension compared to the other two. In the cradled configuration, the middle cable experiences more friction forces than the other two cables. To be conservative, assume that the total required tension is distributed equally between two of the three cables.

Weight correction factors for cradled and triangular configurations are calculated using the formulas shown in Eqs. (6.8) and (6.9), respectively.

$$w_c = 1 + \frac{4}{3}\left(\frac{OD_{cable}}{ID_{conduit} - OD_{cable}}\right)^2 \quad \text{(cradled)} \qquad (6.8)$$

$$w_c = \frac{1}{\sqrt{1 - \left(\dfrac{OD_{cable}}{ID_{conduit} - OD_{cable}}\right)^2}} \quad \text{(triangular)} \qquad (6.9)$$

Weight correction factors for more than three cables in a conduit or for cables of different size being pulled together are complicated to determine since the configuration of the cables is difficult to ascertain. For such complicated pulls, the cable manufacturer should be consulted.

6.2.1 Pulling Tension

Pulling tension is the tensile force that must be applied to the cable to overcome friction as the cable is pulled through the raceway. The tension required to pull a cable or group of cables through a straight section of conduit is expressed in Eq. (6.10).

$$T = w_c \mu L W \qquad (6.10)$$

where T is the pulling tension in pounds, w_c is the weight correction factor (dimensionless), μ is the coefficient of friction (dimensionless), L is the length of straight section of conduit in feet, and W is the weight of cable in pounds per foot.

The tension required to pull a cable through a horizontal bend is shown in Eq. (6.11).

$$T_o = T_i e^{w_c \mu \theta} \tag{6.11}$$

where T_o is the tension out of the bend in pounds, T_i is the tension coming into the bend in pounds, w_c is the weight correction factor (dimensionless), μ is the coefficient of friction (dimensionless), and θ is the bend angle in radians.

When pulling cable through a vertical bend, the tension is calculated as for a horizontal bend, then the cable weight in the vertical section is either added (if the cable is pulled uphill) or subtracted (if the cable is pulled downhill) from the required tension. When pulling cable downhill, a negative tension can be calculated. This indicates that brakes must be applied to the cable supply reel to prevent the cable from spooling too quickly.

Sometimes physical constraints require that the cable be pulled in one specific direction. If no such constraints exist, an attempt should be made to pull the cable in the direction requiring the lesser pulling tension.

6.2.2 Sidewall Pressure

Sidewall pressure is the crushing force applied to the cable by the conduit in the radial direction as the cable is pulled around a bend. It is a function of pulling tension and bending radius of the conduit and differs according to the arrangement of the cables in the conduit. Three-conductor cables, as shown in Figure 6.9, behave like three single conductor cables lying in a triangular configuration.

The formulas to calculate sidewall pressure for various cable arrangements are shown in (Eqs. 6.12) through (6.14):

$$P = \frac{T_o}{r} \text{ (one single-conductor cable)} \tag{6.12}$$

$$P = \left(\frac{3w_c - 2}{3}\right)\frac{T_o}{r} \text{ (three single-conductor cables cradled)} \tag{6.13}$$

$$P = \left(\frac{w_c}{2}\right)\frac{T_o}{r} \text{(three single-conductor cables triangular or}$$
$$\text{one three-conductor cable)} \tag{6.14}$$

Sidewall pressure is typically the controlling factor in raceway design for large power cables. Sufficiently large conduit bending radii must be used, and pulling tension may need to be limited to values well below the maximum allowable pulling tension for the cable to keep the sidewall pressure below the maximum allowed by the cable.

6.2 CONDUIT AND DUCT SYSTEMS

Figure 6.9 Three-Conductor Cable. (*Photo courtesy of The Okonite Company.*) A, Conductor; B, Inner Semiconductor Layer; C, Ethylene Propylene Rubber (EPR) Insulation; D, Extruded Semiconducting EPR Insulation Screen; E, Phase Identification Tape; F, Copper Grounding Conductor; G, Uncoated Copper Shield; H, Fillers and Binder Tape; J, Jacket

6.2.3 Design Examples

Consider the conduit raceway layout shown in Figure 6.10.

All bends lie in the horizontal plane. Three single-conductor 500-kcmil tape-shielded 5-kV power cables need to be pulled through the conduit. The cable has a 1.10-inch outside diameter and weighs 1.83 lb/ft. A coefficient of friction of 0.25 is anticipated. The raceway can be designed as follows:

First, the conduit must be sized. The cross-sectional area of the three cables is

$$A_{\text{cables}} = 3\left[\pi\left(\frac{1.10}{2}\right)^2\right] = 2.851 \text{ in}^2. \tag{6.15}$$

The required cross-sectional area of the conduit to yield a 40% fill, which is the maximum fill allowed by the NEC, is

$$A_{\text{conduit (40\%)}} = \frac{2.851}{0.40} = 7.128 \text{ in}^2. \tag{6.16}$$

The corresponding diameter to this cross-sectional area is

$$\text{ID}_{\text{conduit (40\%)}} = \sqrt{\frac{4\,(7.128)}{\pi}} = 3.013 \text{ in}. \tag{6.17}$$

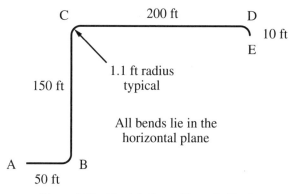

Figure 6.10 Conduit Design Example No. 1

This indicates that a nominal 3-inch rigid steel conduit can be used, as its inside diameter is 3.09 inch, per Table F.1. But before settling on a 3-inch conduit, the jam ratio must be checked. The jam ratio in the 3-inch conduit is

$$j_{3''} = \frac{3.09}{1.10} = 2.809. \tag{6.18}$$

This value is in the "danger zone" of $2.8 \leq j \leq 3.0$ where jamming can occur. In addition to the jam ratio, another issue must be addressed. When a conduit is bent, a slight flattening occurs, transforming the circular cross-section of the conduit into an ellipse. The major axis of the ellipse is typically 5% larger than the diameter of the circular cross-section. This enlarged effective diameter yields a different jam ratio in the bends of the raceway. This jam ratio is

$$j_{3'' = \text{bends}} = \frac{1.05\,(3.09)}{1.10} = 2.950. \tag{6.19}$$

This jam ratio is also in the "danger zone." Since jamming is most likely to occur in a bend, this conduit size should not be selected. The next larger nominal conduit size is $3\frac{1}{2}$ inches. The cross-sectional area of a $3\frac{1}{2}$-inch rigid steel conduit, using the inside diameter from Table F.1, is:

$$A_{\text{conduit}(3-1/2'')} = \pi \frac{(3.57)^2}{4} = 10.010\,\text{in}^2. \tag{6.20}$$

The percent fill for the $3\frac{1}{2}$-inch conduit is

$$\%\text{fill} = \frac{2.851}{10.010} = 28.5\%. \tag{6.21}$$

Next, the jam ratios are checked for the $3\frac{1}{2}$-inch conduit:

$$j_{3-1/2''} = \frac{3.57}{1.10} = 3.245. \tag{6.22}$$

Since jamming cannot occur when $j > 3$, this jam ratio is acceptable. Now, the jam ratio is checked in the bends, allowing for a 5% elongation of the conduit's inside diameter:

$$j_{3-1/2''\text{-bends}} = \frac{1.05\,(3.57)}{1.10} = 3.408. \tag{6.23}$$

This jam ratio is also acceptable, so a $3\frac{1}{2}$-inch conduit will be used.

A minimum bending radius must now be determined for the conduit. The minimum bending radius of the cable is 12 times its outside diameter, or

$$r_{\min} = 12(1.10) = 13.2 \text{ in}. \tag{6.24}$$

This radius will be assumed for all conduit bends, but may have to be increased if the maximum allowable sidewall pressure is exceeded.

The last term to be calculated before the pulling calculations can be performed is the weight correction factor. Because of the high jam ratio, a cradled cable configuration is assumed.

$$w_c = 1 + \frac{4}{3}\left(\frac{1.10}{3.57 - 1.10}\right)^2 = 1.26 \tag{6.25}$$

Now the pulling calculations can be done. First, the calculations will be done to pull the cables from point A to point E. The tension required to pull the cables from point A to point B, using Eq. (6.10), is

$$T_{A-B} = (3)(1.26)(0.25)(50)(1.83) = 86.47 \text{ lb}. \tag{6.26}$$

Note the multiplier of 3 to account for three cables being pulled. The tension required to pull the cables through bend B, using Eq. (6.11), is

$$T_B = 86.47 e^{(1.26)(0.25)(\pi/2)} = 141.83 \text{ lb}. \tag{6.27}$$

The tension required to pull the cables from point B to point C is

$$T_{B-C} = (3)(1.26)(0.25)(150)(1.83) + T_B = 259.40 + 141.83 = 401.23 \text{ lb}. \tag{6.28}$$

The tension required to pull the cables through bend C is

$$T_C = 401.23 e^{(1.26)(0.25)(\pi/2)} = 658.09 \text{ lb}. \tag{6.29}$$

To pull the cables through the next straight section, C to D, the required tension is

$$T_{C-D} = (3)(1.26)(0.25)(200)(1.83) + T_C = 345.87 + 658.09 = 1003.96 \text{ lb}. \tag{6.30}$$

To pull through bend D, the required tension is

$$T_D = 1003.96 e^{(1.26)(0.25)(\pi/2)} = 1646.67 \text{ lb}. \tag{6.31}$$

Pulling through the final straight section, D to E, requires

$$T_{D-E} = (3)(1.26)(0.25)(10)(1.83) + T_D = 17.29 + 1646.67 = 1663.96 \text{ lb}. \tag{6.32}$$

A total tension of 1663.96 lb is required to pull the cables from point A to point E. Next, the pulling calculations are redone, this time pulling from point E to point A. The calculations are as follows:

$$T_{E-D} = (3)(1.26)(0.25)(10)(1.83) = 17.29 \text{ lb} \tag{6.33}$$
$$T_D = 17.29 e^{(1.26)(0.25)(\pi/2)} = 28.36 \text{ lb} \tag{6.34}$$
$$T_{D-C} = (3)(1.26)(0.25)(200)(1.83) + T_D$$
$$= 345.87 + 28.36 = 374.23 \text{ lb} \tag{6.35}$$
$$T_C = 374.23 e^{(1.26)(0.25)(\pi/2)} = 613.80 \text{ lb} \tag{6.36}$$
$$T_{C-B} = (3)(1.26)(0.25)(150)(1.83) + T_C$$
$$= 259.40 + 613.80 = 873.20 \text{ lb} \tag{6.37}$$
$$T_B = 873.20 e^{(1.26)(0.25)(\pi/2)} = 1432.20 \text{ lb} \tag{6.38}$$
$$T_{B-A} = (3)(1.26)(0.25)(50)(1.83) + T_B$$
$$= 86.47 + 1432.20 = 1518.67 \text{ lb} \tag{6.39}$$

Pulling the cables from point E to point A requires only 1518.67 lb of tension. Since less tension is required to pull the cables from point E to point A than in the other direction, the preferred pulling direction is from point E to point A. Sometimes, due to physical constraints, it is more difficult or even impossible to pull in one particular direction. If this is the case, the cables need to be pulled in the more feasible direction. If no pulling constraints exist, the direction requiring the lesser pulling tension should be chosen.

Now it has been determined that the cables will be pulled from point E to point A; sidewall pressures are calculated at each bend. A maximum allowable sidewall pressure of 500 lb/ft of radius is determined to be

$$P_{\max} = (1.10)(500) = 550 \text{ lb/ft}. \tag{6.40}$$

Due to the large jam ratio, a cradled cable configuration is assumed. The following calculations determine the sidewall pressures at each bend:

$$P_D = \frac{3(1.26) - 2}{3} \left(\frac{28.36}{1.10} \right) = 15.30 \text{ lb/ft} \tag{6.41}$$
$$P_C = \frac{3(1.26) - 2}{3} \left(\frac{613.80}{1.10} \right) = 331.08 \text{ lb/ft} \tag{6.42}$$
$$P_B = \frac{3(1.26) - 2}{3} \left(\frac{1432.20}{1.10} \right) = 772.52 \text{ lb/ft} \tag{6.43}$$

The sidewall pressure at bend B is in excess of the maximum allowable. This can be remedied by increasing the bend radius of bend B. The minimum bend radius as to not violate the sidewall pressure limit can be calculated as follows:

$$r_{\min} = \frac{(772.52)(1.10)}{550} = 1.55 \text{ ft} = 18.6 \text{ in} \tag{6.44}$$

So, if a 20-inch bend radius is used at bend B, the sidewall pressure there becomes

$$P_{B(20'')} = \frac{(3)(1.26) - 2}{3} \left(\frac{1432.20}{\frac{20}{12}} \right) = 509.86 \text{ lb/ft}. \tag{6.45}$$

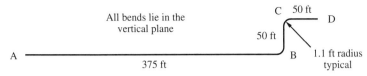

Figure 6.11   Conduit Design Example No. 2

If only bend B is enlarged to a 20-inch bend radius and the other bends remain at a 13.2-inch bend radius, it is critical that the cables be pulled from point E to point A. Pulling in the reverse direction would violate the maximum sidewall pressure limit at the last bend, probably damaging the cables in the process. Also, two different bend radii must be used when the conduit is installed. To safeguard against pulling the cables in the wrong direction and to simplify the conduit installation by requiring only one bend radius, the engineer may choose to use 20-inch bend radii at all bends. If physical constraints allow for the larger bend radii, using a 20-inch bend radius at each bend is advisable.

Next, consider the conduit raceway layout shown in Figure 6.11.

All bends lie in the vertical plane. As in the last example, three single-conductor 500-kcmil tape-shielded 5-kV power cables will be pulled through the conduit. The cable outside diameter, weight, and coefficient of friction are 1.10 inch, 1.83 lb/ft, and 0.25, respectively. The same conduit sizing procedure as used in Example No. 1 determines a $3\frac{1}{2}$-inch conduit size. A minimum bend radius of 13.2 inch and a weight correction factor of 1.26 also will be used.

First, the required tension to pull the cables from point A to point D is calculated:

$$T_{A-B} = (3)(1.26)(0.25)(375)(1.83) = 648.51 \text{ lb} \tag{6.46}$$
$$T_B = 648.51 e^{(1.26)(0.25)(\pi/2)} = 1063.67 \text{ lb} \tag{6.47}$$

To determine the tension required to pull the cable uphill through the vertical section B–C, the weight of the cables in that section is added to the tension as calculated by Eq. (6.10):

$$\begin{aligned} T_{B-C} &= (3)(1.26)(0.25)(50)(1.83) + 1063.67 + (3)(50)(1.83) \\ &= 86.47 + 1063.67 + 274.50 = 1424.64 \text{ lb} \end{aligned} \tag{6.48}$$

Note that the cables must be supported by special hangers in the vertical section of conduit to prevent damage to the cables caused by supporting their own weight:

$$T_C = 1424.64 e^{(1.26)(0.25)(\pi/2)} = 2336.65 \text{ lb} \tag{6.49}$$
$$\begin{aligned} T_{C-D} &= (3)(1.26)(0.25)(50)(1.83) + 2336.65 \\ &= 86.47 + 2336.65 = 2423.12 \text{ lb} \end{aligned} \tag{6.50}$$

A total of 2423.12 lb is required to pull the cables from point A to point D. Next, the tension calculations are redone, this time pulling the cables in the reverse direction:

$$T_{D-C} = (3)(1.26)(0.25)(50)(1.83) = 86.47 \text{ lb} \tag{6.51}$$
$$T_C = 86.47 e^{(1.26)(0.25)(\pi/2)} = 141.83 \text{ lb} \tag{6.52}$$

The weight of the cables in vertical section C–B is subtracted from the tension as calculated by Eq. (6.10), since the cables are being pulled downhill:

$$T_{C-B} = (3)(1.26)(0.25)(50)(1.83) + 141.83 - (3)(50)(1.83)$$
$$= 86.47 + 141.83 - 274.50 = -46.20 \, \text{lb} \tag{6.53}$$

The negative tension indicates that the cable reel must be braked during installation. Assume a braking force sufficient to provide the equivalent of 50 ft of horizontal pulling tension (86.47 lb) developed at the entrance to bend B:

$$T_{\text{entering B}} = (3)(1.26)(0.25)(50)(1.83) = 86.47 \, \text{lb} \tag{6.54}$$
$$T_B = 86.47 e^{(1.26)(0.25)(\pi/2)} = 141.83 \, \text{lb} \tag{6.55}$$
$$T_{B-A} = (3)(1.26)(0.25)(375)(1.83) + 141.83$$
$$= 648.51 + 141.83 = 790.34 \, \text{lb} \tag{6.56}$$

When pulling the cables from point D to point A, only 790.34 lb of tension is required.

The last step of the design is to verify that the sidewall pressure is less than the 550 lb/ft of bend radius maximum. The tension out of each bend is the same (141.83 lb), so the sidewall pressure, assuming a cradled cable configuration, is

$$P_C = P_B = \frac{3(1.26) - 2}{3} \left(\frac{141.83}{1.10} \right) = 76.50 \, \text{lb/ft}. \tag{6.57}$$

This value is well below the maximum allowable sidewall pressure, so a 1.10-ft bending radius is acceptable.

6.3 CABLE TRAY SYSTEMS

As defined in the NEC, a cable tray system is "a unit or assembly of units or sections and associated fittings forming a rigid structural system used to securely fasten or support cables and raceways." Being part of the facility's structural system, *structural* as well as *electrical* considerations must be considered when designing cable tray systems.

Several standards and guidelines exist for the design of cable tray systems. Articles 250, 318, and 800 of the NEC address various aspects of cable tray systems. In addition, the National Electrical Manufacturers Association (NEMA) has published three documents regarding cable trays.

NEMA standard VE1 covers general cable tray definitions, manufacturing standards, performance standards, test standards, and application information. NEMA standard FG1 specifically addresses fiberglass cable tray systems, and NEMA standard VE2 is an installation guideline for cable trays.

The Cable Tray Institute (CTI) is a trade association comprised of major manufacturers in the cable tray industry. CTI provides specifiers, designers, and installers with advice as to which type of cable tray system is best suited to a specific application. Types of cable trays available include ladder, solid-bottom, trough, channel, wire mesh, and single rail.

Ladder tray provides solid side rail protection and system strength with smooth radius fittings and a wide selection of materials and finishes. It is the best type of tray for long span applications because of its strength. It comes in standard widths of 6, 12, 18, 24, 30, and 36 inches, and standard depths of 3, 4, 5, and 6 inches. Ladder tray sections are typically available in 10-, 12-, 20-, and 24-ft lengths. Rung spacings of 6, 9, 12, and 18 inches are available.

Solid-bottom tray provides nonventilated continuous support for delicate cables. Solid metal covers can be provided for use with nonplenum-rated cable in environmental air areas. Standard widths of 6, 12, 18, 24, 30, and 36 inches and standard depths of 3, 4, 5, and 6 inches are available. Solid-bottom tray usually comes in 10-, 12-, 20-, and 24-ft lengths. The most common applications of solid-bottom tray are for minimal heat-generating electrical or telecommunication applications with short to intermediate support spans of 5 to 12 feet.

Trough cable tray provides moderate ventilation with added cable support frequency. The bottom of the tray provides cable support every 4 inches. Trough tray is available in both metal and nonmetallic materials. It commonly comes in standard widths of 6, 12, 18, 24, 30, and 36 inches and standard depths of 3, 4, 5, and 6 inches. Trough cable tray is available in 10-, 12-, 20-, and 24-ft lengths. Trough cable tray is generally used for moderate heat-generating applications with short to intermediate support spans of 5 to 12 feet.

Channel cable tray provides an economical support for cable drops and branch cable runs from the backbone cable tray system. It comes in standard widths of 3, 4, and 6 inches in metal systems and up to 8 inches in nonmetallic systems. Standard depths vary from $1\frac{1}{4}$ to $1\frac{3}{4}$ inches in metal systems and from 1 to over $2\frac{1}{4}$ inches in nonmetallic systems. It commonly comes in 10-, 12-, 20-, and 24-ft lengths. Channel cable tray is used for installations with limited numbers of tray cable when conduit is undesirable. Support requirements allow for short to medium support spans of 5 to 10 feet.

Wire mesh cable tray is used for communication and fiber-optic cables. Most manufacturers use a zinc-plated steel mesh although other materials may be available. Mesh tray comes in standard widths of 2, 4, 6, 8, 12, 16, 18, 20, and 24 inches, and standard depths of 1, 2, and 4 inches. Standard length for mesh tray is a rather unusual 118 inches. Wire mesh cable tray is installed on short support spans of 4 to 8 feet.

Single rail cable trays provide ease of cable installation and maximum freedom for cables to enter and exit the raceway. These trays are wall mounted and are often configured in multiple tiers. Standard widths are 6, 9, 12, 18, and 24 inches, and standard depths are 3, 4, and 6 inches. Commonly available lengths are 10 and 12 ft. Single rail cable tray is generally used for low-voltage and power cable installations where maximum cable freedom, side fill, and speed to install are factors.

Most manufacturers provide cable trays in a variety of materials. *Plain steel* is hot rolled, pickled, and oiled, and is available in both commercial and structural quality. *Pregalvanized* is mill galvanized prior to fabrication. *Hot dip galvanized after fabrication* is plain steel that is hot dipped after fabrication and offers superior corrosion resistance. *Stainless steel*, types 304 and 316L, are fully annealed and suitable for the most corrosive environments. *Aluminum* tray is also available, with the most common alloys being 6063-T6 and 5052-H32, and fiberglass tray is typically

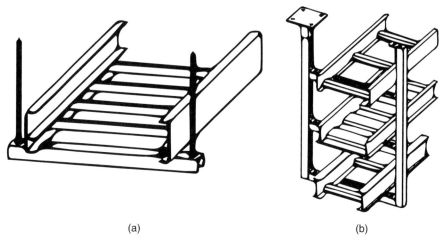

(a)   (b)

Figure 6.12   Trapeze Supports *Reprinted from NEMA Standard VE2-2006 with Permission*

available in polyester and vinyl ester resins and carry Class 1 flame rating and self-extinguishing requirements of ASTM standard D-635, published by the American Society for Testing and Materials International.

After the type of cable tray is selected, the support method must be determined. *Trapeze supports*, shown in Figure 6.12, are perhaps the most commonly used type of support method. Trapeze hangars can be built from a variety of components including threaded rod, channel, angle, and other structural shapes. They can hold single trays or be multitiered.

Trapeze supports are strong and perform well under seismic conditions. Because of the vertical support members on both sides of the tray, cables must be pulled into the tray, similar to the way cables are pulled into conduit.

Hanger rod clamps, shown in Figure 6.13, can be used to support single trays or multiple trays at the same elevation. While offering less strength than trapeze supports, hanger rod clamps are a less-expensive option.

If ease of installation outweighs the need for structural strength, *center hanger supports*, shown in Figure 6.14, may be considered. While being less robust than two-support hangers, cables can be added or removed easily after raceway installation by accessing them from the sides. Note that the cable jackets must be protected from the sharp threads of the threaded-rod support with a smooth bushing slipped over the threads.

If hanging the cable trays from the ceiling is not convenient, *wall support* systems, such as the one shown in Figure 6.15, can be used. While allowing sideloading of the cables and easy access to the cables after raceway installation, support spacing decreases dramatically as cable weight increases, especially for wide trays.

When attaching cable trays to floors or roofs, *elevated strut supports*, as shown in Figure 6.16, can be used. Covers, either *ventilated* or *nonventilated*, are often used in outdoor applications. If accumulation of snow or airborne particulate such as fly ash is a concern, *peaked* covers can be used instead of *flat* covers.

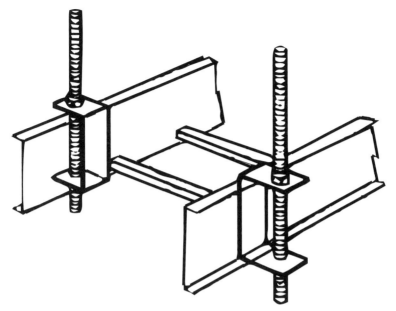

Figure 6.13 Hanger Rod Clamps. (*Reprinted from NEMA Standard VE2-2006 with Permission.*)

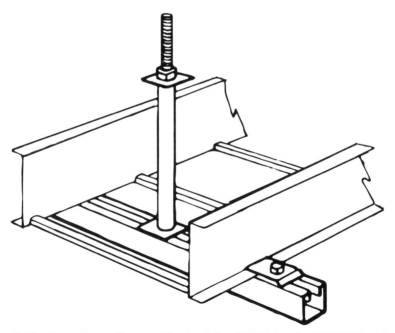

Figure 6.14 Center Hanger Support. (*Reprinted from NEMA Standard VE2-2006 with Permission.*)

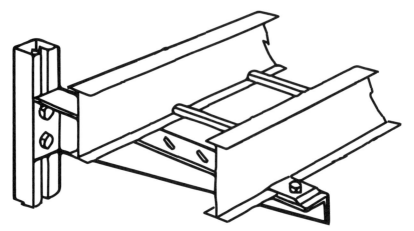

Figure 6.15 Cantilevered Wall Support. (*Reprinted from NEMA Standard VE2-2006 with Permission.*)

In addition to structural requirements, electrical requirements must also be carefully considered. Bonding cable tray sections with flexible *jumpers*, either *insulated*, *bare*, or *braided*, is a requirement for metallic trays. Nonmetallic trays must carry a ground wire, typically bare copper or copper-clad. A typical bonding jumper installation is shown in Figure 6.17.

Installing cable in trays is similar to pulling cable into conduit in that maximum allowable tension and sidewall pressure must not be exceeded. Installation is similar to that for conduit, as trapeze hangers, the most common type for suspending

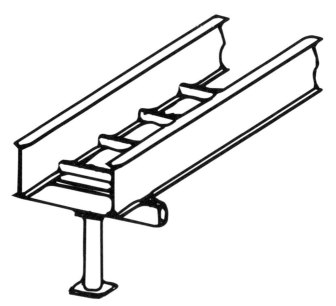

Figure 6.16 Elevated Strut Support. (*Reprinted from NEMA Standard VE2-2006 with Permission.*)

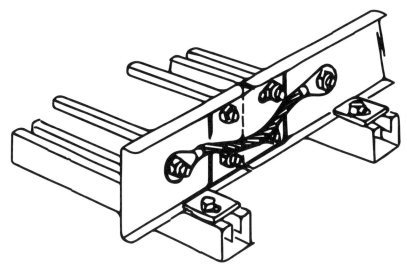

Figure 6.17   Bonding Jumpers. (*Reprinted from NEMA Standard VE2-2006 with Permission.*)

cable tray, prevent cables from simply being laid into the tray from the side. If a cantilever-type hanger is used to mount the tray to a wall, installing the cable becomes much simpler.

Different types of cables (medium-voltage power, low-voltage power, control, instrumentation, communications, etc.) may not be mixed in a single tray unless they are separated by metallic barriers. NEC Section 318-9 defines the requirements for installing multi-conductor cables in tray. For power cables, multiconductor cables larger than AWG 4/0 must be placed side by side and cannot be stacked more than one layer deep. Smaller cables can be stacked, but the total cross-sectional area of the cables cannot exceed 40% of the cross-sectional area of the tray. The total cross-sectional area of control (nonpower) cables in trays with a depth of 6 inches or less cannot exceed 50% of the cross-sectional area of the tray.

Article 318 of the NEC defines the requirements for installing cables in cable tray. The allowable fill, in square inches, is determined by the type of cable installed (single conductor or multiple conductor) and by the type of tray used (solid-bottom or ladder/ventilated trough). Table 6.1 summarizes the allowable cable tray fill in square inches for cables rated 2000 V or less.

For cables rated 2001 V or greater, the sum of the cable diameters cannot exceed the width of the tray, and the cables must be laid in a single layer. If the trays are covered for more than 6 ft, the NEC stipulates derating factors to decrease the ampacity of the cables.

For example, consider 12 single-conductor, 500-kcmil, 600-V cables with an outside diameter of 1.15 inch. To size a ladder-type tray to carry these cables, the total cross-sectional area for the cables must be calculated:

$$A_{\text{CABLES}} = 12 \left[\pi \left(\frac{1.15}{2} \right)^2 \right] = 12.46 \, \text{in}^2 \qquad (6.58)$$

TABLE 6.1 Allowable Cable Tray Fill in Square Inches for Cables Rated 2000 V or Less

Inside Width of Tray	Multi-Conductor Cables				Single-Conductor Cables	
	Solid-Bottom Trays		Ladder or Ventilated Trough Trays			
	All conductors smaller than AWG 4/0	Some conductors AWG 4/0 or larger	All conductors smaller than AWG 4/0	Some conductors AWG 4/0 or larger	All conductors smaller than 1000 kcmil	Some conductors 1000 kcmil or larger
6"	5.5	5.5 – sd[a]	7.0	7.0 –(1.2 × sd[a])	6.5	6.5 –(1.1 × sd[a])
9"	8.0	8.0 – sd[a]	10.5	10.5 –(1.2 × sd[a])	9.5	9.5 –(1.1 × sd[a])
12"	11.0	11.0 – sd[a]	14.0	14.0 –(1.2 × sd[a])	13.0	13.0 –(1.1 × sd[a])
18"	16.5	16.5 – sd[a]	21.0	21.0 –(1.2 × sd[a])	19.5	19.5 –(1.1 × sd[a])
24"	22.0	22.0 – sd[a]	28.0	28.0 –(1.2 × sd[a])	26.0	26.0 –(1.1 × sd[a])
30"	27.5	27.5 – sd[a]	35.0	35.0 –(1.2 × sd[a])	32.5	32.5 –(1.1 × sd[a])
36"	33.0	33.0 – sd[a]	42.0	42.0 –(1.2 × sd[a])	39.0	39.0 –(1.1 × sd[a])

[a] "sd" denotes the sum of the diameters of all cables in the tray larger than AWG 4/0 (for solid-bottom trays) or 1000 kcmil (for ladder or ventilated trough trays).

The column of Table 6.1 labeled "all conductors smaller than 1000 kcmil" under "single conductor cables" in "ladder or ventilated trough trays" indicates that a 12-inch tray can accommodate 13.0 in² of fill, so a 12-inch tray would be selected.

If three of the cables from the previous example were replaced by three single-conductor, 1000-kcmil, 600-V cables with an outside diameter of 1.48 inch, the sum of the diameters of all cables in the tray *1000 kcmil or larger* must be determined:

$$sd = 3(1.48) = 4.44 \text{ in} \quad (6.59)$$

The total cross-sectional area of the cables must be recalculated:

$$A_{\text{CABLES}} = 9\left[\pi\left(\frac{1.15}{2}\right)^2\right] + 3\left[\pi\left(\frac{1.48}{2}\right)^2\right] = 14.51 \text{ in}^2 \quad (6.60)$$

The product of 1.1 times sd, or 4.88, must be subtracted from the fill area allowed for smaller cables, as shown in the far right column of Table 6.1. This means that an 18-inch tray, with an adjusted allowable fill of 19.5–4.88 = 14.62 in², must be selected.

Once the tray is sized, the required pulling tension must be found. A simplified calculation method can be used if the installation rollers or sheaves have a low coefficient of friction and are positioned above the tray, spaced such that the cables do not sag to touch the bottom of the tray. If these conditions are met, the pulling tension for horizontal sections of tray can be calculated using Eq. (6.61):

$$T = \mu w L \quad (6.61)$$

where

μ = coefficient of friction (0.15 is a typical value)
w = total cable weight in pounds per foot
L = length of horizontal section in feet

Equation (6.61) does not account for the tension developed as the cable is pulled off the supply reel. An estimate of this tension is 25 times the total cable weight per foot.

When pulling cable into vertical sections of tray, like those shown in Figure 6.18, the total cable weight in the vertical section is either added to (if pulling uphill) or subtracted from (if pulling downhill) the calculated tension, as when pulling cable through vertical sections of conduit.

Spacing of the installation rollers is critical to prevent cable damage during pulling. Near the end of the pull where the tensions are higher, roller spacing can be increased, since the cables will sag less. An approximation of the spacing requirement between rollers can be determined using Eq. (6.62):

$$d = \sqrt{\frac{8hT}{w}} \qquad (6.62)$$

Figure 6.18   Power Cables Run in Vertical Ladder-Type Cable Tray. (*Photo courtesy of Tampa Electric Company.*)

where

d = distance between rollers in feet
h = height from bottom of tray to top of roller
T = tension in pounds
w = total cable weight in pounds per foot

It is a good practice to provide additional rollers to assure that the cables do not sag to the bottom of the tray while being pulled.

6.3.1 Design Example

A three-conductor 500 kcmil copper 15 kV armored cable with an outside diameter of 3.60 inches and weight of 8.67 pounds per foot must be pulled into the cable tray shown in Figure 6.19. First, the tension off the supply reel is estimated at 25 times the cable weight per foot.

$$T_A = 25\,(8.67) = 216.75 \text{ lb} \tag{6.63}$$

Next, the tension at point B is determined by adding the weight of the cable from the reel to point B:

$$T_B = (3025)(8.67) + 216.75 = 433.50 \text{ lb} \tag{6.64}$$

The tension at point C is found using Eq. (6.58):

$$T_C = (0.15)(8.67)(150) + 433.50 = 628.58 \text{ lb} \tag{6.65}$$

The tension at point D is the tension at point C increased by the weight of the cable in the vertical tray section:

$$T_D = (25)(8.67) + 628.58 = 845.33 \text{ lb} \tag{6.66}$$

Finally, the tension at point E is

$$T_E = (0.15)(8.67)(250) + 845.33 = 1170.46 \text{ lb} \tag{6.67}$$

The estimated spacing between the installation rollers can be found using Eq. (6.62). A height of 3 ft is assumed from the top of the roller to the bottom of the tray. For the 150-ft section of tray,

$$d_{150 \text{ ft. section}} = \sqrt{\frac{(8)(3)(433.50)}{8.67}} = 34.64 \text{ ft} \tag{6.68}$$

This estimate indicates that six installation rollers will be needed, spaced 30 ft apart. Note that the tension used in Eq. (6.68) was the tension at the beginning of

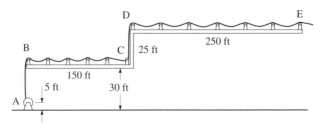

Figure 6.19 Cable Tray Design Example

the horizontal section. This leads to a very conservative result. An average tension of $(T_B + T_C)/2$ could be used, which would lead to an estimate of five rollers spaced 37.5 ft apart. This spacing could cause excessive sag near point B where the tension is less than the average tension in the horizontal section. The conservative approach, which uses the minimum tension, is recommended.

Continuing to use the conservative assumption of minimum tension, the installation roller spacing for the 250-ft horizontal tray section is

$$d_{250\,\text{ft. section}} = \sqrt{\frac{(8)(3)(845.33)}{8.67}} = 48.37\,\text{ft.} \qquad (6.69)$$

This estimate requires seven rollers spaced 41.67 ft apart.

Also, the minimum bend radius of the cable must be obeyed. As with installation into conduit, 12 times the cable outside diameter is assumed as the minimum bending radius. Therefore, all bends must have a radius no less than

$$r_{\min} = (12)(3.60) = 43.2\,\text{in.} \qquad (6.70)$$

Tray sections with 48-inch radii should be used.

SUMMARY

Cables are contained in some type of *raceway*, typically *conduit*, *duct*, or *cable tray*. The raceway protects the circuits from mechanical damage and influences the reactance and ampacity of the cables it contains.

When designing conduit and duct systems, sizing of the conduits and pull boxes are governed by the NEC. NEMA and CTI also provide publications useful for the design and construction of raceway systems. *Pulling tension* and *sidewall pressure* must be calculated to assure that the cables will not be damaged during installation. Pulling tension is often the controlling factor for control and instrumentation cable, while sidewall pressure typically controls for power cables.

Cable tray provides more versatility than conduit or duct systems, but installation challenges can still be faced. The NEC also governs the fill and stacking method of cables in tray. Cable installation in tray may be determined by the type of tray and tray hangers used.

FOR FURTHER READING

Cable Tray Installation Guidelines, National Electrical Manufacturers Association, NEMA Standard VE 2, 2013.
Cable Tray Type Selection, CTI Technical Bulletin Number 4, Cable Tray Institute, 1994.
Engineering Handbook, The Okonite Company, Ramsey, New Jersey, 2010.
Fiberglass Cable Tray Systems, National Electrical Manufacturers Association, NEMA Standard FG 1, 1993.
Installation Practices for Cable Raceway Systems, The Okonite Company, Ramsey, New Jersey, 2011.
Metal Cable Tray Systems, National Electrical Manufacturers Association, NEMA Standard VE 1, 2009.

National Electrical Code, National Fire Protection Association, NFPA 70, 2014.
Standard Test Method for Rate of Burning and/or Extent and Time of Burning of Plastics in a Horizontal Position, ASTM International, ASTM D-635, 2014.

QUESTIONS

1. What are the primary factors that determine the tension required to pull cables through a raceway?
2. What is the purpose of the weight correction factor used in cable-pulling calculations?
3. Explain the purpose of calculating a jam ratio when pulling cables into conduit?
4. In general, when is pulling tension the controlling parameter in determining whether the cable safely cn be pulled through a raceway? When is the sidewall pressure the controlling parameter?
5. Why does reversing the direction of a cable pull usually change the required pulling tension?
6. The NEC defines maximum percent fill values for both conduit and cable tray. In the case of conduit, exceeding the NEC-stipulated value can make installation very difficult, if not impossible. The maximum fill value for cable tray, however, is not an installation issue. What is the basis for it?
7. What is the purpose of shielding a power cable? Why does shielding change the minimum bending radius of the cable?
8. What are some advantages and disadvantages of using three-conductor power cable as opposed to three single-conductor cables?
9. What are some of the criteria that could be used to determine whether to use conduit or cable tray as a raceway?
10. Why is it necessary to bond and ground metallic raceways?

PROBLEMS

1. The NEC sets 40% as the maximum allowable fill for conduit. What is the theoretical maximum percent fill if three cables of identical diameter are installed in a conduit? Why is this value substantially higher than 40%?
2. What is the smallest rigid steel conduit that can be used to contain three cables, each with an outside diameter of 1.65 in?
3. How many control cables, each with an outside diameter of 0.78 in, can be pulled in a 2-in rigid steel conduit?
4. A pull box must be sized to accommodate four 2-in conduits and two 1-in conduits. What size pull box must be used for?
   a. a straight-through pull?
   b. a 90° turn?

5. Make a dimensioned sketch showing the minimum dimensions and conduit locations of a pull box necessary to turn three 4-in and three 3-in conduits through 90°.

6. Three 15-kV cables with an outside diameter of 1.35 in and a weight of 2.10 lb/ft must be pulled through the conduit raceway shown below. The bends at C and D are 45°.

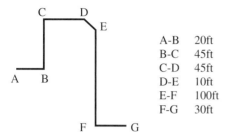

A-B	20ft
B-C	45ft
C-D	45ft
D-E	10ft
E-F	100ft
F-G	30ft

All bends lie in the horizontal plane.

a. Size the conduit (use rigid steel).
b. Calculate the percent fill.
c. Determine the minimum bend radius.
d. Calculate the pulling tension from A to F, and the sidewall pressure at B, C, D, and E.
e. Recalculate the pulling tension from F to A, and the resulting sidewall pressures at B, C, D, and E.

7. Repeat Problem 6 assuming that all bends lie in the vertical plane.

8. The raceway shown in Problem 6, with all bends lying in the horizontal plane, needs to be designed using ladder-type cable tray.
   a. Size the cable tray (width and depth).
   b. Calculate the percent fill.
   c. Determine the minimum bend radius.
   d. Calculate the pulling tension from A to E, and the sidewall pressure at B, C, and D. Use Eq. (6.12) to approximate the sidewall pressure.
   e. Recalculate the pulling tension from E to A, and the resulting sidewall pressures at B, C, and D. Use Eq. (6.12) to approximate the sidewall pressure.

9. Three 5-kV cables with an outside diameter of 0.93 in and a weight of 1.34 lb/ft must be pulled through the conduit raceway shown below. The bends at B and C are 45°.

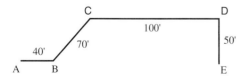

All bends lie in the horizontal plane.
a. Size the conduit (use rigid steel).
b. Calculate the percent fill.

c. Determine the minimum bend radius.
   d. Calculate the pulling tension from A to E, and the sidewall pressure at B, C, and D.
   e. Recalculate the pulling tension from E to A, and the resulting sidewall pressures at B, C, and D.
10. Repeat Problem 9 assuming that all bends lie in the vertical plane.

CHAPTER 7

SWITCHGEAR AND MOTOR CONTROL CENTERS

OBJECTIVES

- Understand the purpose, basic construction, and configuration of switchgear and motor control centers
- Be able to specify switchgear and motor control centers with proper ratings
- Understand the basics of circuit breaker and motor starter operation and application, as well as protection requirements for switchgear and motor control centers
- Understand the hazards associated with and know the requirements for identifying arc flash hazards per National Fire Protection Association (NFPA) Standard 70E and IEEE Standard 1584

7.1 OVERVIEW

In industrial environments, electricity is provided to loads from *load centers*, which contain the equipment necessary to protect and control the loads. Many different types of load centers exist, with their selection based primarily on the electrical requirements and installation environment. Several types of load centers are housed in metal enclosures to protect enclosed equipment, nearby objects, and personnel in the event of equipment malfunction. Load centers supplying large motors and smaller load centers will be referred to as *switchgear*, while smaller load centers specialized to supply small to midsize motors will be called *motor control centers* (MCCs). Auxiliary equipment such as control switches, indicator lights, and metering equipment are sometimes installed in smaller metallic enclosures whose capabilities to withstand the external environment are determined by Standard 250 of the National Electrical Manufacturers Association (NEMA).

Industrial Power Distribution, Second Edition. Ralph E. Fehr, III.
© 2016 The Institute of Electrical and Electronics Engineers, Inc. Published 2016 by John Wiley & Sons, Inc.

7.2 NEMA ENCLOSURES

NEMA standard 250-2003 defines criteria for electrical enclosures based on their ability to withstand external environments. The types of enclosures defined by the standard are summarized in Table 7.1.

7.3 SWITCHGEAR

Switchgear is a specialized type of load center primarily designed to supply relatively large loads, such as large motors and smaller load centers. Switchgear comes in two varieties: *metal enclosed* and *metal clad*. Metal-clad switchgear is the more robust design, as it requires shutters between the bus and the front of the equipment, compartmentalization of live parts, and insulation of the bus and primary components. Also, all switching and interrupting devices must be *drawout mounted*, so they can be removed from the switchgear without unwiring. Metal-enclosed switchgear does not need to meet these criteria and often has lower interrupting ratings, lower breaker duty cycle and may use fused or nonfused switches instead of circuit breakers. Metal-enclosed switchgear is governed by IEEE standard C37.20.3, while metal-clad switchgear is covered by multiple standards. In short, metal-clad switchgear meets or exceeds the requirements for metal-enclosed switchgear, but *not* vice versa.

A switchgear lineup is made up of multiple *sections* or *cubicles* joined side by side, as shown in Figure 7.1. Most manufacturers use a standard 36" section width.

The front of the switchgear is hinged, and opening the doors exposes the circuit breakers. When a breaker is removed from metal-clad switchgear, an insulated barrier separating the cubicle from the energized buswork that runs the length of the switchgear is visible. This barrier is an important safety feature, since metal-clad switchgear commonly is maintained while the main bus is energized.

Although some metal-enclosed switchgear has bare main bus bars, metal-clad switchgear bus is insulated. In addition to increased safety for maintenance personnel, insulated bus greatly reduces the likelihood of the destructive effects of a fault propagating through the switchgear. Most faults occur as line-to-ground faults. The resulting arcing ionizes the air in the vicinity of the fault. The ionized air has a much lower dielectric strength than nonionized air and provides paths for the fault current to follow. This can result in the line-to-ground fault developing into a double line-to-ground or three-phase fault before it can be cleared. Consequently, when a fault occurs in noninsulated bus switchgear, the extent of the damage is much greater than in the case of insulated bus equipment. With insulated bus switchgear, the damage caused by the fault often is contained in a single switchgear section, or in the section where the fault originated and one or two adjacent sections.

Where the consequences of a switchgear fault are exceptionally high, *arc-resistant switchgear* can be specified. Arc-resistant switchgear minimizes the danger to personnel in the vicinity of the switchgear in the event of a fault. If switchgear is installed in a dedicated room that is normally unoccupied, specifying arc-resistant switchgear may not be as critical as when the switchgear is installed in the areas that are typically occupied.

TABLE 7.1 NEMA Enclosure Types

NEMA Standard Type	Description
Type 1	Enclosures constructed for indoor use to provide a degree of protection to personnel against access to hazardous parts and to provide a degree of protection of the equipment inside the enclosure against ingress of solid foreign objects (falling dirt).
Type 2	Enclosures constructed for indoor use to provide a degree of protection to personnel against access to hazardous parts, to provide a degree of protection of the equipment inside the enclosure against ingress of solid foreign objects (falling dirt), and to provide a degree of protection with respect to harmful effects on the equipment due to the ingress of water (dripping and light splashing).
Type 3	Enclosures constructed for either indoor or outdoor use to provide a degree of protection to personnel against access to hazardous parts, to provide a degree of protection of the equipment inside the enclosure against ingress of solid foreign objects (falling dirt and windblown dust), to provide a degree of protection with respect to harmful effects on the equipment due to the ingress of water (rain, sleet, snow), and that will be undamaged by the external formation of ice on the enclosure.
Type 3R	Enclosures constructed for either indoor or outdoor use to provide a degree of protection to personnel against access to hazardous parts, to provide a degree of protection of the equipment inside the enclosure against ingress of solid foreign objects (falling dirt), to provide a degree of protection with respect to harmful effects on the equipment due to the ingress of water (rain, sleet, snow), and that will be undamaged by the external formation of ice on the enclosure.
Type 3S	Enclosures constructed for either indoor or outdoor use to provide a degree of protection to personnel against access to hazardous parts, to provide a degree of protection of the equipment inside the enclosure against ingress of solid foreign objects (falling dirt and windblown dust), to provide a degree of protection with respect to harmful effects on the equipment due to the ingress of water (rain, sleet, snow), and for which the external mechanism(s) remain operable when ice laden.
Type 3X	Enclosures constructed for either indoor or outdoor use to provide a degree of protection to personnel against access to hazardous parts, to provide a degree of protection of the equipment inside the enclosure against ingress of solid foreign objects (falling dirt and windblown dust), to provide a degree of protection with respect to harmful effects on the equipment due to the ingress of water (rain, sleet, snow), that provides an additional level of protection against corrosion, and that will be undamaged by the external formation of ice on the enclosure.
Type 3RX	Enclosures constructed for either indoor or outdoor use to provide a degree of protection to personnel against access to hazardous parts, to provide a degree of protection of the equipment inside the enclosure against ingress of solid foreign objects (falling dirt), to provide a degree of protection with respect to harmful effects on the equipment due to the ingress of water (rain, sleet, snow), that will be undamaged by the external formation of ice on the enclosure that provides an additional level of protection against corrosion, and that will be undamaged by the external formation of ice on the enclosure.

(continued)

TABLE 7.1 *(Continued)*

NEMA Standard Type	Description
Type 3SX	Enclosures constructed for either indoor or outdoor use to provide a degree of protection to personnel against access to hazardous parts, to provide a degree of protection of the equipment inside the enclosure against ingress of solid foreign objects (falling dirt and windblown dust), to provide a degree of protection with respect to harmful effects on the equipment due to the ingress of water (rain, sleet, snow), that provides an additional level of protection against corrosion, and for which the external mechanism(s) remain operable when ice laden.
Type 4	Enclosures constructed for either indoor or outdoor use to provide a degree of protection to personnel against access to hazardous parts, to provide a degree of protection of the equipment inside the enclosure against ingress of solid foreign objects (falling dirt and windblown dust), to provide a degree of protection with respect to harmful effects on the equipment due to the ingress of water (rain, sleet, snow, splashing water, and hose directed water), and that will be undamaged by the external formation of ice on the enclosure.
Type 4X	Enclosures constructed for either indoor or outdoor use to provide a degree of protection to personnel against access to hazardous parts, to provide a degree of protection of the equipment inside the enclosure against ingress of solid foreign objects (windblown dust), to provide a degree of protection with respect to harmful effects on the equipment due to the ingress of water (rain, sleet, snow, splashing water, and hose directed water), that provides an additional level of protection against corrosion, and that will be undamaged by the external formation of ice on the enclosure.
Type 5	Enclosures constructed for indoor use to provide a degree of protection to personnel against access to hazardous parts, to provide a degree of protection of the equipment inside the enclosure against ingress of solid foreign objects (falling dirt and settling airborne dust, lint, fibers, and flyings), and to provide a degree of protection with respect to harmful effects on the equipment due to the ingress of water (dripping and light splashing).
Type 6	Enclosures constructed for either indoor or outdoor use to provide a degree of protection to personnel against access to hazardous parts, to provide a degree of protection of the equipment inside the enclosure against ingress of solid foreign objects (falling dirt), to provide a degree of protection with respect to harmful effects on the equipment due to the ingress of water (hose directed water and the entry of water during occasional temporary submersion at a limited depth), and that will be undamaged by the external formation of ice on the enclosure.
Type 6P	Enclosures constructed for either indoor or outdoor use to provide a degree of protection to personnel against access to hazardous parts, to provide a degree of protection of the equipment inside the enclosure against ingress of solid foreign objects (falling dirt), to provide a degree of protection with respect to harmful effects on the equipment due to the ingress of water (hose directed water and the entry of water during prolonged submersion at a limited depth), that provides an additional level of protection against corrosion, and that will be undamaged by the external formation of ice on the enclosure.

(continued)

TABLE 7.1 *(Continued)*

NEMA Standard Type	Description
Type 12	Enclosures constructed (without knockouts) for indoor use to provide a degree of protection to personnel against access to hazardous parts, to provide a degree of protection of the equipment inside the enclosure against ingress of solid foreign objects (falling dirt and circulating dust, lint, fibers, and flyings), and to provide a degree of protection with respect to harmful effects on the equipment due to the ingress of water (dripping and light splashing).
Type 12K	Enclosures constructed (with knockouts) for indoor use to provide a degree of protection to personnel against access to hazardous parts, to provide a degree of protection of the equipment inside the enclosure against ingress of solid foreign objects (falling dirt and circulating dust, lint, fibers, and flyings), and to provide a degree of protection with respect to harmful effects on the equipment due to the ingress of water (dripping and light splashing).
Type 13	Enclosures constructed for indoor use to provide a degree of protection to personnel against access to hazardous parts, to provide a degree of protection of the equipment inside the enclosure against ingress of solid foreign objects (falling dirt and circulating dust, lint, fibers, and flyings), to provide a degree of protection with respect to harmful effects on the equipment due to the ingress of water (dripping and light splashing), and to provide a degree of protection against the spraying, splashing, and seepage of oil and noncorrosive coolants.

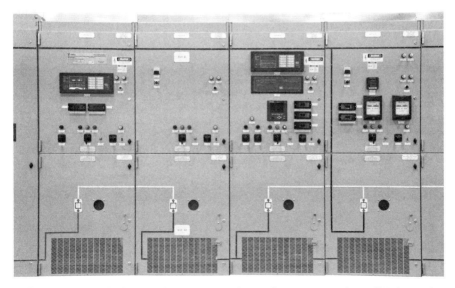

Figure 7.1 Metal-Clad Switchgear—Front View. (*Photo courtesy of Powell Industries.*)

Arc-resistant equipment must withstand two different phenomena: *internal pressure* and *burnthrough*. During a fault, there is a sudden increase in pressure inside the switchgear. The intense heat generated by the arc can vaporize metal very quickly. When subjected to an intense electric arc, one cubic inch of copper can expand to one cubic meter of vapor in less than 4 ms. This generates a pressure wave that can devastate the switchgear. Doors and panels can be blown off the switchgear, resulting in severe damage to anything in the vicinity.

Burnthrough is the loss of integrity of the metal-clad enclosure due to the intense heat inside the switchgear during a fault. Burnthrough requires seconds as opposed to milliseconds. When enclosure integrity is lost, the hot gases produced by the fault are vented through the breaches in the switchgear enclosure, posing a serious threat to nearby personnel and equipment.

While IEEE standard C37.20.7 defines testing requirements for arc-resistant switchgear, no industry standards exist for the construction of arc-resistant switchgear. Each manufacturer offers proprietary versions of the equipment. Most designs use additional reinforcement to strengthen the lower 2 m of the enclosure and provide a more-easily rupturable panel in or near the roof of the switchgear, which would fail during a fault and direct the hot gases upward and away from and personnel or equipment in front of or behind the switchgear.

Devices such as switches, indicator lamps, meters, and protective relays often are mounted on the doors of the switchgear. Opening the door provides easy access to the backs of these devices. When electromechanical relays are mounted on the door, care must be taken when opening and closing the door because the relays are very sensitive to shock and vibration. Slamming a door could easily actuate the relay contacts. Figure 7.2 shows a switchgear cubicle with the door opened.

Figure 7.2 Metal-Clad Switchgear—Front Opened. (*Photo courtesy of Powell Industries.*)

The life of metal-clad switchgear is a function of its maintenance and the environment in which it is installed. A thorough preventive maintenance program, which includes cleaning and lubrication of the switchgear inside and out, is vital to the reliability and longevity of the equipment. Maintenance becomes even more critical when the switchgear is installed in harsh environments. Heat, dust, moisture, and corrosive chemicals are harmful to switchgear. A design modification as simple as air conditioning the switchgear room can add years of life to the switchgear, as well as improve the equipment's reliability and extend the time between maintenance cycles.

Atmospheres that contain sulfur compounds, particularly hydrogen sulfide (H_2S), sulfur dioxide (SO_2), and sulfur trioxide (SO_3), are especially harmful to switchgear. The sulfur encourages the formation of silver sulfide on the silver-plated portions of the switchgear bus and breaker components. The silver sulfide is black and has a whisker-like appearance. It has a substantially higher resistance than pure silver, which causes a significant temperature rise and ultimately failure of the plated components. The whiskers can actually grow to a length where they will cause faults to ground or other energized surfaces. In facilities where sulfur compounds are present, chemical filtration using activated carbon filters can be used in the switchgear room. Another option is to use gold- or tin-plated surfaces instead of silver. The gold-plating option is quite expensive, and both gold and tin have lower conductivity than silver.

An incoming circuit or circuits supply power to the bus, typically through a main circuit breaker. If two supply circuits are used, the switchgear is said to be *double-headed*, as shown in Figure 7.3.

7.3.1 Source Transfer

In Figure 7.3, Sections A and F accommodate the feeds to the main bus. Typically with double-headed switchgear, one supply breaker is normally closed (in Figure 7.3, this is the breaker in Section A), while the second supply breaker (Section F) remains normally open. In the event that Source No. 1 fails, the breaker in Section A trips. The breaker in Section F must close to reenergize the switchgear bus from Source No. 2. Several means of closing the breaker in Section F exist.

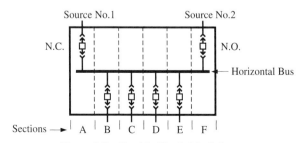

Figure 7.3 Double-Headed Switchgear

The simplest means of switching to Source No. 2 is to manually close the breaker in Section F after the breaker in Section A trips. This method of source transfer is very slow and rather unreliable since it depends on manual intervention. Consequently, manual source transfers are seldom implemented when speed and reliability are critical.

A *fast transfer* can be implemented where the Type B auxiliary contacts on the breaker in Section A actuates the close coil of the breaker in Section F (see Section 8.2). Care must be taken when designing a fast transfer scheme because if the transfer is not completed within five to six cycles, running motors driving high-inertia loads such as fans will slow enough to lose synchronism and generate very high torques when re-energized. Motors driving low-inertia loads such as pumps will lose synchronism in even less time. Since re-energizing the motors out of synchronism with the system can be damaging to the equipment, completing the transfer quickly is critical. Typically, a synchronism check relay is incorporated in fast transfer schemes.

If a fast transfer cannot be implemented, a *slow transfer* can be performed. With a slow transfer, the Source No. 1 breaker trips and a timer starts. After a predetermined amount of time elapses (enough time for the internal voltage on the running motors to decay to less than 25% of normal voltage), the Source No. 2 breaker closes. By using a slow transfer, dangerous torques are avoided, but a perceptible interruption in service is experienced.

Another means of switching sources is a *parallel transfer* where the Source No. 2 breaker closes before the Source No. 1 breaker opens. This results in no interruption of service. However, the protection system must be designed to prevent closing into a fault that may be causing the Source No. 1 breaker to trip. Although a breaker with an adequate close-and-latch rating can close safely into a fault, paralleling the two sources into the fault increases the short-circuit current that the Source No. 1 breaker must interrupt and increases resultant equipment damage.

In Figure 7.3, Sections B through E contain circuit breakers to supply loads. Each section contains provisions for terminating cables. The circuit breaker is of the *drawout* type, meaning it can be physically removed or *racked out* of the switchgear. This is an important feature for easy maintenance and quick replacement.

7.3.2 Configuration

Sections can be *full height*, meaning one circuit breaker per horizontal section, or *half height*, where two circuit breakers are stacked vertically in a single horizontal section. Figures 7.1 and 7.2 depict half-height switchgear. Half-height switchgear reduces cost and space requirements but is more difficult to install and maintain than full-height switchgear. Also, with half-height switchgear, a jacking device must be provided to remove the upper circuit breakers from their cubicles. Since a circuit breaker weighs hundreds of pounds, using jacking devices can be physically strenuous and cumbersome.

7.3.3 Ratings

The standards controlling the design and testing of metal-clad switchgear were developed by the American National Standards Institute (ANSI) in conjunction with the Institute of Electrical and Electronics Engineers (IEEE). IEEE standard C37.20.2 stipulates metal-clad switchgear-rating criteria. Low-voltage switchgear (600-V class and below) is governed by IEEE standard C37.20.1, while IEEE standards C37.04, C37.06, and C37.09 specify medium-voltage circuit breaker rating and testing criteria.

Several important ratings are given to the main buswork in metal-clad switchgear. Both copper and aluminum bus carry these ratings. A *continuous current rating* is assigned to limit the temperature rise of the bus bars to a value that will not compromise the bus insulation.

The bus bars typically are insulated with an epoxy-type material that can withstand fairly high-operating temperatures. Insulators, either porcelain or polymeric composite, support the bus bars. The insulators also have a maximum operating temperature.

It is important to determine the continuous current rating of switchgear bus by temperature rise. While some equipment specifications cite a maximum allowable current density to determine the bus rating, this practice is not allowable. A commonly used current density is 1000 A/inch^2 for copper bus. Standard 1200 A copper bus measuring $1/4'' \times 4''$ experiences a 60°C temperature rise at a current density of 1200 A/inch^2. Standard 2000 A copper bus measuring $1/2'' \times 6''$ rises 59.7°C at a current density of 667 A/inch^2, while 3000 A bus consisting of two $1/2'' \times 6''$ copper bus bars per phase rises 59.5°C at a current density of 500 A/inch^2. The temperature rises for these three scenarios are virtually identical, but the current densities vary tremendously. Specifying a current density of 1000 A/inch^2 would result in underutilization of the 1200 A bus and would overheat substantially the higher capacity buses. While current density may be significant for single conductors in free air, it should not be used for determining switchgear bus ratings.

Short-circuit currents exert tremendous mechanical forces on the buswork. During a short circuit, very large currents flow through the buswork. These currents produce intense magnetic fields that tend to deflect the buswork, forcing the parallel bus sections either toward or away from each other depending on the direction of the current flow. A fault current of 50 kA will exert a force of nearly 17,000 lb/ft of length on $1/4'' \times 4''$ bus bar spaced on 6'' phase centers. The bus insulators and bracing system must withstand these forces to prevent failure of the switchgear during the occurrence of a short circuit. These mechanical forces give rise to a *short-circuit current rating*.

For many years, the fault withstand capability of switchgear and circuit breakers was rated in MVA. This meant that if the equipment was applied at a voltage lower than its rated voltage, an increase in fault current capability was implied. For example, 15-kV switchgear rated at 350 MVA has a fault current capability of

$$I_{SC} = \frac{350{,}000\,\text{kVA}}{\sqrt{3}(15\,\text{kV})} = 13.472\,\text{kA} \tag{7.1}$$

If this switchgear is installed on a 13.2-kV system, its implied fault current capability would be

$$I_{SC} = \frac{350,000\,\text{kVA}}{\sqrt{3}(13.2\,\text{kV})} = 15.309\,\text{kA} \tag{7.2}$$

Installing the switchgear at a voltage slightly below its rating requires an additional 13.6% in fault current capability. This increase in fault current withstand represents a substantial increase in bus bracing and breaker interrupter capability. Stronger bus bracing can be accomplished easily at an increased financial cost. Interrupter capability, however, poses technical problems. With the older bulk oil and air-blast circuit breaker technologies, additional fault current capability with decreased voltage was inherent to the technology. The newer sulfur hexafluoride (SF_6) and vacuum circuit breaker technologies do not provide additional fault current capability with decreased voltage.

Because of the changes in circuit breaker technology, switchgear no longer is manufactured with increased fault current capability voltages lower than rated voltage. Switchgear now is rated in kiloamperes instead of MVA. A point of confusion has been created as a result of the change in rating methodology. The MVA-class nomenclature has remained in the switchgear industry. Today, one can still purchase "350-MVA" switchgear, but the increased fault current capability at voltages less than rated voltage no longer exists. Today's 350-MVA, 15-kV switchgear can withstand no more than 13.472 kA regardless of the voltage at which it is operated.

The electrical properties of the insulation system define a *basic impulse level*, or *BIL rating*, which indicates the maximum voltage that can be tolerated without insulation failure. The BIL rating is tested by performing either a *full-wave* or *chopped-wave* test. The waveform used for the full-wave test has a 1.2 µs rise time to crest voltage and a 50 µs decay to 50% of crest voltage. The chopped-wave test uses the same basic waveform, but with the waveform abruptly interrupted (chopped) 1–3 µs after the crest, depending on the insulation class.

Table 7.2 shows common switchgear nominal voltages and their associated BIL ratings.

The environment in which the switchgear is installed can dictate changes to the required ratings. The effects of altitude on insulation strength can be seen in Table 7.3.

TABLE 7.2   Switchgear Operating Voltages and BIL Ratings

Nominal kV Class	BIL Rating (kV)
5	60
15	95
27	125
38	150

TABLE 7.3 Voltage and Current Derating Factors for Switchgear as a Function of Altitude

Altitude Above Sea Level	Voltage Derating Factor	Current Derating Factor
3300 ft (1000 m) and below (medium-voltage switchgear) 6600 ft (2000 m) and below (low-voltage switchgear)	1.00	1.00
5000 ft (1500 m) (medium-voltage switchgear) 8500 ft (2600 m) (low-voltage switchgear)	0.95	0.99
10,000 ft (3000 m) (medium-voltage switchgear) 13,000 ft (3900 m) (low-voltage switchgear)	0.80	0.96

Note: Intermediate values may be determined by linear interpolation.

Installing switchgear at high altitudes (3300 ft/1000 m above sea level or higher for medium-voltage switchgear and 6600 ft/2000 m above sea level or higher for low-voltage switchgear) requires that both voltage and current ratings be adjusted, since the thinner air at high altitudes has a lower dielectric strength and poorer cooling properties than the denser air at sea level.

Installing switchgear in the areas with high ambient temperatures can require an increase in continuous current rating. Continuous current ratings for switchgear with class 105 insulation are based on a 65°C temperature rise over a 40°C ambient temperature. If the temperature of the environment exceeds 40°C, natural cooling may not be adequate to keep the insulation temperature within acceptable limits. Two methods of dealing with high ambient temperatures are to derate the equipment's continuous current rating until an acceptable temperature rise is achieved or to install fans to force cooling air through the switchgear.

7.3.4 Circuit Breakers

Circuit breakers are devices used to interrupt current flow, either during normal conditions (switching load) or abnormal conditions (clearing a fault). All circuit breakers employ a device called an *interrupter* to stop the flow of current. At a minimum, interrupters consist of a stationary contact (usually the upper contact), a movable contact (usually the lower contact), and some means of extinguishing the arc that results when the contacts part while current is flowing.

The two basic functions of a circuit breaker, interrupting faults and switching load, are very different in nature, and each can pose problems to the circuit breaker designer. The obvious issue with interrupting faults is the tremendous amount of energy that must be withstood by the breaker during the interruption process. As the contacts in the interrupter part, an arc is drawn. When the current waveform crosses the zero axis, the arc stops. As the current magnitude moves away from zero in either the positive or the negative direction, the arc could restrike. This occurs when the voltage across the breaker contacts exceeds the voltage withstand of the gap between the contacts. It is not uncommon for a restrike to occur, particularly if the breaker is switching an inductive or capacitive load. If a restrike does occur, it is imperative that the next current zero crossing results is successful arc extinction.

If the arc remains when the breaker contacts are fully opened, the interrupter has no way of extinguishing the arc and will be destroyed by the thermal energy of the fault.

In addition to the thermal energy produced by the fault, the large currents also produce intense magnetic fields, which give rise to tremendous mechanical forces. These forces cause the current pathways in the breaker to try to move toward or away from each other. The stabs that connect the breaker to the buswork are especially vulnerable to these forces. Since these forces can be calculated by the breaker manufacturer, the design of the breaker can be made to accommodate the magnetic forces experienced during a fault.

Switching low currents (on the order of a few percent of the breaker's continuous current rating) is generally not a problem for vacuum circuit breakers, but can be problematic for air-magnetic breakers. Although the current being switched is only a small fraction of the current produced by a fault, an arc is still drawn and must be extinguished. Air-magnetic breakers rely on a magnetic field generated by the current in the power circuit flowing through auxiliary coils in the interrupter to deflect the arc into arc chutes for extinction. For small load currents, the magnetic field may not be strong enough to deflect the arc, so arc extinction may not occur. This problem can be overcome by adding an air piston to the interrupter unit to blow the arc into the arc chutes. These *puffer* devices are common on newer air-magnetic breakers but may be absent on older models. Additionally, air-magnetic breakers utilize arcing contacts to draw the arc away from the primary contacts and into the arc chute.

Closing the contacts of a circuit breaker is a very involved process. The speed at which the contacts come together is critical to reduce arcing prior to contact. Contact closing speeds up to 75 ft/s are typical in medium voltage breakers. These high speeds are usually accomplished by using very stiff springs, both for opening and closing the contacts. The springs that close the contacts also charge the tripping springs that open the contacts. This assures that a closed circuit breaker always has its tripping springs charged. If this were not the case, it would be possible to close the breaker contacts without having stored energy available to open the contacts.

When the circuit breaker contacts are commanded to open, speed is again very important. Many modern breakers are rated at two cycles meaning that approximately two 60-Hz cycles or 33.33 ms are required for the breaker contacts to part. If the breaker opens to clear a fault, approximately one-half cycle after the inception of the fault is necessary for the protective relays to detect the fault and operate. The breaker operating time is added to that value giving about 2.5 cycles, or 41.67 ms, which is the amount of time the fault exists on the system. Older circuit breakers had operating times of up to eight cycles, which could allow the fault to persist for over 150 ms. The longer a fault persists, the more severe the damage caused by arcing and heating, and the greater the arc flash hazard, as discussed in Section 7.5.

In addition to high contact speed, the impact of the contacts' meeting must be sufficiently damped to avoid bouncing. If the contacts bounce after meeting, the resulting arcing could weld the contacts together. Each manufacturer has methods of preventing contact bounce.

Various types of circuit breakers can be used in metal-clad switchgear. *Air-magnetic circuit breakers* have been used for many years. As the breaker contacts

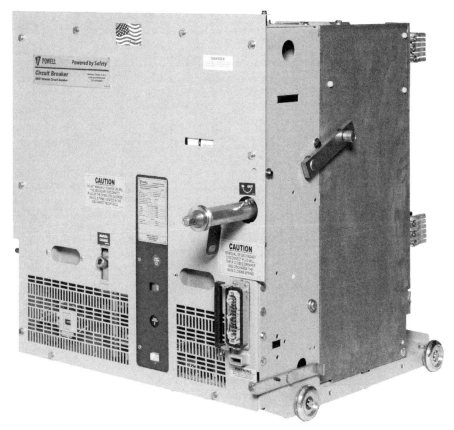

Figure 7.4   Vacuum Circuit Breaker. (*Photo courtesy of Powell Industries.*)

part, a blast of compressed air directs the arc into arc chutes where the arc is extended and subsequently cooled. When insufficient temperature exists for ionization of the surrounding air, the arc is extinguished.

Gas (SF_6) circuit breakers have widespread application at higher voltages and are commonplace in transmission substations. Their use at distribution levels is limited.

A common technology for modern circuit breakers at distribution voltages is the *vacuum circuit breaker*, shown in Figure 7.4, where the contacts separate in a vacuum bottle.

SF_6 interrupters rated at 15 kV have a stroke on the order of 5–6 inch, meaning that the movable contact travels 5–6 inch (12.7–15.2 cm) when they separate from the stationary contact. In contrast, a vacuum interrupter with similar ratings has a stroke on the order of one-half inch (1.3 cm). This large variation in contact separation is due to the large difference in dielectric strength of SF_6 gas (89 kV/cm at 1 atm) versus a vacuum (over 200 kV/cm).

All of the circuit breaker technologies used in metal-clad switchgear are safe and reliable, provided they are maintained and operated as recommended by the manufacturer. Since circuit breakers contain many moving parts, periodic operation of the tripping and closing mechanisms is critical to assure proper performance. Periodic operation of circuit breakers is often not performed, because in many cases, a breaker cannot be operated without de-energizing load. De-energizing certain loads can be objectionable, even during plant outages. Consequently, many circuit breakers operate in the closed position for many years at a time.

When a circuit breaker remains in one state for an extended period of time, several problems can occur. The most common lubricant used on the moving parts of a circuit breaker is soap-based grease. As this grease ages in a warm environment, such as inside a circuit breaker, it hardens and loses its lubricating characteristics. Synthetic greases and dry lubricants can be used to help alleviate this problem. Some manufacturers also use self-lubricating components coated with fluoropolymer resins such as Teflon®.

Bushings are sometimes subjected to asymmetrical loading while the breaker contacts are closed. After years of asymmetrical loading, permanent deformation, or "flats," may occur in the bushings. Also, when a circuit breaker is in the closed position, the springs for tripping the breaker are charged. Keeping these springs compressed for a very long period may cause them to "take a set," which changes their spring constant and thereby the operating speed of the breaker. Care should be taken when designing electrical facilities to provide means for maintaining circuit breakers. Maintenance capability is often omitted from the system design to minimize cost but only *construction* cost is being reduced—*maintenance* cost, which occurs over the entire life of the facility, is being increased, sometimes substantially, in the process.

Circuit breakers should be tested periodically using a power frequency withstand (hi-pot) test. This test will diagnose faulty interrupters prior to failure. Many hi-pot testers utilize direct current (DC) as opposed to 60 Hz current. DC hi-pot tests should not be done on interrupters because the DC can damage the interrupter by severely stressing the insulation within the device. A DC hi-pot test will diagnose all bad interrupters, but may improperly indicate that a good interrupter is bad. If a 60-Hz hi-pot tester is not available, a low-frequency tester can be used. Also, care must be taken when testing vacuum interrupters with voltages in excess of 25 kV, as this can produce X-rays during the test.

The interrupters in vacuum circuit breakers have such a high dielectric strength that they actually can interrupt the current flow before the current waveform crosses zero. When this happens, the chopped current waveform results in high transient voltages. These transient voltages are most severe when the current waveform is chopped 4–7 ms prior to a zero crossing.

The magnitude of the transient voltage resulting from chopping the current waveform is a function of the material of which the breaker contacts are made. Older breaker designs used copper–silver alloys, which chopped the current waveform when it dropped to 18–20 A. This resulted in transient voltages magnitudes up to 20 times the system operating voltage. Advances in metallurgy developed copper–bismuth alloys that chop the current waveform at about 5 A causing a transient voltage on the order of five times the system operating voltage. Copper–chromium alloys chop the current waveform at about 3.5 A and produce transient voltage magnitudes on the order of 2.5–3 times the system operating voltage.

Voltage transients can be very problematic for motors and generators, since rotating machines are not designed with BIL ratings to withstand such transients. Due to reflection of the travelling wave, rotating machines can experience transient voltage magnitudes of 200% of the peak-to-peak transient voltage magnitude. Most engineers specify surge-suppressing capacitors and/or surge arresters on higher voltage motors, particularly 13.2-kV motors, when they are switched with vacuum devices. Lower voltage motors can also be surge protected—in practice, about half the 4-kV motors rated 4000 hp and above installed in industrial facilities have surge protection installed, but relatively few motors below these ratings do. Surge protection should always be provided on dry-type transformers and capacitor banks that are switched with vacuum devices.

Metal-clad switchgear circuit breakers are *drawout devices*, meaning they can be easily removed or *racked out* of the switchgear for maintenance. Six stabs protrude from the back of the circuit breaker to penetrate shutter-covered holes in an insulated safety barrier and connect with the horizontal bus. Three stabs are the line-side connections for the breaker and plug into the switchgear bus. The other three stabs are the load-side connections for the breaker and plug into short bus sections to which cables to supply the load are terminated. The stabs can be seen in Figure 7.5.

The breaker is built of a steel frame equipped with heavy-duty casters that allow the breaker to be *racked out* of the switchgear for maintenance. To prevent current

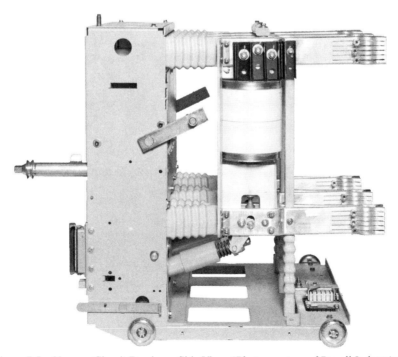

Figure 7.5   Vacuum Circuit Breaker—Side View. (*Photo courtesy of Powell Industries.*)

from flowing through the breaker while it is being racked out, a mechanical interlock prevents the breaker from being racked out unless the breaker contacts are open.

Switchgear breakers are readily available in frame sizes up to 3000 A. If higher continuous current ratings are required, two options exist. Cooling fans can be installed on the breakers to increase their continuous current rating upward of 4000 A. Paralleling circuit breakers can also be done. Two 2000-A breakers in parallel will have a continuous current rating of about 3500 A, while paralleling two 3000-A breakers will result in a continuous current rating of about 5000 A. These two upgrading options increase only the continuous current rating of the circuit breaker—they do *not* increase the interrupting rating. Before applying either of these upgrade options, the circuit breaker and switchgear manufacturers should be consulted.

7.4 MOTOR CONTROL CENTERS

An MCC is a load center customized to serve small to midsize motors. Circuit breakers are replaced with *combination motor starters*. Since the starters are much smaller than switchgear breakers, stacking more than two units per horizontal section and reducing the width of the horizontal sections from 36″ to 20″ can save considerable space. A typical MCC is pictured in Figure 7.6.

Figure 7.6   Motor Control Center. (*Photo courtesy of Siemens Industry, Inc.*)

7.4.1 Configuration

The MCC is made up of *units* which are stacked to form *sections*. One or more sections make up the MCC. The MCC pictured in Figure 7.6 contains eight sections fully visible in the photo. The height of a unit depends on its contents. If it contains a small motor starter, the height can be as small as 6″. As the size of the unit contents increases, the unit height is increased in multiples of 6″.

Most manufacturers allow stacking of units up to 90 inch. Blank units can be specified to fill a horizontal section. Standard MCC depth is 21″, although some manufacturers provide a 16″ depth option if the units are not installed back to back.

7.4.2 Ratings

A main (horizontal) bus similar to that used in switchgear is implemented in MCCs. Since the loads supplied by an MCC are smaller than the loads fed from switchgear, the available bus ratings tend to be lower in an MCC. A system of vertical buswork must be incorporated into the design to provide power to each unit in each horizontal section. Commonly available bus ratings range from 600 to 2500 A for the horizontal bus, and 300–1200 A for the vertical bus although some manufacturers may offer other ratings. Interrupting ratings of 65 kA and 100 kA are common.

The incoming section of an MCC can be equipped with a fusible switch or circuit breaker. Fusible switches and circuit breakers can also be installed in the other units to serve nonmotor load, such as lighting panels. Continuous ratings from 15 A to over 200 A can be specified. Units can also be equipped with starter modules, protected by either fusible disconnect switches or molded case circuit breakers.

Although MCCs are usually implemented at low voltages (240 or 480 V), medium-voltage MCCs are available. Starters for medium-voltage MCCs are specified by type, for example, a reversing autotransformer starter for a 1000-hp squirrel cage motor. Starters up to 8000 hp with 570,000 kVA of interrupting capability are available for 7200-V MCCs.

7.4.3 Starters

Motor starters provide the ability to start and stop a motor, as well as *overload protection* to protect the motor windings against thermal damage due to excessive mechanical, and in turn, electrical load on the motor. A starter alone does not provide protection against short circuits.

Starters are sized using NEMA standard sizes, based on the horsepower of the motor. The International Electrotechnical Commission (IEC) also developed a set of standard starter sizes, where the full-load current for each starter size is adjustable over a specific range. While IEC starters are used throughout much of the world, the United States uses NEMA starters, and this section will focus on the NEMA standard.

Starters up to NEMA size 7 (300 hp at 230 V or 600 hp at 460 V) are commonly available from most manufacturers. Some vendors also provide fractional NEMA size

TABLE 7.4 NEMA Starter Sizes and Continuous Current and Horsepower Ratings at 60 Hz

NEMA Size	Continuous Current Amperes	Horsepower at 200–208 V	Horsepower at 230–240 V	Horsepower at 460–480 V	Horsepower at 575–600 V
00	9	1.5	1.5	2	2
0	18	3	3	5	5
1	27	7.5	7.5	10	10
2	45	10	15	25	25
3	90	25	30	50	50
4	135	40	50	100	100
5	270	75	100	200	200
6	540	150	200	400	400
7	810	—	300	600	600
8	1215	—	450	900	900
9	2250	—	800	1600	1600

starters (such as a NEMA size $2^1/_2$) to increase cost-effectiveness and increase design options. Table 7.4 shows some standard NEMA starter sizes and their continuous current and horsepower capabilities based on full-voltage starting.

The *contactor* is the heart of the starter. It contains the power contacts that open and close the electrical supply to the motor. Depending on the type of starter, there can be one, two, or three contactors per phase. A *full-voltage, nonreversing* (FVNR) starter is the most common and basic type of starter. It contains one contactor per phase and works essentially like a switch to energize and de-energize the motor.

A *full-voltage reversing* (FVR) starter provides two contactors per phase which are both electrically and mechanically interlocked to prevent both sets of contactors operating simultaneously. One set of three contactors applies full-line voltage to the motor in positive (a–b–c) phase sequence. The second set of contactors is cross wired to apply full voltage to the motor in negative (a–c–b) phase sequence. The interlocks are necessary to prevent a line-to-line fault between phases b and c which would result if both contactors close simultaneously. Figure 8.16 illustrates the necessity to interlock the contactors in a reversing starter.

Multispeed starters are available to control squirrel-cage induction motors built to operate at two different speeds. It is common, for example, for a fan motor to have a low and a high speed. Multiple contactors per phase are required for multispeed starters. The need for interlocking is determined by the type of transition desired from one speed to another.

Reduced voltage starters are available for the motors that cannot satisfactorily be started at full voltage, or *across the line*. The starting current of a motor is typically five to six times the rated full-load current of the motor. This is because no counter-EMF is present to decrease the voltage across the motor windings when the rotor is not turning. As the rotor comes up to speed, a counter EMF is generated in the motor, and this voltage subtracts from the terminal voltage to reduce the voltage across the motor windings, and consequently the current drawn by the motor. Figure 9.1 illustrates this concept.

Reduced voltage starters come in different varieties for the different reduced voltage starting methods. Chapter 9 presents several reduced voltage starting methods.

Motor starters usually come in three different duty ratings: *standard*, *jogging*, and *plugging*. *Jogging* is a process of multiple starts and stops over a short period of time. *Plugging* is the sudden reversal of a motor's direction of rotation to facilitate stopping the shaft. Jogging and plugging and methods of implementing them are discussed more thoroughly in Chapter 8.

A *standard duty starter* should not be required to stop and start its motor more than five times per minute or ten times in 10 minutes. The thermal design of the starter is based on these starting criteria. A *jogging duty starter* is designed for frequent stops and starts, such as required by tool positioning, and can provide frequent stops and starts without overheating. As a safeguard, most standard duty starters cannot be wired for jogging applications. A *plugging duty starter* is designed to quickly open the forward contactors and close the reverse contactors long enough to suddenly stop the motor.

7.4.4 Protection

Two protection issues regarding MCCs must be addressed. First, protection must be provided against short circuits, either in the cable feeding the motor, or in the motor itself. This is done using thermal-magnetic circuit breakers, motor circuit protectors, or fusible disconnect switches. Also, protection must be provided for the insulation on the motor windings in the event of an overload condition. As the mechanical load on the motor increases, the current drawn by the motor and subsequently the heating in the motor windings increase. At some amount of overload, the thermal rating of the motor insulation will be exceeded. Before this happens, the motor should be de-energized. This is done using *overloads* or *heaters*.

Thermal magnetic circuit breakers and *motor circuit protectors* are similar interrupting devices capable of opening safely while fault current is flowing. Each device has both a continuous current rating and an interrupting current rating. The continuous current rating should be no less than 115% of the motor's full-load current rating. The interrupting current rating must be adequate to withstand the highest fault magnitude available from the system. *Fusible disconnect switches* open during fault conditions based on the fuse element's time–current characteristic and also provide a visual open point in the power circuit. Recognizing an obvious open point in the power circuit is necessary to safely maintain the motor. The NEC requires a disconnection device be provided within sight of the motor if the motor is more than 50 ft from the power source.

Overloads or *heaters* are thermally sensitive devices, often bimetallic strips, that cause contacts to change state when the heater reaches a specific temperature. Normally, the heater is configured as a normally closed contact that opens when the critical temperature is reached. Application of overload contacts in control circuits can be found in Chapters 8 and 9. Heaters are available in a wide variety of sizes based on the full-load current of the motor. The NEC governs the methods of heater sizing. The criteria used to design both circuit and winding protection can be found in Chapter 9.

7.5 ARC FLASH HAZARD

For many years, the relative hazard of working on an energized industrial power system was determined by the fault current magnitude alone. But over time, it was determined that most injuries incurred while working on energized electrical systems arose from burns, which result from the release of thermal energy during a short-circuit fault. The thermal energy released during a fault depends not only on fault current magnitude but also on the duration of the fault, since energy is the fault current magnitude squared times the time for which that current flows ($I^2 t$).

The NEC has historically addressed the issue of electric shock and leads the engineer toward a design that minimizes the chance of electric shock occurring. Arc flash hazard, however, was not specifically addressed by the NEC until the 2002 edition, when warning labels describing incident energy levels were first required. Two additional standards have been developed to provide guidance for establishing workplace safety practices to reduce the hazards of arc flash incidents to workers and to assist the engineer with properly calculating arc flash parameters. These standards are NFPA standard 70E-2015, *2015 Standard for Electrical Safety in the Workplace* by the NFPA (also the publisher of the NEC), and IEEE standard 1584-2002, *IEEE Guide for Performing Arc Flash Calculations*.

NFPA 70E defines four categories of arc flash hazard based on the calculated incident energy at the defined working distance. These categories are summarized in Table 7.5.

NFPA 70E also defines three protection boundaries to quantify the degree of hazard to which a worker would be exposed. The boundaries are *flash protection boundary*, *limited approach boundary*, and *restricted approach boundary*. Figure 7.7 depicts these boundaries graphically.

NFPA 70E mandates the use of personal protective equipment (PPE) for any worker who comes within the flash protection boundary. PPE includes arc-rated clothing made of natural fibers such as cotton. Synthetic materials such as polyester are prohibited, as these materials melt when exposed to high temperatures, causing severe burns. Suitable headgear, such as hard hat, hearing protection, safety glasses, face shield, balaclava, and hood, must also be worn to protect the face, ears, and eyes. Only about 5.0 J/cm^2 or 1.2 cal/cm^2 can be tolerated by bare skin without incurring second-degree burns.

NFPA 70E requires clear labeling of electrical equipment which can subject workers to arc flash hazard. The incident energy at a specific working distance, the minimum arc rating for clothing, the required PPE level, and the flash hazard

TABLE 7.5 Hazard Classification per NFPA 70E-2015

Category	Incident Energy (cal/cm^2)
1	4
2	8
3	25
4	40

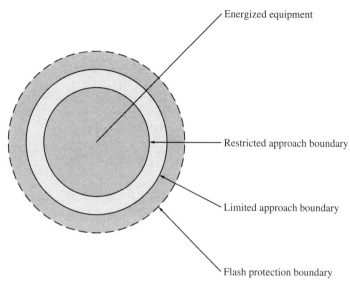

Figure 7.7 Protection Boundaries per NFPA 70E-2015

category must be clearly stated on the label as well as the arc flash boundary and nominal system voltage. A typical warning label is shown in Figure 7.8.

To perform an arc flash analysis, a calculation method must be selected. The three acceptable choices are defined by NFPA 70E Table 137(c)(15)(A and B), IEEE 1584, or specifically developed arc flash software. The NFPA method tends to be conservative while the IEEE method is based on statistical probabilities. Software-based analysis is the most accurate method, but justifying specialty software may not be possible if only a few number of buses need to be analyzed. Ideally, the three methods could be compared to determine the best means of protection.

NFPA 70E Table 137(c)(15)(A and B) are task-based methods. The required PPE requirements can be determined easily if the maximum fault current and clearing time are known. The table footnotes are critical for the proper application of the table.

IEEE 1584 uses a detailed multistep procedure based on empirical data. First the arcing current is estimated, then the normalized incident energy corresponding to an arc duration of 0.2 seconds and a distance of 610 mm can be approximated. Next the incident energy at a given working distance can be calculated. Finally, the arc flash hazard category can be determined.

For voltages in excess of 1000 V, the empirical formula for arcing current is

$$I_a = 10^{(0.00402 + 0.983 \log I_{bf})} \tag{7.3}$$

where

I_a = arcing current in kA

I_{bf} = three-phase bolted fault current in kA.

For lower voltages, the formula is much more complicated:

$$I_a = 10^{[K + 0.0966 V + 0.000526 G + \log I_{bf}(0.662 + 0.5588 V - 0.00304 G)]} \tag{7.4}$$

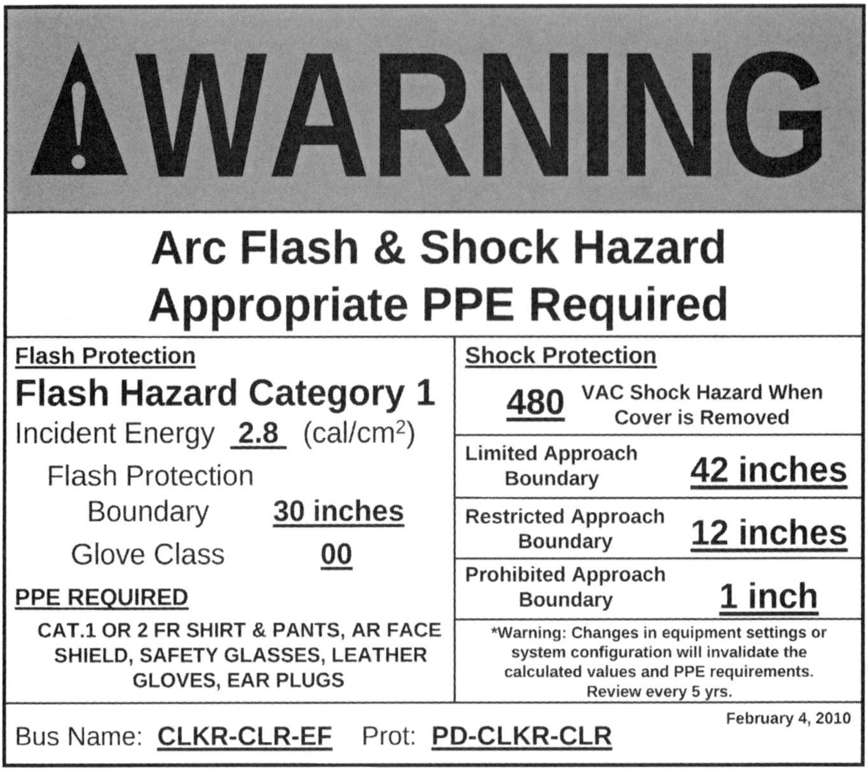

⚠ WARNING

Arc Flash & Shock Hazard
Appropriate PPE Required

Flash Protection	Shock Protection	
Flash Hazard Category 1 Incident Energy **2.8** (cal/cm²) Flash Protection Boundary **30 inches** Glove Class **00**	**480** VAC Shock Hazard When Cover is Removed	
	Limited Approach Boundary	**42 inches**
	Restricted Approach Boundary	**12 inches**
PPE REQUIRED CAT.1 OR 2 FR SHIRT & PANTS, AR FACE SHIELD, SAFETY GLASSES, LEATHER GLOVES, EAR PLUGS	Prohibited Approach Boundary	**1 inch**
	*Warning: Changes in equipment settings or system configuration will invalidate the calculated values and PPE requirements. Review every 5 yrs.	
Bus Name: **CLKR-CLR-EF** Prot: **PD-CLKR-CLR**	February 4, 2010	

Figure 7.8 Typical Arc Flash Warning Label

where

I_a = arcing current in kA

I_{bf} = the three-phase bolted fault current in kA

K = −0.153 for open construction, or

−0.097 for enclosed (box) construction

V = the system voltage in kV

G = gap between conductors in mm.

For voltages below 1000 V, a second arc current equal to 85% of I_a should be calculated and a corresponding arc duration determined. The worst case incident energy of the two cases should be used.

For example, if a lineup of 4160-V switchgear has a maximum calculated fault current of 47.6 kA, the arcing current would be

$$I_a = 10^{[0.00402+0.983 \log(47.6)]} = 45 \text{ kA} \tag{7.5}$$

7.5 ARC FLASH HAZARD

If a 480-V MCC (enclosed construction) has a maximum calculated fault current of 27 kA and an arcing gap of 32 mm, the arcing current would be

$$I_a = 10^{\{-0.097+0.0966(0.48)+0.000526(32)+(\log 27[0.662+0.5588(0.48)-0.00304(32)]\}} = 14.4 \text{ kA} \tag{7.6}$$

Since the nominal voltage is below 1000 V, the fault clearing time corresponding to 85% of 14.4 kA, or 12.2 kA, should also be checked.

Next, the incident energy can be calculated. The normalized incident energy is based on an arc duration of 0.2 seconds and a distance of 610 mm (2 ft) from the arc. The normalized incident energy, per IEEE 1584, is calculated using Eq. (7.7):

$$E_n = 10^{(K_1+K_2+1.081 \log I_a+0.0011 G)} \tag{7.7}$$

where

E_n = the normalized (0.2 second, 610 mm) incident energy in J/cm²
K_1 = −0.792 for open construction, or
−0.555 for enclosed (box) construction
K_2 = 0 for ungrounded or high impedance grounded systems, or
−0.113 for grounded systems
G = gap between conductors in millimeters.

Then the actual incident energy at a specific distance and arc duration can be calculated.

$$E = 4.184 C_f E_n \left(\frac{t}{0.2}\right) \left(\frac{610}{D}\right)^X \tag{7.8}$$

where

E = the incident energy at distance D and arc duration t in J/cm²
C_f = calculation factor (1.0 for V > 1 kV, 1.5 otherwise)
t = arc duration in seconds
D = working distance from arc in mm
X = distance exponent based on enclosure type
  Cable: 2
  Open air: 2
  MCCs and Panelboards: 1.641
  Switchgear > 1 kV: 0.973
  Switchgear < 1 kV: 1.473.

A common value for the working distance from the arc is 460 mm, or about 18 inch.

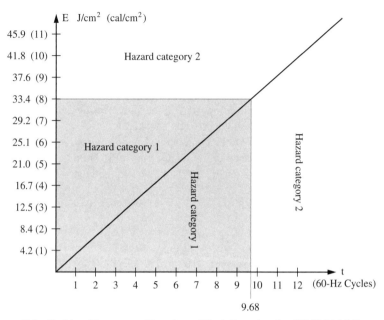

Figure 7.9 Incident Energy as a Function of Fault Duration for 480-V MCC Example

So, assuming a solidly grounded system and using the calculated arc current from Eq. (7.6), the normalized incident energy would be

$$E_n = 10^{(K_1+K_2+1.081 \log I_a+0.0011\,G)}$$
$$= 10^{[-0.555-0.113+1.081 \log 14.4+0.0011(32)]} = 4.16\,\text{J/cm}^2 \quad (7.9)$$

This value leads to an actual incident energy at 460 mm of

$$E = 4.184(1.5)(4.16)\left(\frac{t}{0.2}\right)\left(\frac{610}{460}\right)^{1.641} = 207t\,\text{J/cm}^2 \quad (7.10)$$

where t is the fault duration in seconds.

Graphing this relationship between incident energy and fault duration leads to Figure 7.9, which clearly shows how the arc flash hazard level increases with fault interruption time.

The arc flash protection boundary, as defined by IEEE 1584, specifies the distance from uninsulated parts at which a person could receive second-degree burns. Since second-degree burns can occur at an incident energy level of 5 J/cm², E in Eq. (7.8) can be set to 5, then solving for D specifies the flash protection boundary, denoted D_b:

$$D_b = 610 \sqrt[x]{4.184\,C_f E_n t} \quad (7.11)$$

where

C_f = calculation factor (1.0 for V > 1 kV, 1.5 otherwise)
E_n = the normalized (0.2 second, 610 mm) incident energy in J/cm^2
t = arc duration in seconds
X = distance exponent based on enclosure type
  Cable: 2
  Open air: 2
  MCCs and Panelboards: 1.641
  Switchgear > 1 kV: 0.973
  Switchgear < 1 kV: 1.473

Commercial software is readily available to perform arc flash calculations. The disadvantage of a software-based approach is not only the expense of the software and the training required to use it properly but also the additional data requirement posed by the software. But it is this additional data requirement that produces superior results by considering relay and breaker operating times, fuse characteristics, so on. While the NFPA and IEEE methods provide useful approximations of the hazards posed by arc flash events, a detailed software-based calculation is the preferred approach to assure personnel safety.

The level of hazard posed by arc flash events can be reduced by reducing the fault current magnitude, reducing the arc duration, or both. Reducing the fault current magnitude probably involves increasing the impedance of the system, which could have adverse effects on system performance, especially the ability to start large motors. Reducing the duration of the fault can be addressed by detecting and interrupting the fault more quickly. Replacing electromechanical relays with microprocessor-based relays and changing out old circuit breakers, which tend to be slower than modern circuit breakers, to faster breakers can have a profound effect on reducing arc duration. The use of current-limiting fuses, which operate in less than $1/4$ cycle, is also effective in reducing arc flash hazard.

SUMMARY

In industrial environments, electricity is provided to loads from *load centers*, including *switchgear* and *MCCs*. Both of these load centers are having hinged doors for easy internal access. Switchgear can be *metal enclosed* or *metal clad*. A system of *buswork* runs inside the load center to furnish power to each *unit*, *section*, or *cubicle*. Switchgear with two power sources (*double-headed switchgear*) requires a method of *source transfer* to switch between sources. Examples of source transfer methods include *fast transfer*, *slow transfer*, and *parallel transfer*.

Smaller electrical components can be housed in enclosures that are classified by *NEMA Type*, which defines the ability of the enclosure to resist environmental factors.

Several important ratings apply to both switchgear and MCCs. A *continuous current rating* limits the temperature rise of the load center components, particularly the insulation. A *short-circuit current rating* determines both the interrupting capability of the circuit breakers and fuses and the mechanical strength of the bracing and support systems which must resist the severe mechanical forces exerted by the intense magnetic fields present during a fault. The basic impulse level, or BIL rating, determines the electrical strength of the insulation.

Circuit breakers can be of different types depending on the rating requirements. *Air-magnetic* and *vacuum* breakers are common in switchgear, while MCCs often contain *thermal magnetic*, *molded case breakers*, or *fused disconnect switches*. MCCs utilize *motor starters* to provide control to motor loads. Starters come in different types and NEMA sizes to accommodate a wide variety of motor sizes and types. IEC-rated starters are also available and are commonly used outside the United States.

Circuit protection and motor-winding protection are two different issues accomplished by different types of devices. Circuit breakers or fuses protect circuits from short-circuit conditions, while overloads or heaters protect the motor windings from high temperatures caused by overload conditions.

Arc flash hazard must be addressed to determine safe working procedures around energized equipment. NFPA standard 70E and IEEE standard 1584 address the arc flash hazard by providing guidance for performing calculations to determine the incident energy during an arc flash event. The incident energy is used to determine the distance from the arcing point to the flash protection boundary, as well as the types of PPE needed to assure personnel safety when personnel are within the arc flash boundary of energized equipment.

FOR FURTHER READING

AC High-Voltage Circuit Breakers Rated on a Symmetrical Current Basis—Preferred Ratings and Related Required Capabilities, IEEE Standard C37.06, 1997.

Enclosures for Electrical Equipment (1000 V Maximum), National Electrical Manufacturers Association, NEMA 250, 2011.

IEEE Guide for Performing Arc Flash Hazard Calculations, IEEE Standard 1584, 2002.

IEEE Guide for Testing Metal-Enclosed Switchgear Rated Up to 38 kV for Internal Arcing Faults, IEEE Standard C37.20.7, 2007.

IEEE Standard for Metal-Clad Switchgear, IEEE Standard C37.20.2, 1999.

IEEE Standard for Metal-Enclosed Interrupter Switchgear, IEEE Standard C37.20.3, 2001.

IEEE Standard for Metal-Enclosed Low-Voltage Power Circuit Breaker Switchgear, IEEE Standard C37.20.1, 2002.

IEEE Standard Rating Structure for AC High-Voltage Circuit Breakers, IEEE Standard C37.04, 1999.

IEEE Standard Requirements for Overhead, Pad Mounted, Dry Vault, and Submersible Automatic Circuit Reclosers and Fault Interrupters for AC Systems, IEEE Standard C37.60, 1981.

IEEE Standard Test Procedure for AC High-Voltage Circuit Breakers Rated on a Symmetrical Current Basis, IEEE Standard C37.09, 1999.

National Electrical Code, National Fire Protection Association, NFPA 70, 2014.

Standard for Electrical Safety in the Workplace, National Fire Protection Association, NFPA 70E, 2015.

QUESTIONS

1. Mechanical bracing requirements for switchgear and MCCs are determined by what calculated value?
2. Current interruption capability of a circuit breaker is determined by what calculated value?
3. In addition to fault current ratings, what other current rating is assigned to switchgear and MCCs, and why is it important?
4. Why was the nomenclature for fault withstand capability of circuit breakers changed from MVA to kA?
5. What environmental factor can be harmful to silver-plated components? Describe the problem and suggest a means of mitigating the problem.
6. What are the main functions performed by a motor starter?
7. What devices can be used to protect a motor from
   a. overloads?
   b. short circuits?
8. Why are drawout-type components often used in metal-clad switchgear?
9. What options exist for correcting the following problems:
   a. a circuit breaker serving a load in excess of its continuous current rating
   b. a circuit breaker installed on a bus with fault current in excess of the breaker's interrupting rating
10. What factors limit how many sections or units can be installed in a switchgear lineup or MCC?
11. Why must arc flash calculations, and not just short-circuit calculations, be used to determine safe working procedures around energized equipment?
12. What are the two standards that address arc flash hazard?
13. How do the calculation procedures used in the two arc flash standards differ?
14. What measures can be taken to reduce the level of arc flash hazard?
15. What units are used to measure incident energy to determine arc flash hazard category, and what is the maximum incident energy the bare skin can tolerate without incurring second-degree burns?

PROBLEMS

1. Which NEMA enclosure types cannot be used outdoors?
2. Switchgear must be installed in a facility 12,000 ft (3660 m) above sea level. If the 4.16-kV switchgear bus has a nameplate rating of 2000 A and the 480-V switchgear bus has a nameplate rating of 1200 A, what are the adjusted voltage and current capabilities of the switchgear after installation?
3. A 25-hp motor can be installed either on a 480-V bus or on a 240-V bus. What size starter would be required for these two installation options?

4. What is the interrupting rating in amperes of the following switchgear when installed below 3300 ft (1000 m) above sea level?

   a. 500 MVA, 5 kV
   b. 750 MVA, 5 kV
   c. 750 MVA, 15 kV
   d. 1000 MVA, 15 kV

5. Sketch the waveforms used for the full-wave and chopped-wave test for BIL verification.

6. Calculate the arc current for

   a. a 24.9 kV system with a maximum three-phase fault current of 43.5 kA.
   b. a 600 V system in open air with a maximum three-phase fault current of 87.5 kA and a gap of 15 mm.
   c. a 480 V system in an enclosed panel with a maximum three-phase fault current of 112 kA and a gap of 10 mm.

7. Calculate

   a. the normalized (0.2 second, 610 mm) incident energy for a grounded 12.47-kV system in open air with an arc current of 32 kA and a gap of 25 mm.
   b. the actual incident energy for the conditions in Part a for a fault duration of 8 cycles (at 60 Hz) and a working distance of 460 mm.
   c. the actual incident energy for the conditions in Part a for a fault duration of 5 cycles (at 60 Hz) and a working distance of 460 mm.

8. Calculate

   a. the normalized (0.2 second, 610 mm) incident energy for a grounded 24.9-kV system in metal-clad switchgear with an arc current of 42.5 kA and a gap of 35 mm.
   b. the actual incident energy for the conditions in Part a for a fault duration of 8 cycles (at 60 Hz) and a working distance of 460 mm.
   c. the actual incident energy for the conditions in Part a for a fault duration of 5 cycles (at 60 Hz) and a working distance of 460 mm.

9. Calculate

   a. the normalized (0.2 second, 610 mm) incident energy for ungrounded 480-V switchgear with an arc current of 65 kA and a gap of 10 mm.
   b. the actual incident energy for the conditions in Part a for a fault duration of 8 cycles (at 60 Hz) and a working distance of 460 mm.
   c. the actual incident energy for the conditions in Part a for a fault duration of 5 cycles (at 60 Hz) and a working distance of 460 mm.

10. Determine the arc flash protection boundary for a normalized incident energy of 2.4 J/cm^2 and a clearing time of 8 cycles (at 60 Hz)

    a. for a 600-V system in open air.
    b. for a 480-V panelboard.
    c. for a 480-V switchgear.
    d. for a 4160-V switchgear.
    e. for a 4160-V cable termination.

CHAPTER 8
LADDER LOGIC

OBJECTIVES

- Identify and apply the basic components used to implement ladder logic
- Understand the basic concepts of designing ladder logic
- Explain the operation of a control circuit
- Design a control circuit to perform specific tasks

8.1 FUNDAMENTALS

Motor control circuits are often implemented using *ladder logic*. It is so called because the schematic representation of the control circuit resembles a ladder with two vertical rails and a number of horizontal rungs, as shown in Figure 8.1.

The vertical rails are energized at two different potentials, often 120 V and ground. Any available low-voltage source, either AC or DC, can be used for the control circuit.

If the motor operates at 240 V or less, it is not uncommon to obtain the control voltage directly from the power circuit, as shown in Figure 8.2.

If the motor operates at a higher voltage that is not suitable for the control circuit, a *control power transformer* can be used to feed the control circuit, as shown in Figure 8.3.

Current will flow from the left-hand rail to the right-hand rail if a closed path is provided. Some impedance is required in this closed path to prevent a short circuit. Impedance can be provided by relay coils and indicator lamps. Fusing protects the control circuit against heavy current caused by a short circuit in the control circuit. If the control circuit is left ungrounded, both rails of the ladder can be fused, but only the left rail is fused if the right rail is grounded.

Closing the path between the rails is done by contacts placed in the rung. The contacts, which can be normally open or normally closed, are placed in series to implement a logical *and* function or in parallel to implement a logical *or* function. The contacts can be actuated by a relay coil or by automatically or manually controlled switches. These devices will be discussed in the next section.

Industrial Power Distribution, Second Edition. Ralph E. Fehr, III.
© 2016 The Institute of Electrical and Electronics Engineers, Inc. Published 2016 by John Wiley & Sons, Inc.

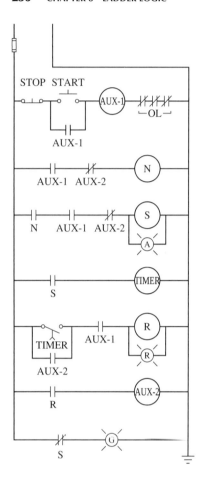

Figure 8.1 Ladder Diagram

8.2 CONSIDERATIONS WHEN DESIGNING LOGIC

Some terminology must be understood before proceeding with implementing ladder logic. When a relay coil is energized, that relay is said to be *picked up*. When a relay picks up, its contacts change state: the normally open contacts close and the normally closed contacts open. Normally open contacts are called *Type A* contacts, and normally closed contacts are called *Type B* contacts.

When a relay coil is de-energized, that relay is said to *drop out*. Care must be exercised so relay coils do not inadvertently drop out due to inadequate voltage. This can be a concern because the starting of a large motor can significantly depress the system voltage and thereby the control circuit voltage. This is one reason why voltage drop during motor starting must be kept within specific parameters. If a running motor experiences a voltage drop in excess of the coil dropout voltage of its contactor, the running motor will shut down. This problem can be mitigated in many ways, ranging from changing the motor starting method to reduce the voltage drop, to using contactors capable of withstanding larger voltage drops without dropping out,

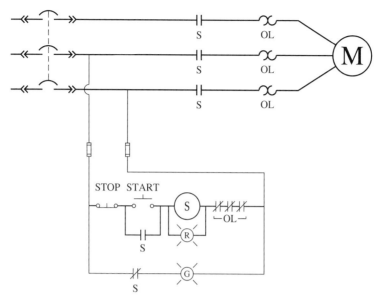

Figure 8.2 Control Circuit—Line Voltage

or to supplying the control circuit from a power source independent of the starting motor. A direct current control system supplied from batteries would not experience voltage fluctuations while motors are starting.

Symbols are used to represent commonly used control circuit devices on schematic diagrams. Although variations in drafting standards exist, the symbols shown in Table 8.1 are fairly standard throughout the industry.

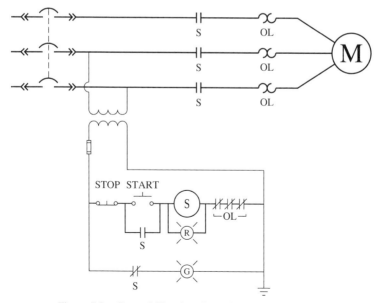

Figure 8.3 Control Circuit—Control Power Transformer

TABLE 8.1 Device Symbols

Device	Normally open	Normally closed	Device	Symbol
			Relay coil	—(S)—
			Type "A" contact	—| |—
Momentary pushbutton	o⊥o	o⊥o	Type "B" contact	—|/|—
Limit switch			Form "C" contacts	
Pressure switch			Indicator lamp letter indicates color	—(R)—
Level switch			Thermal overload	—⌒⌒—
Flow switch			Control circuit fuse	—▭—
Temperature switch				
Timer contacts — Delay after coil pickup			Three-phase drawout-type circuit breaker	
Timer contacts — Delay after coil dropout				

Accidental grounds in the control circuit should never pick up a relay coil. In Figure 8.4, an accidental ground at point F would short out the three Type A contacts and pick up the coil. This could result in a very dangerous situation, such as a motor at rest suddenly starting without warning and with all electrical interlocks defeated. An accidental ground in the control circuit should instead blow the control circuit fuses. Blowing the fuses would safely shut down the motor if it is running, prevent it from starting if it is at rest, and indicate to maintenance personnel that a problem exists in the control circuit.

Control circuits should always be designed in such a way that they fail in a *safe mode*. In other words, the controlled motor should safely shut down in the event of a control circuit failure. Under no circumstances should a control circuit failure cause a de-energized motor to start.

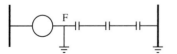

Figure 8.4 Poorly Designed Logic

8.3 LOGIC IMPLEMENTATION

After the control logic is designed, it can be implemented in one of the two ways. Traditionally, *discrete components* were hardwired together to build logic circuits. Relays with various contact configurations aided in the implementation of the logic. Electromechanical timers provided delay functions. Contacts from remote devices such as limit switches and circuit breaker auxiliary contacts were wired into the circuit to implement the desired logic.

When implementing logic using discrete devices, a problem was frequently encountered. Often, a device such as a limit switch has only one set of Type A contacts, but the logic circuit may require say two sets of Type A contacts and a set of Type B contacts. Contacts can be multiplied using an *auxiliary relay* such as the one shown in Figure 8.6. The auxiliary relay diagrammed in Figure 8.5 multiplies a single set of Type A contacts to two sets of Type A and two sets of Type B contacts.

One side of the Type A contacts from a device such as a limit switch is connected to the left rail of the ladder diagram. The other side of the Type A contacts is wired to Terminal 1 of the auxiliary relay. Terminal 2 of the auxiliary relay is connected to the right rail of the ladder diagram. Now, when the Type A contacts of the limit switch close, two other Type A contacts (Terminal 3–Terminal 9 and Terminal 5–Terminal 7) also close. In addition, two Type B contacts (Terminal 4–Terminal 10 and Terminal 6–Terminal 8) open. Any unused contacts on an auxiliary relay are left open circuited.

A more versatile method of constructing logic circuits is to use a *programmable logic controller*, or *PLC*. A PLC is a specialized microcomputer housed in a rugged enclosure to withstand the harsh conditions frequently encountered in an industrial environment such as temperature extremes, high humidity and moisture, vibration, and dirt. The software is designed to let the user define the logic by constructing a ladder diagram. Devices such as timers (Figure 8.7), counters, and auxiliary relays are mimicked by the PLC.

A great deal of flexibility is provided by the software, including troubleshooting modes. For example, if the control logic needs to be totally redesigned, instead of replacing discrete components, the changes can be made via software. Reliability also increases because of the reduced part count. Remote control and monitoring capabilities are easily accomplished by a simple telephone line connection. Logic changes can be made offline and downloaded to the PLC by modem. PLCs are designed in a modular fashion, so in the event of hardware failure, the bad components can easily be isolated and replaced.

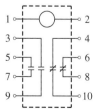

Figure 8.5 Auxiliary Relay Schematic Diagram

240 CHAPTER 8 LADDER LOGIC

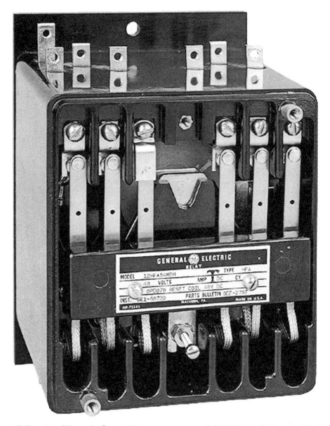

Figure 8.6   Auxiliary Relay. (*Photo courtesy of GE Digital Energy Multilin.*)

Because a PLC is a digital computer, it can also be miniaturized to conserve valuable space and minimize power consumption. The PLC shown in Figure 8.8 measures 19-inch wide by 14.5-inch high by 20-inch deep and is designed to be either panel mounted or installed in a standard 19″ instrument rack.

8.4   SEAL-IN CIRCUITS

It is often desirable to have the normally open contacts of a momentary pushbutton switch remain closed after the pushbutton is released. Using a nonmomentary or latching, switch would accomplish this but would also cause other problems. For example, using nonmomentary switches, it would be possible to have both the stop and start pushbuttons for a single motor depressed simultaneously. Also, if the power supply to a running motor is interrupted, the start pushbutton would remain depressed, allowing the motor to restart unexpectedly when the power supply is restored. The use of momentary switches eliminates these problems, but a means of latching the contacts must be developed. This is the purpose of a *seal-in circuit*.

Figure 8.7 Timing Relay. (*Photo courtesy of IS-Rayfast, Ltd.*)

With a seal-in circuit, such as that shown in Figure 8.9, pressing a momentary pushbutton picks up a relay coil. A Type A contact from that relay is paralleled across the momentary pushbutton. When the relay picks up, a path is closed across the pushbutton switch. When the switch is released and its contacts open, a closed path around the switch still exists. This is equivalent to holding the momentary switch contacts closed.

Another reason to use seal-in circuits is to provide undervoltage dropout for contactors. This prevents the motor from automatically restarting when the voltage returns to normal after dropping below the voltage required to keep the relays picked up. If a nonmomentary pushbutton was used instead of the momentary pushbutton and seal-in circuit, the motor would automatically restart when the voltage returns to normal. While this could be a desirable mode of operation, it could be unacceptable from a safety standpoint. To restart the motor with the seal-in circuit, the pushbutton must be pressed again after the voltage returns to normal.

Figure 8.8   Programmable Logic Controller. (*Photo courtesy of Rockwell Automation.*)

Momentary pushbuttons can be sealed-in using a *three-wire control* scheme. Three separate wires are run to the pushbutton station; one providing the voltage source, one to energize the start coil, and the third to provide the seal-in function. The three-wire control scheme is shown in Figure 8.10.

When a device not requiring seal-in, such as a level switch, is used for motor control, a *two-wire control* scheme can be used, where the switch is simply wired in series with the start coil. When power is restored after a power failure, motors controlled by a two-wire control scheme will restart automatically.

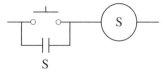

Figure 8.9   Seal-in Circuit

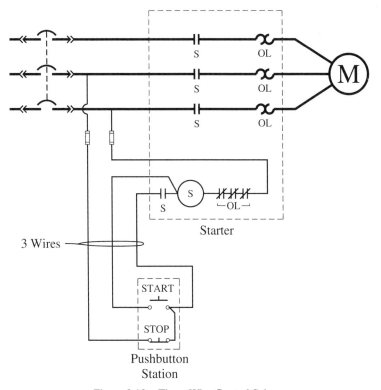

Figure 8.10   Three-Wire Control Scheme

8.5   INTERLOCKS

It is good design practice to allow parts of a control circuit to function only if certain conditions are met. Put another way, sometimes parts of a control circuit should be disabled unless certain conditions are met. This is the concept of an *interlock*.

Perhaps a fan should be started only if a damper is in a particular position. To assure this, *a limit switch* can be installed on the damper. The limit switch has Type A contacts that are closed only when the damper is in the position it needs to be for starting the fan. If the damper is in any other position, the limit switch contacts will be open.

To accomplish the interlock function, the Type A contacts from the limit switch can be wired in series with the start and stop pushbuttons, as shown in Figure 8.11.

The fan motor control circuit pushbuttons can be operated in any way, but the motor will be allowed to start if and only if the damper is in the appropriate position.

Other types of devices can be used to provide interlock functions. If the level of a liquid in a tank must be verified prior to allowing a pump to start, a *level switch* can be used. Level switches work on a principle similar to that of the float mechanism in a toilet tank. When the liquid in the tank reaches a certain level, instead of closing a water valve like the float mechanism in a toilet tank, the level switch's

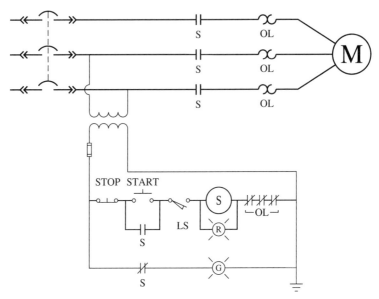

Figure 8.11 Limit Switch Interlock

electrical contacts change state. Other specialty switches include *pressure switches*, *temperature switches*, and *flow switches*.

Electrical switching devices such as power circuit breakers, circuit switchers, and motor-operated air break switches can be purchased with *auxiliary contacts* which are Type A or Type B contacts that can be used in control circuits to provide interlock functionality by indicating the status (closed or open) of the switching device.

If more than one interlock is desired for starting a motor, the contacts from the additional interlock devices can be wired in series (if an AND function is desired) or in parallel (if an OR function is desired). For example, a large pump motor can only be started if (a) the liquid in a storage tank is at a certain level, (b) the DC lubricating oil pump for the large pump is running, and (c) at least one of the three valves is open. A level switch (LEV) is installed on the storage tank, a Type A auxiliary contact (LUB) from the lubricating oil pump motor is utilized, and position switches (PS1, PS2, and PS3) are installed on each of the three valves. LEV will close when the tank level is adequate for motor starting. LUB will be closed when the lubricating oil pump is running, and the position switches on the valves will be closed when the valves are open. The interlock circuit is designed as shown in Figure 8.12.

When the START pushbutton is pressed, in order for the S coil to be picked up, the level switch LEV and the lubricating oil pump Type A auxiliary contacts LUB must be closed. Also, at least one of the pressure switches must be closed. If any of these conditions are not met, the S coil will not pick up when the START pushbutton is pressed, and the large pump motor will not start.

Sometimes the interlock logic is designed to allow the motor to continue running in addition to allowing the motor to start. In this case, the interlock logic could be wired in series with the start pushbutton seal-in contact instead of in series with the start pushbutton.

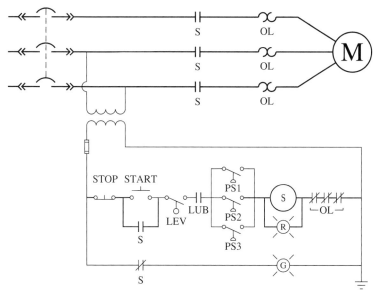

Figure 8.12 Interlock Example

Motor overload relays provide Type B contacts that open when the stator winding temperature exceeds a certain limit. These contacts are typically wired in series with the run coil, so if any one overload opens, the motor will shut down.

If a motor must be able to run in both the forward and reverse directions, a *reversing starter* is required. When reversing starters are used, electrically interlocking the forward and reverse contactors is essential to prevent the possibility of a line-to-line fault in the motor's power circuit (see Section 8.7).

8.6 REMOTE CONTROL AND INDICATION

It is often desirable to have control over and indication of motor status from a location remote from the motor control center. This remote location may be at the motor itself or in a control room.

Multiple START pushbuttons can be implemented by wiring the switches in *parallel*. Parallel switch contacts form an OR logic gate, meaning that pushbutton one *or* pushbutton two *or* pushbutton n must be pressed to start the motor.

Multiple STOP pushbuttons can be implemented by wiring the switch contacts in *series*. Series switch contacts form an AND logic gate, meaning that pushbutton one *and* pushbutton two *and* pushbutton n must be in their normal state for the motor to run. So pressing any one pushbutton switch will de-energize the motor. Multiple START and STOP pushbuttons are shown in Figure 8.13.

Colored lamps are used to indicate the status of a motor. In most applications, a red lamp indicates a running motor and a green lamp indicates a de-energized motor. This color convention is consistent with circuit breaker status indication (red is closed, green is open), but may seem counterintuitive. An easy way to remember

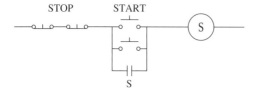

Figure 8.13 START and STOP Pushbuttons

this convention is to *think safety first*. If a motor circuit needs to be maintained, a red light means STOP (do not work on the motor circuit because the motor is running) and a green light means GO (the motor is not running, so carefully proceed and work on the circuit). It must be noted that this red/green color convention is not always followed, particularly outside the United States. For this reason, and to accommodate workers with color blindness, labeling the indicator lamps with engraved escutcheons is recommended.

Variable speed motors and motors with reduced voltage starters may have more complex lamp color designations. For example, an amber and an orange lamp may represent a two-speed motor running on low and high speed, respectively. An amber lamp could also be used on a motor equipped with a reduced voltage starter to indicate that the motor is running at reduced voltage (during the starting process).

Indicator lamps to indicate a running motor are paralleled across the coils of the relays that operate the contactors. Lamps should not be put in series with relay coils because a burned out bulb would disable the coil from picking up. The series connection would also reduce the voltage across the coil due to the voltage drop across the lamp. The green STOP lamp is usually put in series with a Type B contact from the run relay coil. Indicator lamp connections can be seen in Figure 8.14.

Schematically, all START and STOP pushbuttons and indicator lamps are drawn adjacent to each other to facilitate reading the diagram. Physically, these pushbuttons and lamps may be a considerable distance apart. A note or wire mark may be used on the schematic to indicate the location of a remote device.

8.7 REVERSING STARTERS

Many times, a motor must be able to run in either a forward or a reverse direction. This type of operation is common for conveyors, cranes, hoists, and other such applications. Bidirectional operation is accomplished by using a *reversing starter*, as shown in Figure 8.15.

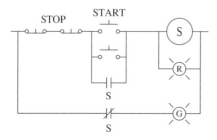

Figure 8.14 Indicator Lamps

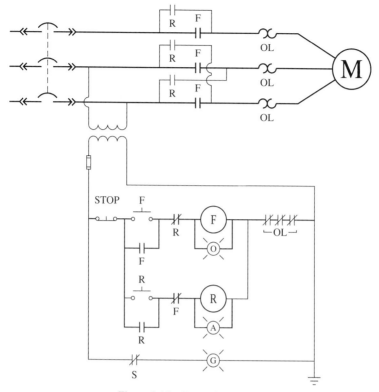

Figure 8.15 Reversing Starter

Reversing the direction of rotation of a three-phase motor requires that a negatively sequenced set of voltages be applied to the motor. Transposing any two of the three line connections to the stator windings will provide negative sequencing. This is precisely how a reversing starter works.

Two separate contactors are required for a reversing starter—one for running the motor in the forward direction and the other in the reverse direction. Both electrical and mechanical interlocks are provided to prevent the simultaneous operation of both the forward and reverse contactors. Energizing both contactors at the same time would create a line-to-line fault in the power circuit, as can be seen in Figure 8.16.

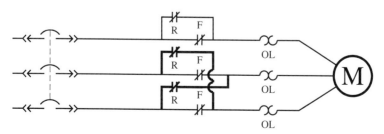

Figure 8.16 Line-to-Line Fault

The Type B contacts in Figure 8.15 make up the electrical interlock. When the F coil is picked up, the Type B contacts of the F relay are open, thereby preventing the R coil from picking up. When the R coil is picked up, a similar interlock is accomplished with the F coil. This type of electrical interlock is a very good method of preventing two coils from being energized at the same time. The interlock forms a logical *exclusive or*, or *xor*, function.

8.8 JOGGING

Certain types of loads such as presses, cranes, and hoists require their motors be started and stopped repeatedly for very short periods of time to bring the machinery into a specific position. This is known as *jogging*. From a control circuit standpoint, jogging is essentially a start operation without a seal-in function. To provide jogging capability for a nonreversing starter, an auxiliary relay can be utilized as shown in Figure 8.17.

When the JOG pushbutton is pressed, the S coil picks up. This coil closes the contactor for the motor, so the motor starts. When the S relay's Type A contacts close, the seal-in circuit is not complete because the AUX relay's Type A contacts are still open. The motor will run only as long as the JOG pushbutton is held closed.

When the START pushbutton is pressed, the AUX coil picks up. When the AUX relay's Type A contacts close, the S coil picks up. At this point, both Type A contacts connected in series to form the seal-in circuit are closed, so the start pushbutton is

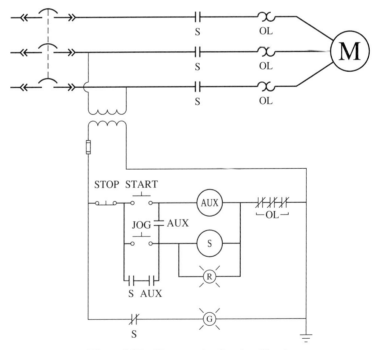

Figure 8.17 Nonreversing Jogging Circuit

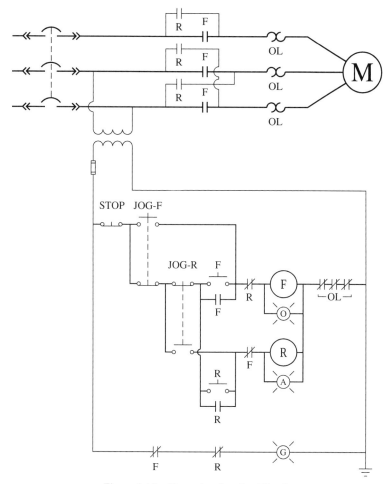

Figure 8.18 Reversing Jogging Circuit

sealed-in. Releasing the START pushbutton will not stop the motor. Pressing the STOP pushbutton, however, will cause both the AUX and S coils to drop out thereby opening the contactors and stopping the motor.

Providing jogging capability for a reversing starter is considerably more complicated, as is evident in Figure 8.18.

Two contactors are provided for the motor: the F coil closes the forward contactor and the R coil closes the reverse contactor. The jogging function requires gang-operated pushbuttons. When the JOG-F pushbutton is pressed, the F coil picks up and remains picked up as long as the JOG-F pushbutton is held closed. Releasing the pushbutton stops the motor. Similarly, pressing the JOG-R pushbutton picks up the R coil, which remains picked up as long as the pushbutton is held closed.

Pressing the F pushbutton also picks up the F coil, but when the Type A contacts of the F relay close, the F pushbutton is sealed in. The motor will remain running in the forward direction until the STOP pushbutton is pressed to drop out the F coil.

Pressing the R pushbutton picks up and seals in the R coil. The motor will remain running in the reverse direction until the STOP pushbutton is pressed to drop out the R coil.

8.9 PLUGGING

Sudden reversal of a motor's direction is called *plugging*. Sometimes plugging is desired, for example, to suddenly stop a motor. In some applications, plugging subjects the driven equipment to unacceptably high torques and must be prevented.

Using a timer in the control circuit, as shown in Figure 8.19, can prevent plugging. Timers can be purchased with a wide variety of operating characteristics. The timer used in this example is of the *delay after pickup* type. This means that

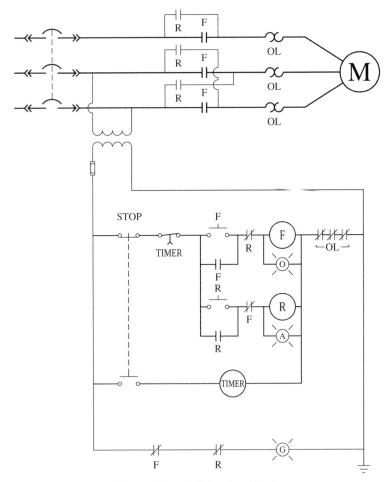

Figure 8.19 Antiplugging Circuit

when the timer coil is picked up, the preset delay time elapses and the timer contacts change state. The coil must remain picked up until the delay time elapses or the timer contacts will not change state.

SUMMARY

Ladder logic is often used to implement motor control circuits. Two vertical "rails" are energized at different voltages. Horizontal "rungs" connect the two rails with a combination of *relay coils, indicator lamps*, and *relay and switch contacts*. When a completed path is formed between the rails, current flows through that rung. Relay coils and indicator lamps are the elements that provide impedance, which is necessary to prevent a short circuit from rail to rail.

Contacts can be *sealed in*, so current can continue to flow through the rung after the contacts open. *Auxiliary contacts* can be used to form *interlocks*, which allow a function to happen only under certain specific conditions. Indication and control components can be installed in multiple physical locations although they are typically shown adjacent to each other on the ladder diagram.

A *reversing starter* uses two sets of contactors, one which allows the motor to run in the forward direction, and another which transposes two of the three phases causing the motor to run backwards.

Special control circuits can be designed to allow *jogging* (brief energization of the motor to adjust its positioning) or to prevent *plugging* (sudden direction reversal).

Ladder logic can be implemented using conventional relays and discrete components, or using the solid-state circuitry of a PLC.

FOR FURTHER READING

Electrical Standard for Industrial Machinery, NFPA 79, 2015.

QUESTIONS

1. What determines whether multiple contacts are connected in series versus parallel in a ladder logic circuit?
2. Why should an indicator lamp never be put in series with a relay coil?
3. What is the purpose of mechanically and electrically interlocking the forward and reverse contactors in a reversing starter?
4. Why are control circuits fused?
5. Using two momentary pushbuttons labeled "A" and "B", design a circuit to light an indicator lamp when the following conditions are met:
    a. A and B

b. A or B

   c. A xor B (Use switches configured like the jog pushbuttons in Figure 8.18)

6. What is the purpose of an auxiliary relay?

7. Why can excessive voltage drop during motor starting or other low-voltage conditions be problematic for control circuits?

8. If nonmomentary pushbutton switches were used instead of momentary pushbuttons to start and stop a motor, what operational problems would arise?

9. Why are all motor overload Type B contacts put in series with the motor contactor coil or with an auxiliary relay that energizes the motor contactor coil?

10. Momentary switches need to be sealed-in so a closed current path remains after the pushbutton is released. Why is it sometimes necessary to seal-in a relay coil?

PROBLEMS

1. Design control circuits for two motors (Motor A and Motor B), which start across the line as shown in Figure 8.3, meet the following criteria. Motor A requires a single START/STOP pushbutton station with one set of indicator lamps. Motor B requires two START/STOP pushbutton stations and two sets of indicator lamps. Motor B can only start if Motor A is running. Also, Motor B must shut down if Motor A stops.

2. A 6900-V-induced draft fan motor can be started across the line, as shown in Figure 8.3. Before it can be started, lube oil pump LO1 must be running. Limit switches LS1 and LS2 also must be closed. When the fan is running, the limit switches may open, but a blue indicator light should be lit if either switch opens. The fan motor should be stopped if LO1 stops.

   Draw the power and control circuits for the induced draft fan. Use two START/STOP/INDICATOR LAMP stations to control the fan. Show all necessary interlocks.

3. Flow switches FS1, FS2, and FS3, and pressure switches PS1, PS2, and PS3 provide interlocks for the start coil S.

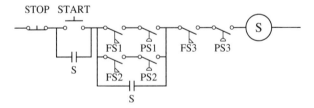

   a. What conditions are necessary to start the motor?

   b. What conditions are necessary to keep the motor running?

4. How does relocating the seal-in contact in Problem 3, as shown, change the answers to parts (a) and (b)?

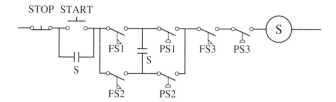

5. Temperature switch TS-A is a normally open switch that closes when the temperature exceeds 80°C. Temperature switch TS-B is a normally closed switch that opens when the temperature drops below 40°C. Design a logic circuit where a STOP/START pushbutton station with red and green indicator lamps controls a motor that can only start when the temperature lies between 40°C and 80°C. Once the motor is running, temperature is not a criterion to keep the motor running.

6. Modify the control circuit in Problem 5 so the motor shuts down if the temperature falls outside the range required for starting.

7. Three lubricating oil pumps (LO1, LO2, LO3) are necessary to support a large compressor. The starting sequence requires that LO1 be started first. LO2 cannot be started until flow switch FS1 closes. LO3 cannot be started until flow switch FS2 closes. When flow switch FS3 closes, the contactor for the compressor can be picked up. Design a control circuit where a single START/STOP pushbutton station with red and green indicator lamps controls the compressor.

8. Modify the control circuit in Problem 7 to include three amber indicating lamps to show that the lubricating oil pumps are running and three blue indicating lamps to show that the three flow switches are closed.

9. Design a reversing jogging control circuit with antiplugging logic included.

10. A hoist motor requires a reversing jogging control circuit. If normally open limit switch LS-U is closed, the hoist motor must not be opened in the RAISE direction. If normally open limit switch LS-D is closed, the hoist motor must not be operated in the LOWER direction. Indicator lamps and sealed-in pushbuttons are not required. Design the control circuit.

CHAPTER 9

MOTOR APPLICATION

OBJECTIVES

- Recall the fundamentals of motor modeling and know the differences between various NEMA motor designs
- Know how to use speed–torque curves to specify a motor for a specific application
- Understand the behavior of induction motors during starting and the reasons for using special motor starting methods
- Recognize the differences between open and closed transitions from reduced voltage to full voltage
- Implement control logic for the various types of reduced voltage starters
- Understand the fundamentals of electronic motor starting methods

9.1 FUNDAMENTALS

Consider a three-phase induction motor connected to a three-phase source by an open circuit breaker. When the breaker is closed, the source voltage will be applied to the motor terminals. If the motor windings are connected in a delta configuration, each winding will have the full source line-to-line voltage applied across it. If the motor windings are connected in a wye configuration, each winding will have $1/\sqrt{3}$ times the source line-to-line voltage applied across it.

Assume that the driven load has a horsepower requirement equal to the horsepower rating of the motor. In other words, the motor is loaded to its nameplate rating. There are two major issues must be considered when starting this motor. First, *is the system adequate to start the motor?* The motor will have a minimum starting torque requirement that is dependent on the driven load. Since torque is proportional to the voltage squared, the system must be able to provide adequate voltage to accelerate the starting motor; otherwise, the motor will stall. The second issue is, *what impact will the starting motor have on the rest of the system?* If the voltage drop during motor starting is excessive, loads already running could be impacted.

When a motor is initially energized, the rotor speed is zero. Consequently, there is no generated electromotive force (counter EMF) produced by the motor to limit

---

Industrial Power Distribution, Second Edition. Ralph E. Fehr, III.
© 2016 The Institute of Electrical and Electronics Engineers, Inc. Published 2016 by John Wiley & Sons, Inc.

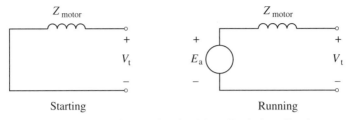

Figure 9.1   Running and Starting Motor Equivalent Circuits

the current flowing into the motor terminals. Only the winding impedance limits the current. When the rotor is spinning, a counter EMF (E_a) is generated. This reduces the voltage across the motor impedance and thus limits the current. The starting and running motor equivalent circuits are compared in Figure 9.1 and the corresponding equations are shown in Eqs. (9.1) and (9.2):

$$I_{\text{motor (running)}} = \frac{V_t - E_a}{Z_{\text{motor}}} \tag{9.1}$$

$$I_{\text{motor (starting)}} = \frac{V_t}{Z_{\text{motor}}} \tag{9.2}$$

The motor impedance Z_{motor} is made up of a winding resistance component in series with the inductive reactance of the windings. The resistance varies slightly with temperature but can be considered to be constant over short periods of time. Since the resistance term is essentially constant during the motor starting process and is much smaller than the reactance term, the resistance term will be neglected in the remainder of this discussion.

The lack of counter EMF at the time of starting effectively decreases the impedance of the starting motor. Since a higher voltage exists across the windings of a starting motor than across the windings of a running motor due to the changes in counter EMF, a higher current flows into the terminals of a starting motor than into the terminals of a running motor for the same terminal voltage. This increase in current appears to the electrical source as a decrease in motor impedance. A reduced effective impedance explains the high starting current experienced—typically five to seven times the motor's full-load current. The actual inrush current is determined by the NEMA code letter assigned to the motor, per NEMA standard MG 1-10.37.2. The NEMA code letters for locked rotor kVA are shown in Table 9.1.

As the rotor begins to rotate, a counter EMF is generated due to the interaction of the magnetic fluxes produced by the armature and the field currents. The counter EMF reduces the voltage across the motor windings, thereby reducing the current flowing into the motor. When the rotor is turning at full speed, the counter EMF will be at its maximum and the current flowing into the motor will be equal to rated full-load current. So, as the counter EMF builds, the effective impedance of the motor appears to increase. Full-load current and effective reactance as a function of rotor speed can be seen in Figures 9.2 and 9.3.

Slip is the speed differential between the rotating magnetic field and the rotor speed of an induction motor. The stator field rotates at *synchronous speed* n_s, which

TABLE 9.1   NEMA Locked-Rotor kVA Code Letters

Code Letter	kVA/HP	Code Letter	kVA/HP	Code Letter	kVA/HP
A	0.0–3.15	H	6.3–7.1	R	14.0–16.0
B	3.15–3.55	J	7.1–8.0	S	16.0–18.0
C	3.55–4.0	K	8.0–9.0	T	18.0–20.0
D	4.0–4.5	L	9.0–10.0	U	20.0–22.4
E	4.5–5.0	M	10.0–11.2	V	>22.4
F	5.0–5.6	N	11.2–12.5		
G	5.6–6.3	P	12.5–14.0		

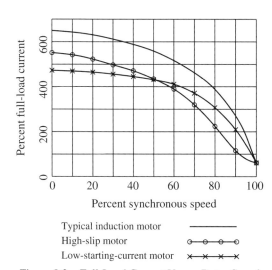

Figure 9.2   Full-Load Current Versus Rotor Speed

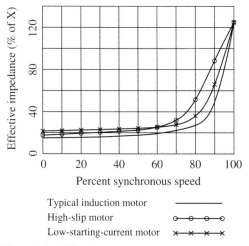

Figure 9.3   Effective Reactance Versus Rotor Speed

can be determined using Eq. (9.3), where f is the system frequency and p is the number of poles the motor contains:

$$n_s = \frac{120f}{p} \tag{9.3}$$

The rotor lags slightly behind the stator field. Slip is calculated as

$$s = \frac{n_s - n_{rotor}}{n_s} \tag{9.4}$$

and is usually expressed in either percent or per unit. Typical induction motor designs have a slip of a few percent.

Starting time also can be an issue for the motors driving high-inertia loads. As starting times increase to tens of seconds, the resulting impact on both the motor and the electrical system can become significant due to the high starting current over a long time period. Sometimes, special motor protection schemes including speed-sensing relays are required to properly protect the motor during the starting process. The motor's load may be decoupled by means of a clutch to facilitate motor starting. Special system criteria and configurations may be required when starting very large motors to assure adequate fault current during the starting process.

High starting currents cause large voltage drops. It is important to realize that a sizable voltage drop can result whenever a large current passes through even a small impedance. This means that significant voltage drops can occur throughout the power system, beginning at the starting motor terminals and continuing upstream toward the source.

Another parameter that influences a motor's current requirements is the operating voltage of the motor. Motor voltage ratings do not correspond exactly to the nominal system voltages for which they are designed. For example, a motor designed to be used on a nominal 480-V system will have a nameplate voltage rating of 460 V. A comparison of motor nameplate voltages and nominal system voltages is shown in Table 9.2.

A range of horsepower ratings is commonly available at each motor operating voltage. Figure 9.4 shows the commonly available motor sizes for several different operating voltages.

A common sizing strategy used by many US companies is shown in Table 9.3.

TABLE 9.2 Comparison of Motor Nameplate and Nominal System Voltages

Motor Nameplate Voltage	Nominal System Voltage
115	120
230	240
460	480
575	600
4000	4160
6600	6900
13,200	13,800

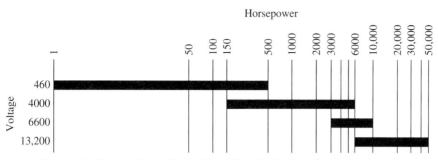

Figure 9.4 Commonly Available Motor Sizes for Various Operating Voltages

9.2 ENERGY CONVERSION AND LOSSES

A motor is an energy conversion device that converts electrical energy to mechanical energy. As with any energy conversion, the law of conservation of energy must be obeyed. This means that the amount of electrical energy required by the motor must equal the amount of mechanical energy required by the driven load, plus the energy required to furnish any losses that occur during the energy conversion process. Losses incurred by a motor vary considerably, but for many motor applications, in the 7–15% range. In general, the higher the loading and the larger the motor, the lower the losses, or the higher the efficiency.

For some types of approximate analyses, neglecting the motor losses may have a relatively minor effect on the accuracy of the answer. If this is the case, the energy conversion process can be modeled as the block diagram in Figure 9.5.

If losses are quantified by quoting an *efficiency* (η), Figure 9.5 can still be used to model the energy conversion process. The efficiency, in per unit, gives the relationship between P_{out} and P_{in} as

$$P_{in} = \frac{P_{out}}{\eta} \tag{9.5}$$

A more detailed model of the energy conversion process, showing losses, is shown in Figure 9.6. The equivalent circuit of an induction motor is shown above a power flow diagram.

TABLE 9.3 Common US Motor Voltage Sizing Strategy

Horsepower Rating	Operating Voltage
1–250	460
250–5000	4000
5000–50,000	13,200

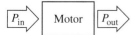

Figure 9.5 Energy Conversion Performed by a Motor Neglecting Losses

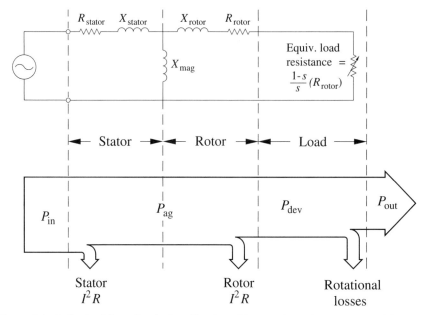

Figure 9.6 Induction Motor Equivalent Circuit and Power Flow Diagram *Adapted from Hambley–Electrical Engineering: Principles and Applications*

It is important to realize that losses in an induction motor occur in three places only: power losses in the stator windings, power losses in the rotor conductors, and mechanical or rotational losses. The driven load is modeled as an electrical resistance although it is actually a mechanical term and not an electrical term. The driven load is modeled as the product of the rotor resistance and a factor $(1-s)/s$, where s in the per-unit slip. Since slip is variable, the load resistance is shown as a variable resistor. The per-unit value of this variable resistance is much larger than the real resistances R_{stator} and R_{rotor}. The power dissipated in this variable "resistor" represents the power (brake horsepower) delivered to the load.

9.3 SPEED–TORQUE CURVES

A *speed–torque curve* shows the amount of torque produced at a given speed (in the case of a motor) or the amount of torque required to maintain a given speed (in the case of a driven load). An induction motor has an interesting speed–torque characteristic, shown in Figure 9.7.

At zero speed, a *locked-rotor torque* is produced. This is the torque magnitude that starts the acceleration of the rotor. The minimum torque produced between zero speed and the breakdown point is called the *pull-up torque*. The maximum torque produced by the motor is the *breakdown torque*. After the breakdown point is reached, the acceleration has essentially ended and the motor is in its normal operating range. The torque characteristic falls abruptly and close linearly. At synchronous speed, where slip is zero, there is no torque produced.

9.3 SPEED–TORQUE CURVES

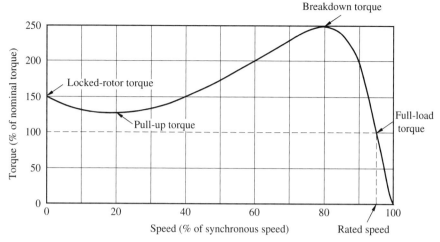

Figure 9.7   Speed–Torque Curve for a Polyphase Induction Motor

The rotor design has a profound effect on the characteristics of the speed–torque curve. By varying the design of the rotor bars, the speed–torque curve can be shaped to better match the speed–torque curve of the load. Rotor bars with a rectangular cross-section set just below the surface of the rotor correspond to NEMA Designs "A" and "B." Design "C" uses a double-cage design, while Design "D" uses high-resistance conductor material sometimes with a nonrectangular cross-section. IEC has developed a similar set of standard designs that approximately parallel the NEMA standards. Examples of the NEMA standard rotor designs are shown in Figure 9.8.

NEMA Design Class A motors are considered the standard induction motor design. The rotor bars are essentially rectangular and are set just beneath the surface of the rotor. This rotor design produces normal starting torque and starting current. Slip at rated torque is less than 5%, and the pull-up torque occurs at a speed of less than 20% of synchronous speed. Starting torque is 100–200% of full-load torque.

The largest concern with Class A motors is starting current, which is typically five to eight times the full-load current. When large inrush currents are not acceptable, Class B designs may be a better choice.

In NEMA Design Class B motors, the rotor bars are buried a bit more deeply beneath the rotor surface. This increases the leakage reactance, thereby reducing the

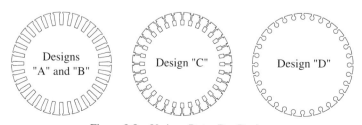

Figure 9.8   Various Rotor Bar Designs

starting current. Starting torque is comparable to the Class A design. Slip at rated torque is less than 5%, and the pull-up torque is less than that of the Class A design.

NEMA Class C designs offer high starting torque with fairly low starting current. The pull-up torque is less than that of Class A or B motors, but the starting torque is up to 250% of the full-load torque. Class C motors use a double-cage rotor bar design, with one set of bars just beneath the rotor surface and the second buried deeply in the rotor. This elaborate rotor design substantially increases the cost of the motor. Class C designs are commonly used for the loads requiring high starting torque.

NEMA Class D motors provide even higher starting torque than Class C designs, with starting torques in excess of 275% of rated torque. The starting current is fairly low, but slip at rated torque is very high. Typical Class D motors have a slip at rated torque of 7–10% although values approaching 20% are sometimes found. High starting torque with low starting current requires a high-rotor resistance. To increase the resistance of the rotor bars, materials such as bronze are often used. Because of the very high slip, Class D motors are often used in conjunction with a mechanical energy storage device such as a flywheel.

NEMA Class E motors are high-efficiency motors with speed–torque characteristics similar to those of Class A and B designs. In the United States, the Energy Independence and Security Act of 2007 (EISA) mandates higher energy efficiency requirements for general purpose motors rated from 1 to 200 hp manufactured after December 19, 2010. NEMA Class E is an industry solution for the requirements stipulated by EISA.

A comparison of the speed–torque curves of NEMA Class A through E motors is shown in Figure 9.9.

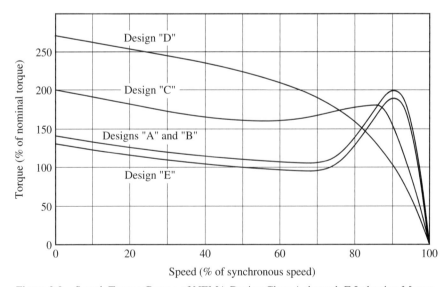

Figure 9.9 Speed–Torque Curves of NEMA Design Class A through E Induction Motors

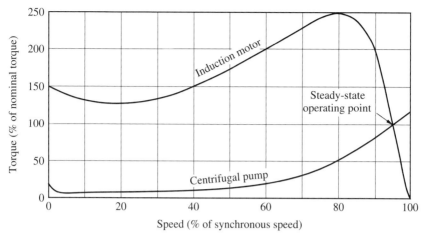

Figure 9.10 Speed–Torque Curves for a Motor and Pump

9.4 MOTOR STARTING TIME

The amount of time required for an induction motor to accelerate its driven load to rated speed can be determined by examining the speed–torque curves for both the motor and the load. The speed–torque curves for the induction motor and its driven load must be plotted on the same graph, as in Figure 9.10.

At any speed, the difference between the torque provided by the motor and the torque required by the load represents the *accelerating torque*:

$$T_{acc} = T_{motor} - T_{load} \tag{9.6}$$

The accelerating torque available is one of the two determining factors that define the amount of time required to accelerate the load from one speed to another. The *moment of inertia* of the mechanical system, meaning the motor and driven load, is the other. If the motor is coupled to the load through a clutch, gearbox, or similar device, the moment of inertia of the coupling device must also be considered.

The time t, in seconds, required to accelerate the mechanical system from one speed to another is given by Eq. (9.7), where WR^2 is moment of inertia of the mechanical system in pound-feet squared (lb-ft^2) and T_{acc} is the average accelerating torque available during the acceleration period in pound-feet (lb-ft). ΔRPM is the speed difference in revolutions per minute, and 308 is a unit correction factor:

$$t = \frac{WR^2 \times \Delta RPM}{308 T_{acc}} \tag{9.7}$$

The dimensional analysis of the unit conversion factor (308) is shown in Eq. (9.8):

$$\frac{32.2 \text{ ft/s}^2}{2\pi \text{ radians/revolution}} = 308 \frac{\text{revolution-ft}}{\text{min-s}} \tag{9.8}$$
$$60 \text{ s/min}$$

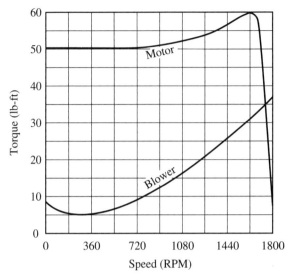

Figure 9.11   Induction Motor and Blower Speed–Torque Curves

Consider an example of an induction motor driving a blower. It is given that the WR^2 values for the blower and motor are 15 lb-ft^2 and 3.26 lb-ft^2, respectively, so the WR^2 of the mechanical system is

$$WR^2_{mech\ sys} = WR^2_{blower} + WR^2_{motor} = 15 + 3.26 = 18.26\ \text{lb-ft}^2. \tag{9.9}$$

The speed–torque curves for the motor and blower are required. Figure 9.11 shows both curves on the same graph.

Since the accelerating torque varies over the acceleration period, the acceleration period is broken into intervals. There is no need for the intervals to be uniform, but we will use equal intervals of 180 RPM, except for the last interval, which is smaller due to the slip speed. For each interval, the average accelerating torque for that interval is calculated by averaging the accelerating torque at the beginning of the interval with the accelerating torque at the end of the interval. Figure 9.12 shows the 10 intervals and notes the average accelerating torque for each interval at the top of the graph.

The total acceleration time can be approximated by applying Eq. (9.6) to each interval of the acceleration period, as shown in Eq. (9.10):

$$t = \frac{18.26}{308}\left(\frac{180}{43} + \frac{180}{45} + \frac{180}{44} + \frac{180}{42} + \frac{180}{40} + \frac{180}{37} + \frac{180}{34} + \frac{180}{32} + \frac{180}{30} + \frac{126}{14.5}\right)$$
$$= 3.06\ \text{seconds} \tag{9.10}$$

9.5   CABLE SIZING

The NEC requires circuits supplying single motors to have an ampacity of not less than 125% of the motor full-load current rating. Exceptions to this requirement may

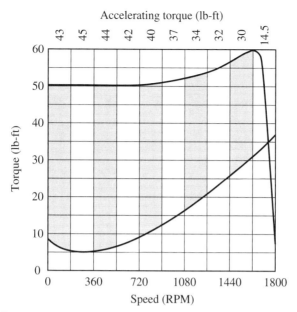

Figure 9.12 Speed–Torque Curves with Accelerating Period Broken into Intervals

be made for short time, intermittent, periodic, or varying duty motors, and for some DC motors. When multiple motors are supplied from a single circuit, that circuit must have an ampacity of at least the sum of all the full-load currents, plus 25% of the full-load current of the largest motor in the group, plus the ampere ratings of any other loads fed from the circuit. Some exceptions can be made to this requirement, including the subtraction of full-load currents of motors that are interlocked such that they cannot run while other motors on the circuit are running.

For example, consider three motors each with a full-load current rating of 24 A designed in such a way that through interlocks, only two of the three motors can run at the same time. In this case, 48 A plus 25% of 24, or 6, would be the full-load current used to determine cable size. If no other load is supplied from this circuit, the required ampacity would be

$$I_{req} = 24 + 24 + (0.25)(24) = 54 \text{ A}. \tag{9.11}$$

9.6 MOTOR PROTECTION

Two different protection schemes must be implemented to protect a motor properly. First, the cable supplying the motor must be protected against damage due to fault current. This is known as *circuit protection* and is dictated by the NEC. In addition to circuit protection, the insulation on the motor windings must be protected from thermal damage due to overloads. This is known as *winding protection* and is also

TABLE 9.4 Motor Circuit Protective Device Maximum Ratings and Settings

Type of Motor	Percent of Full-Load Current			
	Nontime–Delay Fuse	Dual Element (Time-Delay) Fuse	Instantaneous Trip Breaker	Inverse Time Breaker
Single phase	300	175	800	250
Polyphase squirrel cage	300	175	800[a]	250
Polyphase synchronous	300	175	800	250
Wound rotor	150	150	800	150
Constant voltage DC	150	150	250	150

[a] Protection on circuits supplying "Design E" motors can be set to 1100% of full-load current.

governed by the NEC. While these two protection schemes serve different purposes, they must be coordinated with each other to protect properly the motor and its supply circuit.

9.7 CIRCUIT PROTECTION

The NEC defines maximum ratings and settings for motor circuit protective devices. These are summarized in Table 9.4. Consult the current edition of the NEC for specific details and exceptions to these ratings and settings.

9.8 WINDING PROTECTION

Motor windings must be protected from thermal damage due to excessive current caused by a mechanical overload or failure to start. The thermal protection can be integral with the motor or can be provided by an overload relay. The maximum ratings and settings for these devices as stipulated by the NEC are shown in Table 9.5.

Overload relays shall be set to not more than 140% of full-load current for motors with a service factor of 1.15 or greater, or a temperature rise of 40°C or less. Overloads for all other motors shall be set to not more than 130% of full-load current.

TABLE 9.5 Thermal Protection Integral with Motor

Full-Load Current	Thermal Protection Rating in Percent of Full-Load Current
Less than 9 A	Not more than 170%
9–20 A	Not more than 156%
More than 20 A	Not more than 140%

9.9 MOTOR STARTING METHODS

Many methods of motor starting exist. The simplest method, *across-the-line* starting, applies full voltage across the motor windings. If full-voltage starting results in unacceptably high starting current, reduced voltage starting methods may be necessary. These methods include *resistor starting, reactor starting,* and *autotransformer starting.* W*ye–delta starting* and *part-winding starting* also can be utilized but require specially built motors.

9.9.1 Across-the-Line

The simplest means of starting a motor is to apply full-line voltage across the stator windings. Simply closing a contactor or circuit breaker that connects the motor to a power source can do this. This method of starting is called *across-the-line* starting.

Since full voltage is applied to the motor terminals, rated torque is available at starting time. In reality, some voltage drop and resulting torque reduction will occur when the motor starts since torque is proportional to voltage squared. Across-the-line starting not only provides the maximum starting torque of all the starting methods available but also provides the highest starting current—typically from five to seven times rated full-load current. If this starting current produces a voltage drop that is too large for the system to tolerate, a reduced voltage starting method must be investigated. But, unless the motor is very large or the system is very weak, across-the-line starting is usually acceptable and is by far the most common method of starting a motor.

A simple across-the-line starting scheme is shown in Figure 9.13. When the START pushbutton is pressed, the S coil picks up, seals in the START pushbutton, and closes the contactor, thus energizing the motor.

9.9.2 Reduced Voltage Starting

When starting current from across-the-line starting becomes excessive, a reduced voltage starting method can be used. Because a starting motor is a constant impedance device, as the voltage applied to a starting motor decreases, the starting current decreases proportionately. Unfortunately, because torque is proportional to voltage squared, starting torque decreases quadratically with applied voltage. This relationship can limit the application of reduced voltage starters, particularly for the motors driving high-inertia loads. Table 9.6 assumes the motor that has a starting current equal to six times the rated full-load current.

Another issue must be addressed before discussing specific reduced voltage starting methods. Although a starting motor behaves as a constant impedance device, a running motor behaves as a constant kVA device. During the starting process, the motor characteristics change significantly. This change is caused by the production of counter EMF in the motor. When reduced voltage motor starting is used, at some point before the motor begins to behave like a constant kVA device, the applied voltage must be increased, or current will increase to potentially damaging levels.

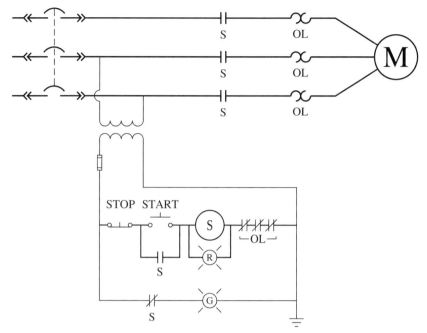

Figure 9.13 Across-the-Line Starter

This increase in current would cause the overload relays to operate, thereby shutting down the motor.

Several methods can be used to determine when to transition from reduced voltage to full voltage. The simplest method is to use a timer. If the approximate starting time is known, the transition can be timed so it occurs when the rotor is at approximately 90% of rated speed. This method can be problematic if the motor takes longer to start than anticipated. Lengthened starting times could be caused by low bus voltage, mechanical problems with the motor or driven load, or abnormal operating conditions. If a timer is used to transition voltage on a motor that is taking longer than anticipated to start, the transition will occur at too low a rotor speed. This premature transition will result in high fundamental frequency and harmonic current, which could disrupt other loads in the plant. A more reliable, but more complex,

TABLE 9.6 Starting Current and Starting Torque as a Function of Voltage

Voltage (%)	Starting Current (% of Full-Load Current)	Starting Torque (%)
100	600	100
90	540	81
80	480	64
70	420	49
60	360	36
50	300	25

transition means is to use a tachometer or speed-sensing relay to determine, based on rotor speed, exactly when to transition to full voltage. This method is used primarily on very large motors because of its relative complexity and expense.

Another transition method is to monitor motor current and transition to full voltage based on the actual current flowing to the motor. This type of transition can be accomplished either with conventional relays or with a programmable logic controller.

When it is necessary to transition to full voltage, two methods of transition are possible—an *open transition* or a *closed transition*. An open transition interrupts (opens) the reduced voltage power circuit then re-energizes the motor at full voltage, while a closed transition applies voltage to the motor at all times. Open transitions can cause serious transient and harmonic disturbances on the electrical system and have the potential to produce very high mechanical torques that can damage the driven equipment. Consequently, open transition starters are seldom used.

9.9.2.1 Resistor Starting One method of providing a reduced voltage for motor starting is to drop voltage across a resistor or a series of resistors. Resistor starters provide smooth acceleration and often are used where abrupt torque changes must be avoided. This is particularly important for the loads driven by a belt or gear train since sudden torque application to these loads could damage the driven equipment. Resistor starters are easy to implement since interlocking of the contactors is not required. Also, resistor starters provide a closed transition from reduced voltage to full voltage while improving the starting power factor by inserting a resistance in series with the inductive motor load. On the other hand, the power loss in the resistors can impact the starting duty cycle of the motor due to excessive heating of the resistors. The torque efficiency is also low, and modification of the starting characteristics cannot be done easily in the field. However, these disadvantages can be mitigated by careful design. Resistor starters are most commonly applied to fairly small motors at relatively low voltages.

Sizing of the resistor is crucial when implementing a resistor starter. The ohmic value of the resistor determines the magnitude of the voltage that will be applied to the motor for starting. Once the ohmic value is known, the power rating for the resistors can be determined by multiplying the starting current (with the starting resistors in the circuit) squared by the ohmic value of the starting resistors. This wattage will be very high (tens or hundreds of kilowatts), but the resistor is only in the circuit until the transition to full voltage. Starting resistors are rated in kilowatts for a specified number of seconds. Liquid resistors, where electrodes are submersed in a conductive brine, are best suited to this type of duty.

Since reduced voltage motor starters are used when the current drawn during an across-the-line start is unacceptably high, a good strategy for sizing the resistor in the resistor starter is to drop as little voltage across the resistors as possible, resulting in the highest acceptable starting current. This approach will lead to a minimum resistor size.

Consider the example of a 230-V, 100-hp motor with a 1550-A locked-rotor current at rated voltage. It is determined that 1050 A is the maximum starting current that can be tolerated by the system. A starting resistor can be sized as follows.

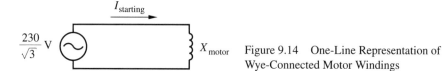

Figure 9.14 One-Line Representation of Wye-Connected Motor Windings

If the motor windings are wye-connected, line-to-neutral voltage is impressed across each winding. This results in a one-line diagram as shown in Figure 9.14.

Neglecting the resistance of the motor, which is a reasonable assumption given the large X/R ratio of an induction motor, the reactance of the motor which limits the starting current to 1550 A is

$$X_{motor} = \frac{230/\sqrt{3}}{1550} = 0.085671 \ \Omega \tag{9.12}$$

The starting resistor will be in series with the motor winding. Solving Eq. (9.12) for impedance and adjusting it to include a resistance R is series with the motor,

$$\frac{(230/\sqrt{3}) \ \underline{/0°}}{1050 \ \underline{/\phi}} = 0.085671 \ \underline{/90°} + R \ \underline{/0°} \tag{9.13}$$

The magnitude of the impedance must equal the magnitude of the phasor on the left side of the equal sign, so

$$\frac{(230/\sqrt{3})}{1050} = \sqrt{0.085671^2 + R^2} \tag{9.14}$$

Solving for R,

$$R = 0.093029 \ \Omega \tag{9.15}$$

If the motor windings are delta-connected, a delta-to-wye transformation can be used to reconfigure the motor windings as wye-connected, allowing the above process to also be used for delta-connected windings.

A control circuit for a resistor starter is shown in Figure 9.15. When the START pushbutton is pressed, the S coil picks up, sealing in the START pushbutton, lighting an amber lamp to indicate that the motor is energized at reduced voltage, and closing the S contactors which energize the motor through a series resistor. The voltage drop across the resistor reduces the voltage at the motor terminals.

A speed-sensing switch on the motor shaft changes state when the motor speed reaches the desired speed to transition to full voltage. This switch is most likely a centrifugal switch, where springs hold the electrical contacts open but when the centrifugal force due to the shaft rotation exceeds the force exerted by the spring, the contacts close. The speed switch contacts are labeled "SS." When SS closes, coil R is picked up and sealed in. The S coil also drops out when the R coil picks up. Now, the resistors are bypassed and full voltage is applied to the motor.

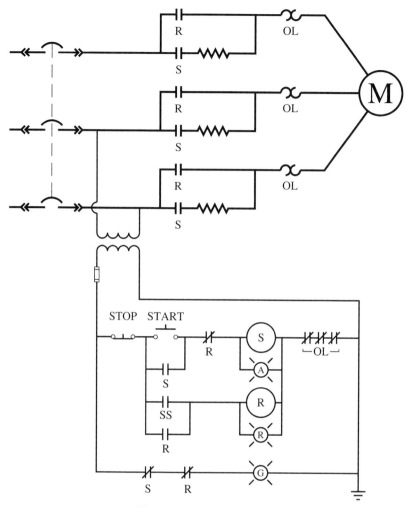

Figure 9.15 Resistor Starter

9.9.2.2 Reactor Starting Another method of providing a reduced voltage for motor starting is to drop voltage across an inductive reactance or reactor. A control circuit for a reactor starter is shown in Figure 9.16.

This starting method is similar to resistor starting with minor differences. The thermal heating of the reactors is much less of a concern than the heating of the resistors in a resistor starter. The starting power factor is not improved as with the resistor starter and is actually worsened by the addition of an inductive reactance in series with the motor. Reactor starters do provide a closed transition from reduced voltage to full voltage and are relatively simple in design. The operation of the control circuit for the reactor starter is identical to that of the resistor starter.

The method for sizing the reactors is similar to the method for sizing the starting resistors in a resistor starter. Using the same example as for the resistor starter, the

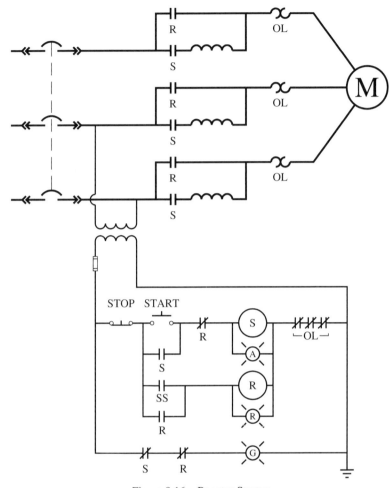

Figure 9.16 Reactor Starter

starting reactor can be sized as follows. In the resistor starter example, the reactance of the motor was found to be 0.085671 Ω.

The starting reactor will be in series with the motor winding. So to limit the starting current to 1050 A,

$$\frac{(230/\sqrt{3})\,\underline{/0°}}{1050\,\underline{/\phi}} = 0.085671\,\underline{/90°} + X\,\underline{/90°} \quad (9.16)$$

Solving Eq. (9.16) for X gives, $X = 0.040796$ Ω. At 60 Hz, the inductance of this reactor is

$$L = \frac{0.040796}{2\pi(60)} = 0.1082\,\text{mH} \quad (9.17)$$

9.9.2.3 Autotransformer Starting

A more efficient means of providing a reduced voltage for motor starting is to use an autotransformer to reduce the line voltage. Autotransformer starters provide the maximum torque per ampere of all of the reduced-voltage starters. If the autotransformer has taps, the starting characteristics can be adjusted easily in the field. Either an open or closed transition from reduced voltage to full voltage can be provided. The autotransformer starter is the most complex of the reduced-voltage starters because of the intricate sequencing of the contactors. It also requires more contactors than the other starting methods. The presence of the autotransformer in the starting circuit can also reduce the number of allowable motor starts per hour.

A control circuit for a closed-transition autotransformer starter is illustrated in Figure 9.17. When the START pushbutton is pressed, the AUX coil picks up, seals in the START pushbutton, and closes the type A contact in the second rung, which picks up the N coil. The N coil closes the N contactors that form the neutral of the autotransformer connection. Note that this neutral is not grounded. Grounding it would provide unacceptably high fault currents if a ground fault occurred during starting. Many times, an N contactor is installed in only two of the phases, as a contactor in the third phase is redundant and unnecessary.

When the N coil picks up, the S coil also picks up, illuminating an amber lamp to indicate that the motor is energized at reduced voltage. When the motor reaches a speed suitable for transitioning to full voltage, the speed-sensing switch closes. This picks up the T coil, which drops out the N coil. This sequence opens the autotransformer neutral, leaving a section of the autotransformer winding in the circuit to act as a reactor for smoothing the current transient produced as full voltage is applied to the motor. Picking up the T coil also picks up the R coil, which closes the R contactor in the power circuit, applying full voltage to the motor. Picking up the R coil also lights the red lamp and causes the S coil to drop out. The autotransformer remains energized, although it carries no current.

9.9.2.3.1 Advantages of the Autotransformer Starter

Because the autotransformer starter produces its reduced voltage using an autotransformer instead of simply dropping voltage across a passive element, as is the case with the resistor and reactor starters, a significant improvement in efficiency should be expected. It turns out that several significant advantages are achieved when using an autotransformer starter. The cost of these advantages is primarily the expense of the autotransformer and the need for the third set of contactors in the power circuit.

Perhaps the greatest benefit of the autotransformer starter is greater torque availability for the same amount of inrush current when compared to resistor or reactor starters. This benefit is due to the fact that the inrush current drawn by the motor (I_m) and the starting current drawn from the system ($I_{starting}$) are of different magnitudes.

Consider the diagram in Figure 9.18, showing one phase of a wye-connected autotransformer, its source, and its load. This figure will be used in the example in the next section.

274 CHAPTER 9 MOTOR APPLICATION

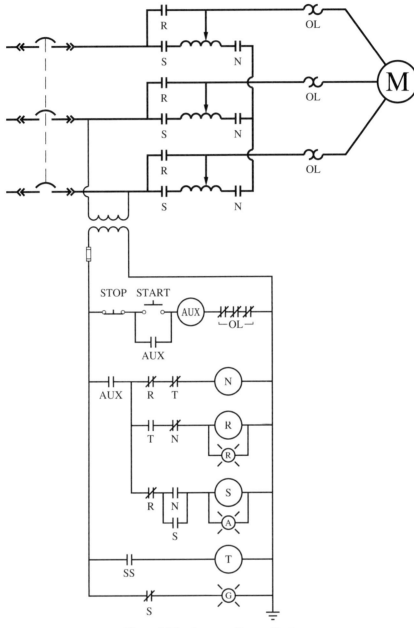

Figure 9.17 Autotransformer Starter

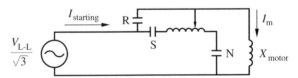

Figure 9.18 One Phase of a Motor Starting Autotransformer Power Circuit

9.9.2.3.2 Determining Autotransformer Tap and Starting Torque Consider the following example: A 100-hp, 460-V motor is unable to start across the line without an unacceptable voltage drop. When the motor is started at reduced voltage ($V_{\text{start}} = 0.35$ p.u.), the motor draws 200% of full-load current at locked-rotor conditions. Its starting torque is 20% of rated full-load torque. Determine the autotransformer tap needed to limit the locked-rotor current to 150% of full-load current, and determine the starting torque.

Since a starting motor is a constant impedance load, the locked-rotor current at full voltage is the locked-rotor current at the reduced voltage times the inverse of the voltage reduction ratio:

$$I_{L(100\% \text{ voltage})} = 2.0 \left(\frac{1.00}{0.35}\right) = 5.71 \text{ p.u.} \tag{9.18}$$

Since torque is proportional to voltage squared, the starting torque at full voltage is

$$T_{S(100\% \text{voltage})} = 0.2 \left(\frac{1.00}{0.35}\right)^2 = 1.633 \text{ p.u.} \tag{9.19}$$

At locked-rotor conditions, the current drawn by the motor is directly proportional to the voltage applied to the motor. This means that the starting torque at 35% voltage is

$$T_{S(35\% \text{voltage})} = k_1 (2.0)^2, \tag{9.20}$$

where k_1 is a proportionality constant. Let the turns ratio of the autotransformer (on the appropriate tap) be represented by the variable a. Since we are limiting the current drawn from the source to 1.5 times the full-load current rating, the current flowing to the motor will be $1.5 \times a$. Then, the starting torque when reduced voltage is applied by the autotransformer is

$$T_{S(\text{auto})} = k_1 (1.5\, a)^2. \tag{9.21}$$

At locked-rotor conditions, torque is also proportional to voltage squared, since voltage and current are directly proportional. Therefore, Eq. (9.20) can be rewritten in terms of voltage as

$$T_{S(35\% \text{voltage})} = k_2 (0.35)^2, \tag{9.22}$$

where k_2 is another proportionality constant. Similarly, Eq. (9.21) can be rewritten in terms of voltage as

$$T_{S(\text{auto})} = k_2 \left(\frac{1}{a}\right)^2. \tag{9.23}$$

Dividing Eq. (9.21) by Eq. (9.20) gives

$$\frac{T_{S(\text{auto})}}{T_{S(35\% \text{voltage})}} = \frac{k_1 (1.5a)^2}{k_1 (2.0)^2} = \frac{9a^2}{16}. \tag{9.24}$$

Similarly, Eq. (9.23) can be divided by Eq. (9.22) to yield

$$\frac{T_{S(auto)}}{T_{S(35\%voltage)}} = \frac{k_2 \left(\frac{1}{a}\right)^2}{k_2(0.35)^2} = \frac{400}{49a^2}. \tag{9.25}$$

Equating Eq. (9.24) and Eq. (9.25),

$$\frac{9a^2}{16} = \frac{400}{49a^2}. \tag{9.26}$$

Solving for a gives

$$441a^4 = 6400 \tag{9.27}$$

or

$$a = 1.95. \tag{9.28}$$

So the percent tap required on the autotransformer is 1/1.95 or 51.3%.

Rearranging Eq. (9.25), we can obtain the following expression for the starting torque when the motor is supplied by an autotransformer set on a 51.3% tap:

$$T_{S(auto)} = T_{S(35\%voltage)} \left[\frac{400}{49\,(a)^2}\right] = 0.20 \left[\frac{400}{49\,(1.95)^2}\right] = 0.429 \tag{9.29}$$

This represents a starting torque of 42.9% of rated torque when a reduced voltage of 51.3% is applied to the motor using an autotransformer.

9.9.3 Wye–Delta Starting

A starting method that requires no special components for reducing voltage is the wye–delta start. A specially built motor, however, is required for this starting method, because access to each end of all three stator windings is required. Most motors have three terminals for connection, but the wye–delta starting method requires a six-terminal motor. When the motor is initially energized, the windings are configured in wye with a floating (ungrounded) neutral. The neutral point of the wye is left ungrounded to limit fault current in the event of a ground fault during starting. Since the windings are connected in wye, each winding will have $1/\sqrt{3}$ times the source line-to-line voltage applied across it. When the rotor reaches the appropriate speed, the windings are reconnected in delta via an open transition. This impresses full line-to-line voltage across the windings. In essence, the power system itself produces the voltage reduction instead of relying on a resistor, reactor, or autotransformer.

Wye–delta starters do not limit the motor's duty cycle, and they provide a high torque efficiency without torque fluctuations or abnormal winding stresses as with a part-winding starter, described in the next section. Wye–delta starters, however, do require a specially built (six-terminal) motor and a rather elaborate control system.

A control circuit for a wye–delta starter is shown in Figure 9.19. When the START pushbutton is pressed, the S coil picks up, which picks up the 1S coil and closes the two 1S contactors that configure the motor windings in wye be connecting motor terminals 4, 5, and 6. The 1M coil also picks up, applying line voltage to motor

9.9 MOTOR STARTING METHODS 277

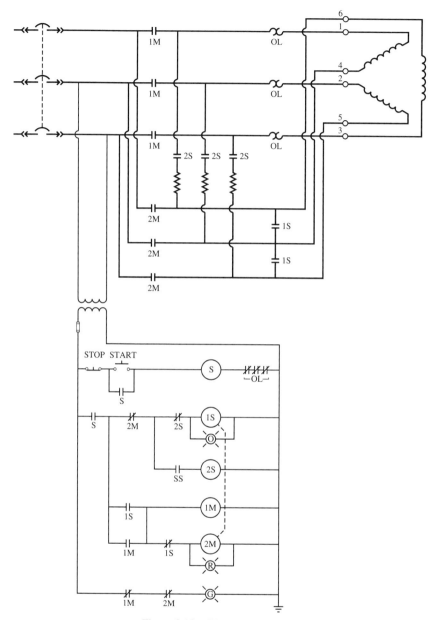

Figure 9.19 Wye–Delta Starter

terminals 1, 2, and 3. The S coil also seals in the START pushbutton. An orange lamp is illuminated to indicate that the motor is energized with its windings in a wye configuration.

When the speed switch closes, coil 2S picks up. This closes the 2S contactors in the power circuit which inserts transient-damping resistors into the circuit and causes the 1S coil to drop out, which opens the neutral of the wye-connected motor windings. Coil 2M also picks up, which connects the motor windings in delta and drops out the 2S coil, thus removing the damping resistors from the circuit. Note that in addition to the electrical interlock between the 1S and 2M coils, a mechanical interlock is also provided, since picking up both coils simultaneously would create a three-phase fault.

9.9.4 Part-Winding Starting

Another method of decreasing the starting current of a motor is to increase the winding impedance during starting. This can be done by building the motor with a stator composed of two sets of parallel windings. At starting, only one of the parallel sets of windings is energized, offering relatively high impedance. As the motor approaches rated speed, the remaining parallel set of windings is energized, thus lowering the winding impedance.

Transition can be accomplished using two sets of contactors and a speed switch to trigger the closed transition when adequate rotor speed is attained. Figure 9.20 shows a control circuit for a part-winding starter.

When the START pushbutton is pressed, the 1M coil picks up and seals in the START pushbutton. The 1M starting contactors close, energizing the starting windings. An amber indicating lamp lights to show that the starting windings are energized. When the speed switch closes, the 2M coil picks up and seals in. This closes the 2M run contactors, energizing the run windings. A red indicator lamp lights to show that the run windings are energized. Note that when the motor is running, both the start and run windings remain energized, and both the amber and red indicator lamps are illuminated.

9.9.5 Solid-State Starting Options

Solid-state electronics provide another means of starting motors in applications where across-the-line starting causes objectionably high starting currents. *Soft starters* use power semiconductors to control the voltage applied to the motor. *Variable-frequency drives (VFDs)*, used primarily to control the speed of an AC motor by varying the frequency of the applied voltage, are also useful to control starting characteristics.

9.9.5.1 Soft Starters A soft starter is another type of reduced voltage starter for induction motors. The soft starter is similar to a primary resistance or primary reactance starter in that it is in series with the supply to the motor. The current into the starter equals the current out. The soft starter uses solid-state devices to control the current flow and therefore the RMS voltage applied to the motor. Soft starters can be connected in series with the line voltage applied to the motor or can be connected inside the delta of a delta-connected motor, controlling the voltage applied to each winding.

Voltage control is achieved by means of solid-state switches in series with each motor winding. These switches can be configured in several ways, as shown in Figure 9.21.

9.9 MOTOR STARTING METHODS 279

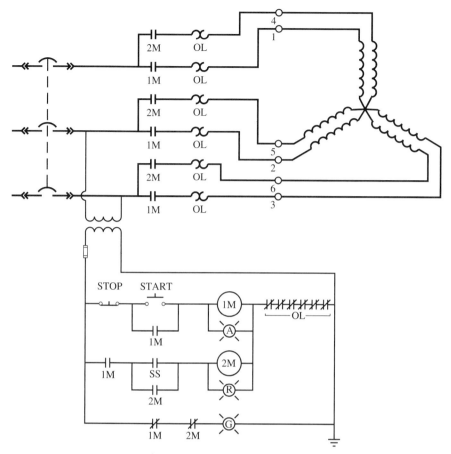

Figure 9.20   Part-Winding Starter

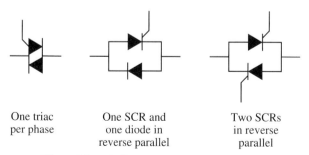

One triac
per phase

One SCR and
one diode in
reverse parallel

Two SCRs
in reverse
parallel

Figure 9.21   Solid-State Switch Configurations

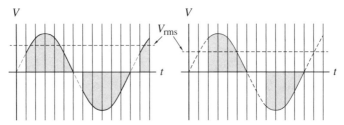

Figure 9.22 RMS Voltage Control by Varying Conduction Angle

The solid-state switches are phase-controlled in a similar manner to that of a light dimmer, in that they are turned on for a part of each cycle. The average voltage is controlled by varying the conduction angle of the switches. Increasing the conduction angle will decrease the RMS output voltage, as shown in Figure 9.22.

Controlling the RMS voltage in this manner has several advantages over the means used by other reduced voltage starters. The efficiency is much higher than the resistance-type starter, since the conduction losses in the solid-state switches will be on the order of 1% of the power dissipated by the resistors in the resistor starter. Perhaps the biggest advantage is the ability to control precisely the RMS voltage by varying the conduction angle. This control can be done dynamically, constantly adjusting the conduction angle during the starting process to produce virtually any speed–torque and inrush current characteristic imaginable.

The disadvantages of the soft starter are the initial cost, which increases dramatically with operating voltage and horsepower, and power quality concerns due to the switching of the power semiconductors. The harmonics produced by the soft starter can be mitigated by good starter design, but these measures further add to the cost. Since the soft starter is bypassed when the motor is up to speed, any power quality concerns only apply during starting, which is typically a very brief period.

9.9.5.2 Variable-Frequency Drives The speed of standard induction motors can be controlled by the variation of the frequency of the voltage applied to the motor. To prevent saturation of the magnetic circuit, the voltage applied to the motor must also change with the frequency to maintain a constant volts-per-Hz ratio.

A VFD is a device put in series with a motor to adjust both the frequency and the voltage. In addition to aiding the starting process much like a soft starter, the VFD also provides speed and torque control. Unlike the soft starter, the VFD remains in the circuit at all times, not just during starting.

The omnipresence of the VFD means that its advantages of speed and torque control are always available. Unfortunately, its disadvantages are also present at all times. Care must be taken when designing the VFD to mitigate harmonic production by incorporating adequate filtering. Reliability of the motor/drive system must also be studied, since the probability of failure of the solid-state components in the drive will be much higher, possibly more than an order of magnitude higher, than the probability of failure of the motor itself. In applications where reliability is critical, the use of a VFD may be called to question.

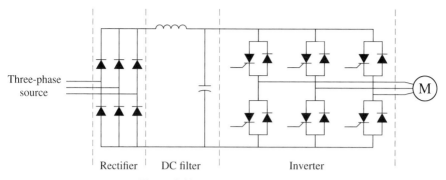

Figure 9.23 VFD Block Diagram

The most serious issue with having the VFD in the circuit at all times may be the efficiency of the drive and its impact on the overall efficiency of the motor/drive/load system. While the power semiconductors will decrease the electrical efficiency due to conduction and switching losses, the ability to control the motor speed will likely increase the efficiency of the mechanical part of the system. If the motor is used to drive a fan, for example, use of a VFD may eliminate the need for input and output dampers in the ducting, which would greatly reduce energy losses in the ductwork. Even though the efficiency of the drive/motor system may be considerably lower than the efficiency of the motor itself, the overall system efficiency (motor/drive/load) must be considered. In many applications, the efficiency reduction caused by using a VFD is offset by the efficiency improvements in the mechanical part of the system.

VFDs contain three major parts: a rectifier, a DC filter, and an inverter. A block diagram depicting these components is shown in Figure 9.23.

The rectifier stage uses diodes to rectify the incoming voltage, which is typically a three-phase voltage. Since the output of the rectifier stage is DC, there is no need for the AC source to be three phase. In fact, a single-phase source would work, but as the number of phases comprising the input voltage decreases, the amount of ripple in the DC output increases, thus increasing the filtering requirement in the second stage. The effect of the number of phases on the DC ripple is shown graphically in Figure 9.24.

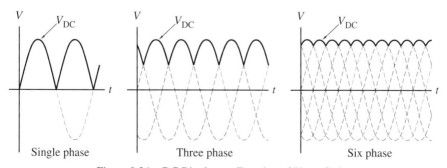

Figure 9.24 DC Ripple as a Function of Phase Order

To reduce the filtering requirement, a high-phase-order input voltage can be provided. Typically, 6- and 12-phase inputs are used. The high-phase-order voltages are produced using special transformer connections as described in Chapter 3.

The DC filter stage uses capacitance to smooth the ripple produced by the rectifier stage. Using only capacitance in the DC filter would result in large currents flowing from the rectifier whenever the capacitor voltage is less than the DC bus voltage. These current surges would result in the production of a rich harmonic spectrum. By adding a choke (inductor) to the filter, the harmonic production can be minimized.

The inverter stage for three-phase output uses six solid-state switches arranged in pairs, each pair making up a *totem pole* configuration. Two totem pole stacks are needed for two-phase output, while a single totem pole stack can produce single-phase output. In small VFDs operating at low voltages, metal-oxide-semiconductor field-effect transistors (MOSFETs) can be used as the switches. As the voltage and horsepower ratings of the drive increase, insulated gate bipolar transistors (IGBTs) and thyristors or silicon-controlled rectifiers (SCRs) are commonly used as switches.

The gating of the switches in each totem pole is often controlled using *pulse-width modulation* (PWM). A train of square waves having varying periods can be generated to represent a sine wave of any frequency by modulating a sine wave produced by an oscillator with a triangle wave of a frequency considerably higher than that of the sine wave. The triangle wave frequency is called the *carrier frequency*. When the carrier has a greater magnitude than the sine wave, a digital zero signal level is generated, and when the sine wave has a greater magnitude than the carrier, a digital one signal level is generated. The resulting square wave is used as a gate input to the totem pole circuit of solid-state switches. A slight gating delay must be used to assure that both switches in the totem pole are never conducting at the same time, as this would short-circuit the DC bus. The PWM carrier and sine wave are shown along with the generated gating signal in Figure 9.25.

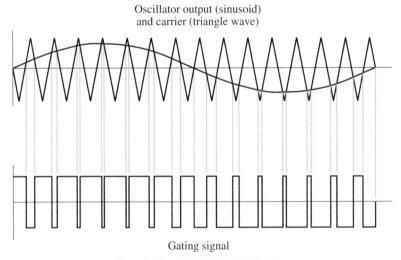

Figure 9.25 Pulse-Width Modulation

In most variable speed drives, a carrier frequency of between 1 and 16 kHz is used. While the higher frequency carriers have the advantage of lying outside the range of human hearing and are therefore quiet, they also subject the solid-state switches to increased stress and losses due to the high switching frequency. To mitigate the effects of the high switching frequency, the switches must be derated. The PWM process also generates high-order harmonics, which can be problematic to both the motor and the rest of the electrical system. To mitigate the effects of these harmonics, additional filtering can be provided on the output of the inverter stage. Surge arresters are commonly installed on motor circuits in conjunction with VFDs, especially at medium voltages. And the motor windings can be specially designed and insulated to better tolerate the high-frequency effects of the harmonics. Such motors are designated as *inverter-duty* motors. When a noninverter-duty motor is used in conjunction with a VFD, insulation failure in the first few turns of the stator windings is common. The high-frequency voltages which cause the failures are attenuated by the inductance of the windings a very short distance into the winding.

SUMMARY

Starting motors behave differently from running motors because of the lack of counter EMF generated in the machine until the machine approaches rated speed. The counter EMF becomes significant at 70–80% of synchronous speed. Circuit breakers or fuses protect the cable circuit supplying the motor, while heaters or protective relays protect the motor windings themselves. *Across-the-line* starting is used whenever possible, but *reduced-voltage* starting methods are sometimes necessary to limit inrush current or starting torque. Reduced voltage starters rely on the fact that a starting motor is a *constant impedance* device. Since a motor running at rated speed is a *constant kVA* device, full voltage must be applied to the motor before rated speed is reached. This can be done through an *open transition* where the motor is briefly de-energized, then re-energized at full voltage, or through a *closed transition* where the motor remains energized as the voltage source is changed. The closed transition is preferred since it provides a smoother torque characteristic and less transient harmonic content than the open transition.

FOR FURTHER READING

Hambley, A. R. *Electrical Engineering: Principles and Applications*, 6th edition. Pearson, 2013. ISBN: 978-0-133-11664-9.

Motors and Generators, National Electrical Manufacturers Association, NEMA Standard MG-1, 10 CFR 431, 2009.

National Electrical Code, National Fire Protection Association, NFPA 70, 2014.

QUESTIONS

1. What causes the current drawn by a motor to decrease as rotor speed increases?

2. How do the electrical characteristics of a starting motor differ from those of a motor running at or near rated speed?

3. Why does reducing the voltage applied to the motor terminals reduce the starting current of the motor?

4. When reduced voltage is used to start a motor, why must a transition to full voltage occur sometime before rated speed is reached?

5. Describe the two methods that can be used to transition from reduced to full voltage.

6. Briefly describe the three reduced voltage starting methods that drop voltage across a circuit element to produce the reduced voltage source.

7. Briefly describe the two special starting methods that require a specially built motor to reduce starting current. Explain how they work without reducing the voltage at the motor terminals.

8. Why is the neutral of a motor-starting autotransformer not grounded?

9. State several methods of determining when the transition from reduced to full voltage should be made when starting a motor.

10. Describe the two components of motor circuit protection, and how they can be implemented.

11. What happens if, at a given speed, the torque required by a driven load exceeds the torque capability of the motor driving the load?

12. What does the difference between the torque capability of the motor and the torque required by the load represent, and how does it affect starting time?

13. What are some advantages of using an autotransformer starting method instead of resistor or reactor starting methods? What are some disadvantages?

14. List some advantages and disadvantages of using a soft starter compared to nonelectronic reduced voltage starting methods.

15. Soft starters and VFDs share many similarities, but have some significant differences when used to supply a motor. List some of these similarities and differences.

PROBLEMS

1. Starting a large fan motor across the line results in an unacceptably high starting current, so a reduced voltage starting method is selected. The motor requires 75% of full-load torque to start. What is the lowest voltage that can be applied to the motor terminals for starting?

2. If a mechanical design change is made to the fan in Problem 1, which reduces the starting torque to 60% of full-load torque, what is the lowest voltage that can be applied to the motor terminals for starting?

3. A new industrial plant is planned. Eight large motors, two at 6000 hp, four at 3500 hp, and two at 2500 hp, are required. Propose an electrical system to supply these motors, specifying bus voltages, current ratings for circuit breakers, and cable sizes. State any assumptions made.

4. A squirrel cage motor having a full-load current rating of 125 A is fed from a motor control center. The 4-kV unshielded 90°C single-conductor cables comprising the circuit are run

in open cable tray. What is the smallest copper conductor that can be used to supply the motor? What is the smallest aluminum conductor that can be used to supply the motor?

5. What is the maximum allowable setting on an inverse-time circuit breaker protecting the circuit feeding the motor in Problem 4?

6. What is the largest time-delay fuse that can be used to protect the circuit feeding the motor in Problem 4?

7. What is the maximum allowable overload size that can be used to protect the windings of the motor in Problem 4?

8. Three motors, each with a full-load current rating of 30 A, are fed from a single circuit breaker. What is the minimum ampacity of the circuit conductors?

9. Four motors, two with a full-load current rating of 25 A and two with a full-load current rating of 15 A, are fed from a single circuit breaker. The circuit also supplies 20 A of lighting load. What is the minimum ampacity of the circuit conductors?

10. Design a control circuit for an autotransformer starter that energizes a 50% voltage tap for 10 seconds, then switches through a closed transition to a 75% voltage tap. After 20 seconds at 75% voltage, a second closed transition switches to full voltage.

11. Size a reactor for a reactor starter to limit the inrush current of a 460-V 450-hp motor with a locked-rotor current of 3900 A to a maximum of 3500 A.

12. Size an autotransformer for an autotransformer starter to meet the same criteria described in Problem 11.

13. Refer to Figure 9.11. If a 15% voltage reduction is applied to the motor, transitioning to full voltage at 1440 rpm, determine the starting time.

14. Repeat Problem 13 for a 20% voltage reduction.

15. Determine the firing angle δ for a soft starter to reduce the rms voltage to 85%.

CHAPTER **10**

LIGHTING SYSTEMS

OBJECTIVES

- Understand the fundamentals of lighting system design, particularly for outdoor areas and basic interior volumes
- Know the various lighting technologies available for commercial and industrial applications, and understand their advantages and disadvantages, and how they work
- Realize the variation in lighting requirements based on the use of the area being lighted
- Be aware of the ballast options that exist for low-pressure and high-intensity discharge lamps

10.1 FUNDAMENTALS

Illuminating spaces, both interior and exterior, is a fundamental engineering function. To perform this function properly, a working knowledge of the characteristics and behavior of light, and a thorough understanding of the use of the space are necessary.

Various activities that will take place within the space will have different illumination requirements. The Illuminating Engineering Society of North America (IESNA) has established criteria for many lighting applications, and the 9th edition of the *IESNA Lighting Handbook* will be used to guide the design examples in this chapter.

A common lighting measurement used in the design of lighting systems is *luminous intensity*. Luminous intensity, measured in *candelas* (cd), is one of the seven basic SI measurement units. The quantity of light emitted by a source is the *luminous flux*, and is measured in lumens, where one lumen equals one candela-steradian. The total luminous flux incident on a surface per unit area is called *illuminance*, and is measured in footcandles, where one footcandle equals one lumen per square foot. Illuminance is often broken into components: horizontal illuminance is a measure of the luminous flux incident on a horizontal surface, and vertical illuminance is the luminous flux incident on a vertical surface.

Industrial Power Distribution, Second Edition. Ralph E. Fehr, III.
© 2016 The Institute of Electrical and Electronics Engineers, Inc. Published 2016 by John Wiley & Sons, Inc.

Color temperature is another consideration when designing a lighting system. Measured in kelvins, color temperature refers to the temperature of a blackbody source that would emit light of that given wavelength. The color temperature of a light source effects the color rendition of objects viewed under that light. Sunlight at the earth's surface has a color temperature between 5500 K and 6500 K. Lower color temperatures (2700–3300 K) produce a reddish-yellow light that accentuates warm hues, so these lower temperatures are often referred to as "warm colors." Higher color temperatures (5000–6500 K) produce a harsher bluish light that is not flattering to the color rendition of human skin. These higher color temperatures are referred to as "cool colors." Note that the use of "warm" and "cool" in reference to color temperature is opposite of the blackbody temperature concept.

In the interest of energy conservation and economics in general, issues such as the efficiency of the lighting system, often expressed as efficacy, power factor, and power quality need also be considered.

Luminous efficacy, or simply *efficacy*, is a measure of the efficiency of producing visible light. It is essentially the ratio of luminous flux produced to the electrical energy consumed.

Because most lighting technologies require ballasting, the ballast design is of interest because of impacts to power factor and power quality. Many ballast designs are commercially available, but before making a selection, their electrical characteristics should be understood.

10.2 LIGHTING TECHNOLOGIES

The lighting technologies covered in this chapter can be categorized as *incandescent*, *low-pressure discharge* (fluorescent and low-pressure sodium), *high-intensity discharge*, or *HID* (mercury vapor, high-pressure sodium, and metal halide), and *solid-state* (light-emitting diode or LED). The general basis of operation, including advantages and disadvantages, lamp construction, and appropriate applications for each, is discussed in the following sections.

10.2.1 Incandescent

The incandescent lamp, shown in Figure 10.1, is the most basic and the oldest type of electric lamp still in general use. The concept was patented in 1879 by Sir Joseph Swan in England and by Thomas Edison in the United States.

The incandescent lamp produces light by passing an electric current through a filament causing the filament to incandesce or glow. In the process, metal from the filament is vaporized which ultimately limits the life of the bulb.

In modern incandescent lamps, a tungsten filament is supported by molybdenum wires inside a glass bulb filled with an inert gas such as argon to reduce the vaporization rate of the filament. Current is carried into the gas-filled bulb by dumet wires, which are made of an alloy designed to form an airtight seal between the wire and the glass surrounding the wire. An exhaust tube, through which the bulb is evacuated and filled with argon during manufacturing, runs through the glass stem.

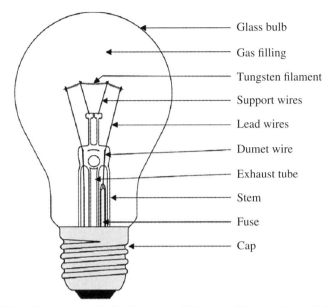

Figure 10.1 Incandescent Bulb. (*Courtesy of Museum of Electric Lamp Technology.*)

When voltage is applied across the filament, a large inrush current flows because of the low filament resistance. As the temperature of the filament rises from ambient to temperatures in excess of 2500°C, where the filament incandesces white-hot, the resistance of the filament rises dramatically. The inrush current, as high as 14 p.u. of the normal operating current of the bulb, diminishes quickly—in less than 50 milliseconds for low-wattage bulbs. High-wattage bulbs may see elevated starting current for over 1 second. Typical incandescent bulb inrush currents are graphed in Figure 10.2.

When a sufficient amount of metal has vaporized from the filament, the filament can no longer carry the required current, and the filament burns in two. Because of the high temperatures present inside the bulb, the gas between the broken pieces of filament could ionize, reestablishing the circuit. If this happens, a very high transient current well in excess of 100 A could result. Currents that are high would cause a very dangerous condition, quite possibly leading to the explosion of the bulb and a fire. To

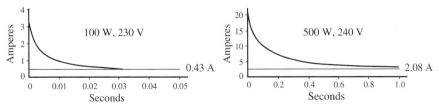

Figure 10.2 Incandescent Bulb Inrush Currents. (*Courtesy of Museum of Electric Lamp Technology.*)

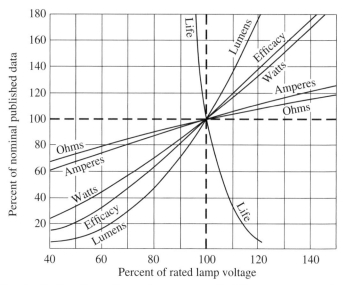

Figure 10.3 Sensitivity of Incandescent Lamps to Supply Voltage. (*Courtesy of Museum of Electric Lamp Technology.*)

prevent dangerously high currents at the end of the bulb's life, fusing is incorporated into the base to safely open the circuit without the possibility of reignition.

Incandescent bulbs are very sensitive to variations in supply voltage. Impedance, current, power consumption, efficacy, light production, and bulb life are all functions of supply voltage, as can be seen in Figure 10.3.

While incandescent lamps are inexpensive to manufacture and operate at line voltage without need for a ballast, unlike the other technologies discussed in this chapter, their shortcoming is very poor efficacy. Efficacy, when used to describe an electric lamp, is the efficiency of converting electrical energy to light. Incandescent lamps have a typical efficacy in the 2–5% range, producing between 14 and 35 lumens of light per watt of electricity. Consequently, 95–98% of the electrical energy consumed by the lamp is converted to heat, not light. Besides being very inefficient in producing light, incandescent lamps have a large disadvantage when used indoors by putting a substantial burden on air-conditioning systems. Many governments, including the federal government of the United States, have adopted efficacy standards that cannot be met by standard incandescent lamps. These standards, as well as an increased focus on energy conservation, are causing many incandescent lighting systems to be replaced by higher efficiency alternatives.

10.2.2 Low-Pressure Discharge

10.2.2.1 Fluorescent Fluorescent lighting dates back to the mid-1800s when the French physicist Alexandre Becquerel began investigating the phenomena of fluorescence and phosphorescence. This led him to conceptualize a fluorescent tube

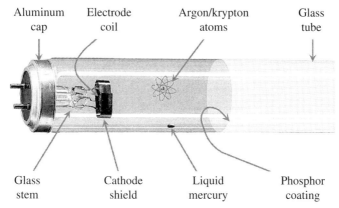

Figure 10.4 Fluorescent Tube. (*Courtesy of Museum of Electric Lamp Technology.*)

design very similar to that used today. The first commercially viable fluorescent lamps became available in the 1930s. Fluorescent lighting became very popular in the 1970s, and today accounts for about 80% of the world's artificial light. A fluorescent tube is shown in Figure 10.4.

Fluorescent lighting is an example of low-pressure mercury discharge technology. A sealed glass tube, coated with a phosphor material on the inside, is filled with a mixture of inert gases, typically argon, krypton, neon, and xenon, at a low pressure. A very small amount of liquid mercury is also inside the glass tube. A coil of tungsten wire at each end of the tube forms a cathode which emits electrons when a voltage is impressed across the electrodes of the cathode. A cathode shield minimizes blackening of the glass tube near the cathode caused by vaporization of the tungsten as well as reduces the visible flicker that occurs at the cathode.

When voltage is applied to the lamp, the cathode emits electrons which move through the glass tube from one end to the other. In the process, some of the liquid mercury transitions to a gas. When electrons collide with the atoms of mercury vapor, electron energy states in the mercury atoms change, producing photons. The photons are in the ultraviolet wavelength range, so cannot be seen as light. But when they collide with a phosphor atom on the inside surface of the glass tube, a photon in the visible light range is produced. By carefully controlling the composition of the phosphor coating, various colors of white light can be produced, ranging from a bluish high color temperature, ironically called "cool white" to a more reddish lower color temperature, referred to as "warm white."

The operating principles of fluorescent lighting require a large distance between the cathodes. This is the reason for the spiral designs of many compact fluorescent lamps (CFLs) intended as incandescent bulb replacements. The most convenient lamp design is a tubular design, with 2-, 4-, and 8-ft length standard sizes in the United States. The letter "T" (for "tubular") followed by a number (indicating the number of eighths of inches of tube diameter) is used to designate different tube sizes. A "T-8" lamp is a $1''$ ($8/8''$) diameter tube, whereas a "T-12" indicates a $1.5''$ ($12/8''$) diameter tube. In general, a smaller tube diameter yields a higher efficiency.

To start a modern fluorescent lamp, a high voltage (usually 400–650 V) is applied across the cathode electrodes to initialize ionization. Once ionization occurs, the cathode voltage is decreased to working voltage (typically 93 V for a 120 V system). A device is needed to control the voltage to allow starting of the lamp. This same device must also provide a second very important function. Unlike current flowing through a solid, increasing current flowing through a gas causes the resistance of the conductor (gas) to decrease. This phenomenon would lead to thermal runaway and destruction of the lamp unless current can be controlled, so current limitation becomes a necessity.

The device which provides both the starting function and current limitation is called a *ballast*. *Magnetic ballasts* are essentially inductors that limit changes in current. The limitation must continue for half of an electrical cycle, until the current goes to zero and reverses direction. The starting function is accomplished using a starter, which is a capacitor that is discharged to provide a high ignition voltage.

Magnetic ballasts have been in use for many years, but have some drawbacks. Being iron-core inductors, they are bulky and heavy. They also produce an audible hum like a transformer due to magnetostriction. Magnetic ballasts are usually oil-insulated, and prior to the 1980s, polychlorinated biphenyls (PCBs) were added to the oil to reduce the risk of fire. Due to toxicity concerns, use of PCBs in the United States was banned by the 1976 Toxic Substances Control Act. Also, being an inductor, the ballast shifts the current to lag the voltage, resulting in a low power factor. Power factor correction should always be considered when magnetic ballasts are used.

An alternative to the magnetic ballast is the solid-state ballast. Solid-state ballasts are used in virtually all new fluorescent lighting systems, as energy conservation requirements imposed by many countries cannot be met using magnetic ballasts. Electronic circuitry is used for current limitation and the production of a starting voltage. A frequency higher than the power system frequency is used to improve efficiency and to reduce flicker and hum. Power factor correction is often implemented into the solid-state ballast design.

The efficacy of fluorescent lamps is considerably higher than that of incandescent lamps, with a typical efficacy of about 22%. Color rendering is also very good, as changing the phosphor composition can produce a wide variety of colors, including hues other than white. Fluorescent tubes have a long expected life, usually in excess of 10,000 hours, and possibly over 50,000 hours with certain triphosphor coatings. Dimming is also possible with ballasts designed for that purpose.

Fluorescent lighting also comes with several disadvantages. All fluorescent lamps contain a small amount of mercury, which is a toxic heavy metal. This is cause for concern when old lamps must be discarded, or if a lamp tube breaks. Fluorescent lamps are also sensitive to ambient temperature, as light output drops significantly with both low and high ambient temperatures, as is shown in Figure 10.5. Flicker can be an issue with magnetic ballasts, and tubes in excess of 4 ft in length can be awkward to handle and store.

10.2.2.2 Low-Pressure Sodium Low-pressure sodium lamps incorporate a mixture of argon, neon, and sodium at a relatively low pressure. The lower pressure improves the efficacy and reduces the operating temperature of the bulb, but

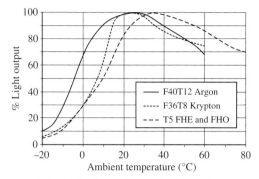

Figure 10.5 Fluorescent Tube Light Output Versus Ambient Temperature. (*Courtesy of Museum of Electric Lamp Technology.*)

requires a longer discharge tube length. Ordinary glass types can be used since the operating temperatures are lower than with the high-pressure discharge technologies, but borate coatings must be used to protect the glass from the corrosive sodium vapor. The longer discharge tube is often accommodated by bending the tube into a U-shape, as seen in Figure 10.6.

A *getter* is used in the evacuated region inside the outer bulb to chemically combine with any stray gas molecules that may find their way into the region, preserving the vacuum. Because of the U-shaped tube design, the cap is typically keyed to assure that the bulb is oriented in the luminaire correctly to maximize light output.

Low-pressure sodium lamps produce a monochromatic light at a wavelength of 589.3 nm, very close to the color to which the human eye is most sensitive. The monochromatic characteristic of the light makes color rendering impossible. Viewing objects illuminated by low-pressure sodium lighting is much like viewing a black-and-white television set while wearing orange-tinted glasses. This eliminates low-pressure sodium lighting for all indoor lighting applications and outdoor applications where

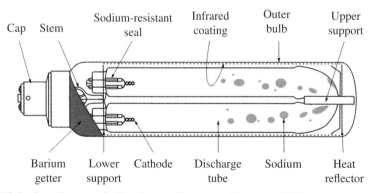

Figure 10.6 Low-Pressure Sodium Lamp. (*Courtesy of Museum of Electric Lamp Technology.*)

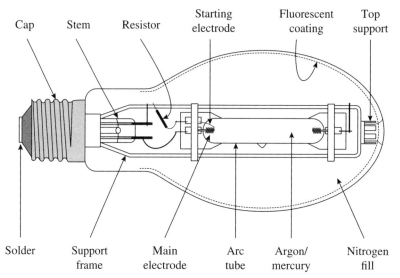

Figure 10.7 Mercury Vapor Lamp. (*Courtesy of Museum of Electric Lamp Technology.*)

color rendering is necessary. Security lighting is the most common application of low-pressure sodium lighting.

Low-pressure sodium bulbs have a typical life on the order of 18,000 hours, and unlike all the other lighting technologies, do not experience a loss of lumen production as the lamp ages. Light output remains constant over the life of the lamp, but electricity consumption increases slightly toward the end of useful life due to electrode degradation.

10.2.3 High-Intensity Discharge

10.2.3.1 Mercury Vapor A mercury vapor lamp, as shown in Figure 10.7, consists of a quartz arc tube with tungsten electrodes at either end. The arc tube contains a small amount of mercury in an argon atmosphere. A starting electrode in series with a current-limiting resistor is located close to one of the main electrodes. The arc tube assembly is surrounded by a phosphor-coated glass envelope, and the volume between the arc tube assembly and the glass envelope is filled with an inert gas, usually nitrogen.

When voltage is applied across the main electrodes, the difference in voltage is inadequate to ionize the gas in the arc tube. But the voltage difference between the starting electrode and the adjacent main electrode is adequate to begin the ionization process. When current begins to flow, the resistor in series with the starting electrode limits the current to a small fraction of the lamp's operating current. A short time after ionization starts, a discharge between the main electrodes is possible, and this current path bypasses the starting electrode essentially removing the starting electrode and resistor from the circuit. At this point, the mercury is fully vaporized and a characteristic greenish-yellow light is produced. The phosphor coating on the glass envelope is used to improve color rendition by filling in the missing red end of the spectrum.

The efficiency of mercury vapor lighting systems is comparable to that of fluorescent lighting, but considerably lower than other HID lighting systems. Long lamp life, on the order of 24,000 hours, and color rendition superior to that of sodium lighting, are the chief advantages of mercury vapor lighting. As metal halide technology continues to advance, mercury vapor lighting declines in popularity.

10.2.3.2 High-Pressure Sodium High-pressure sodium lamps utilize an arc tube made of aluminum oxide ceramic to resist the corrosive effects of the sodium vapor. Xenon is often used as the inert gas in the arc tube to establish ionization, since its low ionization temperature facilitates the starting of the lamp.

An amalgam of mercury and sodium is stored in the amalgam reservoir at the end of the arc tube. After a xenon arc is struck, the heat produced vaporizes the amalgam, and the sodium vapor begins to ionize. A broad-spectrum light is produced, as opposed to the monochromatic light of the low-pressure sodium lamp, because the mercury fills in otherwise missing blue hues and *quasistatic pressure broadening*, made possible by the high pressure inside the arc tube, increases the spectral output from the sodium. A high-pressure sodium lamp is shown in Figure 10.8.

The broad spectral output of the high-pressure sodium lamp combined with its high efficacy and long lamp life make it an attractive option for many outdoor applications, particularly street lighting, as well as some indoor applications such as high-bay and warehouse lighting.

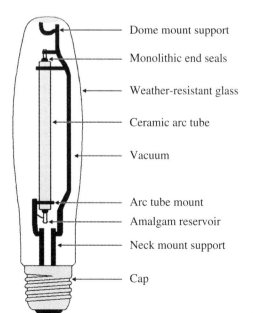

Figure 10.8 High-Pressure Sodium Lamp. (*Courtesy of Commercial Lighting Co.*)

10.2.3.3 Metal Halide Metal halide lamps use arc tubes similar to those used in mercury vapor lamps, but operate at higher temperatures and pressures. Like the mercury vapor lamp, photons in the ultraviolet range are produced in the arc tube. A borosilicate glass envelope, sometimes phosphor-coated to diffuse the light and improve color rendition, surrounding the arc tube reduces the ultraviolet radiation emitted from the lamp.

The metal halide salts in the arc tube dissociate in the heat of the plasma into metal ions and iodine. The metal ions emit photons in the visible spectrum when struck by ultraviolet photons and are the primary source of light. The light characteristics can be altered by controlling the chemical composition of the halide salts.

Metal halide lamps can produce excellent color rendition, making them desirable for photographic applications as well as for athletic field lighting. It can take from 1 to 15 minutes for a metal halide lamp to produce full lumen output after being lit.

When starting, a resistor limits current to the starting electrode, or probe. When the lamp reaches operating temperature, a bimetal switch shorts the probe and applies voltage to the tube through the main thoriated tungsten electrode.

Unfortunately, metal halide lamps have the potential to fail nonpassively. When this occurs, the glass enclosure explodes, posing a significant hazard to people in the vicinity. Another concern with metal halide technology is that ultraviolet radiation can be emitted if the lamp outer jacket cracks. Exposure to ultraviolet radiation is a health concern, as the skin and eyes are particularly sensitive to ultraviolet wavelengths.

These safety issues can be addressed by using luminaires with containment lenses made of acrylic or polycarbonate plastics. Enclosed luminaires should carry an Underwriters Laboratory endorsement (UL standard 1598) to ensure that if an explosion occurs, the fragments will be contained in the fixture. Group relamping at a point between 50% and 70% of rated lamp life is an effective means of reducing the likelihood of explosion. And installing the luminaire in the correct orientation, per the manufacturer instructions, is essential to assure safe operation. Note that the risk of explosion is unique to the metal halide technology, but the ultraviolet hazard also exists with the mercury vapor technology. An external lens with ultraviolet-filtering characteristics can be used with both mercury vapor and metal halide lamps to reduce the risk of ultraviolet radiation emission. A metal halide lamp is shown in Figure 10.9.

Discharge-type lighting systems have a warm-up period that lasts for several minutes, depending on the type of lamp. During this warm-up period, lumen output and color temperature are significantly degraded compared to the lamp at normal operation.

When power is interrupted to a discharge lighting system, the arc extinguishes, and cannot be reestablished until the lamp cools sufficiently to reduce the vapor pressure in the arc tube to a point where restrike is possible. Table 10.1 shows typical warm-up and restrike times for the discharge-type technologies. Actual times can vary significantly based on lamp design.

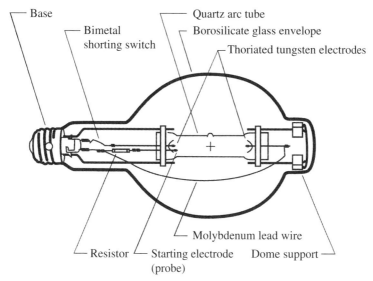

Figure 10.9 Metal Halide Lamp. (*Courtesy of Venture Lighting.*)

10.2.4 Light-Emitting Diode (LED) Lighting

The technology leading to the development of LEDs dates back to the early 1900s when several British and Russian researchers discovered the phenomenon called *electroluminescence*, where certain materials emit light when an electric current is passed through them. In 1962, Nick Holonyak developed the first practical LED at General Electric Company. A typical LED is shown in Figure 10.10.

LEDs emit light of a very specific color, whose wavelength is a function of the bandgap energy of the semiconductor material used to make the LED. This light may or may not be visible to the human eye. The first LEDs produced light in the infrared range. To produce a white light suitable for general lighting purposes, one of two approaches can be taken: a *trichromatic* device can be built, or *phosphor-converted* devices can be used.

Trichromatic LED lamps use separate red, green, and blue LEDs mounted in close proximity to one another. Their light output merges to produce a white color. The intensities of the three colors can be individually controlled to adjust the color

TABLE 10.1 Typical Warm-Up and Restrike Times

Technology	Typical Warm-Up Time	Typical Restrike Time
Fluorescent	$<\frac{1}{2}$–2 minutes	Negligible
Low-pressure sodium	7–10 minutes	3–12 seconds
Mercury vapor	5–7 minutes	3–6 minutes
High-pressure sodium	3–4 minutes	$\frac{1}{2}$–1 minute
Metal halide	2–5 minutes	10–20 minutes

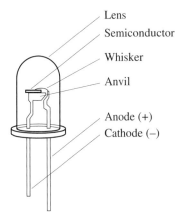

Figure 10.10   Light-Emitting Diode

temperature of the resulting white light. While this seems like a very effective means of controlling the resulting light color, the green and blue wavelengths are relatively close together, whereas the red color is of a much lower wavelength. This results in an uneven spectral density. Adding a fourth color (amber) to form a *tetrachromatic* LED lamp helps the color rendition, but significantly increases the manufacturing cost. In fact, high manufacturing cost is perhaps the largest disadvantage of trichromatic and tetrachromatic LEDs.

A more cost-effective approach to implement LED lighting is to use a single color LED, perhaps in the blue to ultraviolet range in conjunction with a phosphor coating to convert the produced light to the desired color, much like in a fluorescent lamp. With this approach, the output color cannot be adjusted, but the manufacturing cost is much lower.

LED lamps have an efficacy in the range of 65–100 lumens per watt, and a long life of 25,000 to 30,000 hours. Their high initial cost is offset by low energy consumption, about 1/6 that of incandescent lighting, and a reduced maintenance requirement. This puts the life-cycle cost of LED lighting close to that of fluorescent lighting.

All LEDs operate at a relatively low DC voltage, so driver circuitry is needed to reduce voltage, rectify the AC to DC, and hold a constant current output. The driver circuitry can be integrated into the lamp or implemented as a separate device, much like a ballast. A common driver circuit is shown in Figure 10.11.

Capacitor C1 forms a voltage divider with the rest of the circuit which reduces the voltage across capacitor C2 to about 15 V, effectively eliminating the need for a transformer to reduce the line voltage. Resistor R1 prevents the buildup of a DC voltage on capacitor C1. The specifications of resistor R2 depend on the forward voltage drop across the LEDs, the number of LEDs used, and the forward current for which the LEDs are designed.

LED technology can be packaged with both the LEDs and driver circuitry in a single package with a standard E27 Edison base, making it a direct replacement for the incandescent bulb, as shown in Figure 10.12. LED lighting can be cycled on and off frequently, making it well-suited for use with occupancy sensors. And the driver circuit, unlike many HID technologies, is well-suited to dimming.

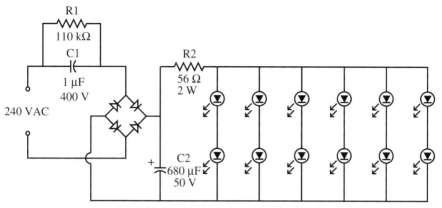

Figure 10.11 LED Driver Circuit

Figure 10.12 LED Lamp. (*Courtesy of Royal Philips Electronics, N.V.*)

10.3 LUMINAIRE DESIGNS

A luminaire is defined by IESNA as *a device to produce, control, and distribute light*. A wide variety of luminaire designs are available to meet virtually any lighting application. In general, considerations when selecting a luminaire for a particular application include construction and installation codes and standards, physical and environmental conditions, electrical and mechanical requirements, thermal characteristics, economics, and most importantly, safety.

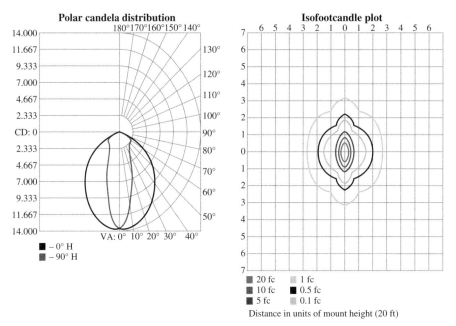

Figure 10.13 Candela Distribution and Isoilluminance Plots. (*Courtesy of Acuity Brands Lighting.*)

Most luminaires are fitted with reflectors, refractors, and/or diffusers to specifically control the distribution of light. Photometric data, as shown in Figure 10.13, including plots of candela distribution and isoilluminance, can be obtained from the luminaire manufacturer.

While this data can be used to perform manual calculations, such calculations tend to be very tedious so software-based solutions are commonly used. Most lighting manufacturers have software available that uses the methods described in this chapter and recommended by IESNA. But as with any software, the engineer needs to understand what the computer is doing prior to depending on the software to help determine a design. Two manual lighting design examples will be done at the end of this chapter. By stepping through the design process manually, an understanding (and appreciation) of the software will be developed.

Luminaires are classified both by application and by photometric characteristics. Application can include residential, commercial, and industrial with subclassification by lighting technology, mounting method, and luminaire construction. Most lighting manufacturers organize their product catalogs using this classification method.

Photometric classification methods have been developed by various engineering organizations, including the International Commission on Illumination (CIE), the National Electrical Manufacturers Association (NEMA), and IESNA. The CIE classification system is based on the ratio of upward-directed light to downward-directed light, and is usually applied to indoor luminaires. Categories include *direct lighting* (90–100% downward light), *semidirect lighting* (60–90% downward light), *general diffuse lighting* (upward and downward components approximately equal),

TABLE 10.2 NEMA Field Angle Classifications and Projection Distances

Beam Type	Beam Spread (degrees)	Projection Distance (ft)
1	10–18	>240
2	18–29	200–240
3	29–46	175–200
4	46–70	145–175
5	70–100	105–145
6	100–130	80–105
7	>130	<80

semi-indirect lighting (60–90% upward light), and *indirect lighting* (90–100% upward light). Outdoor luminaires are sometimes described by cutoff. Three methods are used in photometric reports: *physical cutoff*, *optical cutoff*, and *shielding angle*. Physical cutoff is the angle measured from the downward-directed vertical, or *nadir*, to the point where the lamp is fully occluded. Optical cutoff measures the angle from the nadir to the point where reflection of the lamp in the luminaire's reflector is fully occluded. The shielding angle is the angle measured from the horizontal at which the lamp is just visible.

The NEMA classification system is based on the distribution of luminous flux. It is used primarily for sports lighting and floodlighting. It considers the spread of the light beam in degrees and the projection distance in feet. Seven beam types are defined, as shown in Table 10.2.

The IESNA classification system applies to outdoor luminaires, and is based on the shape of the area illuminated. This classification is commonly used in roadway and parking lot lighting. Six intensity distribution types are defined along with four cutoff classifications, as shown in Tables 10.3 and 10.4.

10.4 ELECTRICAL REQUIREMENTS

All discharge-type lighting systems (all lighting technologies discussed in this chapter except incandescent and LED) require a *ballast* to supply electricity to the lamp. A ballast is a device, either electromagnetic or solid-state, that

1. provides sufficient ignition voltage to start the lamp,
2. acts as a constant current source when the lamp is starting, and
3. acts as a constant power source when the lamp is operating.

TABLE 10.3 IESNA Intensity Distribution Types

Intensity Distribution Type	Description of Illuminance Distribution
Type I	Narrow and symmetric illuminance pattern
Type II	Slightly wider illuminance pattern than Type I
Type III	Wide illuminance pattern
Type IV	Widest illuminance pattern
Type V	Symmetrical circular illuminance pattern
Type VS	Symmetrical nearly square illuminance pattern

TABLE 10.4 IESNA Cutoff Classifications

Cutoff Classification	Description of Intensity Distribution
Full cutoff	Zero candela intensity at angles of 90° and greater above nadir. Additionally, the candela intensity does not exceed 10% of the maximum candela intensity at a vertical angle of 80° above nadir for all lateral angles around the luminaire.
Cutoff	Candela intensity does not exceed 2.5% of the maximum candela intensity at an angle 90° above nadir, and does not exceed 10% of the maximum candela intensity at a vertical angle of 80° above nadir. These limits apply to all lateral angles around the luminaire.
Semicutoff	Candela intensity does not exceed 5% of the maximum candela intensity at an angle 90° above nadir, and does not exceed 20% of the maximum candela intensity at a vertical angle of 80° above nadir. These limits apply to all lateral angles around the luminaire.
Noncutoff	No candela limitation in the zone above maximum candela intensity.

These functions can be implemented using an iron-core reactor (inductor) for current limitation in combination with a capacitor (starter) to provide ignition voltage, or by a variety of solid-state circuit designs. Solid-state ballasting should be used on all new systems, but core-and-coil ballasts have been used for many years, and a significant number remain in service.

During the ignition phase, the ballast must provide a sufficient voltage across the lamp electrodes to initiate and maintain discharge, and must also provide a sufficient current at discharge voltage to force a transition from glow to arc. These voltages and currents are specific to the type of lamp being used, so ballasts must be correctly matched to the lamps they are supplying.

During the warm-up phase, the resistance of the lamp continuously increases. So the ballast must provide a nearly constant current to the lamp, which linearly increases power to the lamp.

In the operating phase, the lamp resistance takes on a value close to the arc impedance. This low resistance requires current limitation to prevent damage to the lamp. As the voltage supplied to the lighting system changes, the lamp will react differently depending on the rate of change. If the change in voltage occurs gradually (over many seconds), a corresponding change in lamp current will try to occur. Again, a constant current supply is needed to keep the lamp operating properly. Sudden changes in voltage could cause extinction of the arc or a sudden increase in current. The constant current characteristic of the ballast will allow proper and safe operation of the lamp through most voltage transients. If the magnitude or duration of the transient exceeds the capabilities of the ballast, the lamp will shut down, and will have to cool before restrike of the arc is possible.

High-pressure sodium lamps experience a rise in lamp voltage over their lifetime. This voltage rise is substantial, often as much as 170% of the lamp voltage experienced when the lamp is new. Therefore, the ballast must keep the lamp power within an acceptable power range over the life of the lamp.

When the voltage applied to the arc tube has a very high frequency (in the kilohertz), standing pressure waves may be created in the arc tube. This phenomenon, known as *acoustic resonance*, can cause a visible distortion of the arc and, in severe cases, even crack the arc tube. Acoustic resonance limits the practical operating frequency of electronic ballasts.

A phenomenon known as *cataphoretic segregation* occurs when a sodium lamp is subjected to DC. The DC voltage causes the sodium atoms to be transported to the cathode side of the arc tube. This renders the lamp useless for producing light. To assure proper discharge, the polarity of the voltage applied across the arc tube must reverse on the order of 10 milliseconds. This means that the DC component of the voltage applied to the arc tube must be kept close to zero.

10.5 LIGHTING SYSTEM DESIGN EXAMPLES

10.5.1 Parking Lot Lighting

A basic application of lighting design involves the lighting of open outdoor areas such as parking lots. In such applications, lighting is necessary for vehicular and pedestrian safety, for protection against crime, and for the convenience of the users. This type of lighting design is based on maintaining specific illuminance levels and keeping the ratio of maximum illuminance to minimum illuminance below a predetermined value. The values for maintained illuminance and *uniformity ratio*, which is the ratio of maximum illuminance to minimum illuminance, are listed in the *IESNA Lighting Handbook* for various applications. According to the 9th edition of the *IESNA Lighting Handbook*, the recommended maintained illuminance values for parking lots are shown in Table 10.5.

Horizontal illuminance is defined as the quantity of luminous flux falling on a horizontal plane. Similarly, *vertical illuminance* is the amount of luminous flux striking a vertical plane. These illuminance values are minimum values whereas the uniformity ratio is a maximum value. Actual requirements for a particular application may surpass these values.

Parking lot lighting is typically designed based on illuminance criteria. The IESNA recommends a minimum horizontal illuminance level of 0.2 footcandles and a minimum vertical illuminance level of 0.1 footcandles for low-activity open parking lots to help assure safety and to deter crime. A uniformity ratio, which compares the

TABLE 10.5 Recommended Maintained Illuminance Values for Parking Lots

Parameter	Basic Applications	Enhanced Security
Minimum horizontal illuminance	0.2 fc	0.5 fc
Uniformity ratio, maximum-to-minimum	20:1	15:1
Minimum vertical illuminance	0.1 fc	0.25 fc

Source: *IESNA Lighting Handbook*, 9th edition.

area with highest illuminance to the area with lowest illuminance, of 20:1 or lower is also specified for parking lots.

A suitable lighting technology must be selected to start the design process. A luminaire style and lamp type must also be selected for the design to proceed. Parking lots are usually lit by HID lighting, although low-pressure sodium can be used. Metal halide is a common choice and will be used in this example. A Type VS luminaire with square distribution designed specifically for parking lot lighting has been chosen. The luminaire utilizes a clear 250-W lamp, and has an initial lumen production of 20,500 lm. The luminous intensity distribution for this luminaire is shown in Table 10.6. A drawing of the parking lot to be illuminated is shown in Figure 10.10. It is decided to use two luminaires, one on a pole at location L1 and the other on a pole at the location L2. A pole height of 35 ft is selected.

To verify adequate illuminance, the horizontal and vertical illuminance levels must be calculated for each point on the parking lot, then the uniformity ratio must be determined. These calculated values then need to be compared to the IESNA recommendations. If any of the recommendations are not met, modifications to the lighting system must be made until the design criteria are satisfied. This process is quite tedious, and can involve an extensive trial-and-error approach. Consequently, computer-based design of lighting systems is usually done. We will show the manual calculation process to illustrate the general methodology.

Instead of calculating the illuminance levels at *every* point on the parking lot, computer software determines the illuminance levels at points on a fixed grid, where the grid spacing can be specified by the user. Since we are doing manual calculations, we will calculate illuminance values at the most likely maximum and minimum illuminance locations. These locations are points A, B, and C on Figure 10.14. If the IESNA illuminance criteria can be met at these locations, we can assume with good certainty that they will be met everywhere in the parking lot.

Being an outdoor application, reflectance is disregarded. Since the light source is more than five times the largest luminaire dimension from the parking lot surface, it can be treated as a point source, and illuminance can be calculated using the inverse square cosine law as shown in Eq. (10.1).

$$E = \frac{I(\theta, \psi) \cos \theta}{D^2} \tag{10.1}$$

Two angles of interest when doing lighting calculations are the vertical angle formed by the incident light beam and the z-axis, and the lateral angle formed by the projection of the light beam onto the x–y plane and the optical axis of the luminaire. The vertical angle θ and the lateral angle ψ can be determined by Eqs. (10.2) and (10.3), respectively.

$$\theta = \tan^{-1}\left(\frac{\sqrt{x^2 + y^2}}{z}\right) \tag{10.2}$$

$$\psi = \tan^{-1}\left(\frac{y}{x}\right) \tag{10.3}$$

10.5 LIGHTING SYSTEM DESIGN EXAMPLES

TABLE 10.6 Parking Lot Luminaire Luminous Intensity Distribution for Type A Photometry

Vertical Angle θ	Intensity (cd) at Lateral Angle ψ									
	0°	5°	15°	25°	35°	45°	55°	65°	75°	85°	90°
0.0°	2886	2886	2886	2886	2886	2886	2886	2886	2886	2886	2886
2.5°	2879	2887	2887	2904	2921	2929	2946	2954	2954	2954	2963
5.0°	2846	2854	2879	2904	2921	2938	2946	2963	2963	2963	2963
7.5°	2829	2829	2862	2879	2896	2896	2904	2912	2921	2929	2929
10.0°	2779	2795	2812	2820	2829	2837	2854	2854	2879	2871	2879
12.5°	2745	2745	2762	2779	2779	2795	2812	2837	2846	2854	2846
15.0°	2678	2678	2695	2728	2753	2762	2787	2812	2820	2812	2812
17.5°	2628	2628	2645	2678	2712	2737	2770	2795	2812	2795	2787
20.0°	2578	2578	2603	2636	2670	2712	2779	2837	2854	2829	2804
22.5°	2561	2553	2553	2611	2653	2728	2812	2904	2921	2837	2804
25.0°	2544	2553	2544	2628	2703	2753	2804	2879	2879	2804	2762
27.5°	2461	2477	2511	2620	2720	2770	2829	2896	2887	2804	2770
30.0°	2427	2435	2461	2611	2720	2812	2904	3021	2963	2829	2812
32.5°	2578	2569	2494	2728	2779	2904	3046	3155	3021	2871	2837
35.0°	2712	2678	2561	2929	2896	3080	3297	3356	3138	3046	2996
37.5°	3046	2921	2653	3122	3063	3406	3599	3691	3373	3306	3339
40.0°	3565	3331	2912	3456	3389	4051	3967	4268	3590	3490	3565
42.5°	3733	3456	3122	3892	3825	4695	4293	4670	3749	3574	3724
45.0°	4118	3599	3189	4067	4101	5030	4729	4812	3984	3833	4109
47.5°	4185	3841	3297	4469	4377	5097	4896	5214	4101	3833	4327
50.0°	3749	3565	3482	4570	4494	5038	4871	5381	4101	3691	4293
52.5°	3314	3097	3465	4327	4243	4829	4645	5281	4385	3716	4770
55.0°	3046	3097	3758	3992	4093	5088	4888	5842	5139	4067	5164
57.5°	2142	2544	3130	4084	5088	6955	5716	5591	4553	3825	4151
60.0°	1741	2134	2410	3281	4461	5323	4385	4143	3808	3197	3448
62.5°	1490	1942	2260	2820	4076	4235	3716	3348	3908	3021	3557
65.0°	1205	1657	2402	2293	3724	3515	3540	3389	3858	3180	4235
67.5°	1038	1398	2193	2151	3574	4009	4452	3816	4921	3222	4143
70.0°	836	1105	1774	1967	3565	4645	5189	4093	4720	3339	3532
72.5°	619	795	1205	1758	2753	3657	4185	3607	3549	2059	2360
75.0°	418	460	518	1138	2025	3021	3063	1891	2092	1013	1364
77.5°	234	192	200	309	535	1029	1071	594	493	326	251
80.0°	117	100	108	142	175	184	209	167	108	92	100
82.5°	50	58	50	50	58	66	50	50	41	33	41
85.0°	33	16	41	16	25	16	16	16	16	16	16
87.5°	0	0	0	0	0	0	0	0	0	0	0
90.0°	0	0	0	0	0	0	0	0	0	0	0

Source: *IESNA Lighting Handbook*, 9th edition.
Total Lumen Output is 20,500 lm.

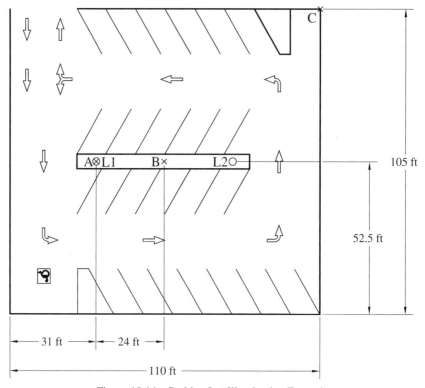

Figure 10.14 Parking Lot Illumination Example

The basic approach will be to determine θ and ψ at the point where the illuminance is desired, then find I as a function of θ and ψ in the luminaire's luminous intensity distribution data (Table 10.6), and finally determine the horizontal and vertical components of the illuminance. Recall the distance formula in three-dimensional space $(D = \sqrt{x^2 + y^2 + z^2})$ and that a lamp height of 35 ft has been chosen.

First, the illuminance at Point A (directly under L1) is determined. The total illuminance is the sum of the illuminance produced by each light source. First, the contribution from source L1 is considered. From the geometry of the problem,

$$D_{(L1-A)} = 35 \text{ ft} \tag{10.4}$$

$$q_{(L1-A)} = 0° \tag{10.5}$$

$$\psi_{(L1-A)} = 0°. \tag{10.6}$$

The luminous intensity of $L1$ at $\theta = 0°$ and $\Psi = 0°$, $I(0°, 0°)$, can be obtained from Table 10.6.

$$I(0°, 0°) = 2886 \text{ cd} \tag{10.7}$$

10.5 LIGHTING SYSTEM DESIGN EXAMPLES

Then the illuminance due to light source L1, $E_{(L1)}$, can be determined at Point A.

$$E_{A(L1)} = \frac{2886 \cos 0°}{35^2} = 2.356 \text{ fc} \tag{10.8}$$

Since the IESNA illuminance criteria are given in terms of horizontal illuminance and vertical illuminance, those components must be obtained by trigonometry.

$$E_{H-A(L1)} = 2.356 \cos 0° = 2.356 \text{ fc} \tag{10.9}$$

$$E_{V-A(L1)} = 2.356 \sin 0° = 0 \text{ fc} \tag{10.10}$$

Next, the contribution from light L2 is determined in a similar fashion.

$$D_{(L2-A)} = \sqrt{48^2 + 35^2} = 59.4 \text{ ft} \tag{10.11}$$

$$\theta_{(L2-A)} = \tan^{-1}\left(\frac{48}{35}\right) = 53.9° \tag{10.12}$$

$$\Psi_{(L2-A)} = 0° \tag{10.13}$$

The luminous intensity at $\theta = 53.9°$ and $\Psi = 0°$, I (53.9°, 0°), is not tabulated in Table 10.4. The closest values listed in Table 10.6, which are I (52.5°, 0°) and I (55.0°, 0°), can be looked up and a reasonable approximation of I (53.9°, 0°) can be made using linear interpolation.

$$I (52.5°, 0°) = 3314 \text{ cd} \tag{10.14}$$

$$I (55.0°, 0°) = 3046 \text{ cd} \tag{10.15}$$

Then, by interpolation,

$$I (53.9°, 0°) = 3164 \text{ cd.} \tag{10.16}$$

Now the illuminance due to L2 can be found at Point A.

$$E_{A(L2)} = \frac{3164 \cos 53.9°}{59.4^2} = 0.5284 \text{ fc} \tag{10.17}$$

$$E_{H-A(L2)} = 0.5284 \cos 53.9° = 0.3113 \text{ fc} \tag{10.18}$$

$$E_{V-A(L2)} = 0.5284 \sin 53.9° = 0.4269 \text{ fc} \tag{10.19}$$

Finally, the total illuminance at Point A can be determined by component.

$$\sum E_{H(A)} = E_{H-A(L1)} + E_{H-A(L2)} = 2.356 + 0.3113 = 2.6673 \text{ fc} \tag{10.20}$$

$$\sum E_{V(A)} = E_{V-A(L1)} + E_{V-A(L2)} = 0 + 0.4269 = 0.4269 \text{ fc} \tag{10.21}$$

These calculated values are well above the minimum criteria set forth by IESNA ($E_H \geq 0.2$ fc and $E_V \geq 0.1$ fc), so the results at Point A are acceptable and the design can continue.

308 CHAPTER 10 LIGHTING SYSTEMS

Next, the same process can be applied to Point B (midway between the two luminaires).

$$D_{(L1\text{-}B)} = \sqrt{24^2 + 35^2} = 42.4 \text{ ft} \tag{10.22}$$

$$\theta_{(L1\text{-}B)} = \tan^{-1}\left(\frac{24}{35}\right) = 34.4° \tag{10.23}$$

$$\psi_{(L1\text{-}B)} = 0° \tag{10.24}$$

The values I (32.5°, 0°) and I (35.0°, 0°) can be found in Table 10.6.

$$I\,(32.5°, 0°) = 2578 \text{ cd} \tag{10.25}$$

$$I\,(35.0°, 0°) = 2712 \text{ cd} \tag{10.26}$$

Then the value I (34.4°, 0°) can be found by interpolation.

$$I\,(34.4°, 0°) = 2680 \text{ cd} \tag{10.27}$$

The illuminance at Point B due to source $L1$ is

$$E_{B(L1)} = \frac{2680 \cos 34.4°}{42.4^2} = 1.230 \text{ fc.} \tag{10.28}$$

Expressed as horizontal and vertical components, the illuminance at Point B due to $L1$ is

$$E_{H\text{-}B(L1)} = 1.230 \cos 34.4° = 1.015 \text{ fc} \tag{10.29}$$

and

$$E_{V\text{-}B(L1)} = 1.230 \sin 34.4° = 0.6949 \text{ fc.} \tag{10.30}$$

The illuminance contribution from source $L2$ is the same as $L1$ by symmetry, so the total illuminance at Point B by component is

$$\sum E_{H(B)} = 2E_{H\text{-}B(L1)} = 2(1.015) = 2.030 \text{ fc} \tag{10.31}$$

and

$$\sum E_{V(B)} = 2E_{V\text{-}B(L1)} = 2(0.6949) = 1.390 \text{ fc.} \tag{10.32}$$

These values exceed the minimum illuminance values in Table 10.5, so we can continue with the design.

The same process is applied a third time to Point C (corner of the parking lot).

$$D_{(L1\text{-}C)} = \sqrt{79^2 + 52.5^2 + 35^2} = 1.390 \text{ fc} \tag{10.33}$$

$$\theta_{(L1\text{-}C)} = \tan^{-1}\left(\frac{\sqrt{79^2 + 52.5^2}}{35}\right) = 69.7° \tag{10.34}$$

$$\psi_{(L1\text{-}C)} = \tan^{-1}\left(\frac{52.5}{79}\right) = 33.6° \tag{10.35}$$

10.5 LIGHTING SYSTEM DESIGN EXAMPLES

The desired luminous intensity value $I\,(69.7°, 33.6°)$ is not listed in Table 10.6, but can be found by looking up the closest listed values and interpolating. For $\psi = 25°$,

$$I\,(67.5°, 25°) = 2151 \text{ cd} \tag{10.36}$$

and

$$I\,(70.0°, 25°) = 1967 \text{ cd.} \tag{10.37}$$

Using linear interpolation,

$$I\,(69.7°, 25°) = 1989 \text{ cd.} \tag{10.38}$$

Then, for $\psi = 35°$,

$$I\,(67.5°, 35°) = 3574 \text{ cd} \tag{10.39}$$

and

$$I\,(70.0°, 35°) = 3565 \text{ cd.} \tag{10.40}$$

Using linear interpolation,

$$I\,(69.7°, 35°) = 3572 \text{ cd.} \tag{10.41}$$

The luminous intensity at $\psi = 33.6°$ can be found by interpolating between values found in Eq. (10.38) and Eq. (10.41).

$$I\,(69.7°, 33.6°) = 3350 \text{ cd} \tag{10.42}$$

Now the illuminance at Point C due to source $L1$ can be found.

$$E_{C(L1)} = \frac{3350 \cos 69.7°}{101.1^2} = 0.1137 \text{ fc} \tag{10.43}$$

The horizontal and vertical components of $E_{C(L1)}$ are

$$E_{H\text{-}C(L1)} = 0.1137 \cos 69.7° = 0.03945 \text{ fc} \tag{10.44}$$

and

$$E_{V\text{-}C(L1)} = 0.1137 \sin 69.7° = 0.1066 \text{ fc.} \tag{10.45}$$

Next, the illuminance contribution of $L2$ can be calculated at Point C.

$$D_{(L2\text{-}C)} = \sqrt{31^2 + 52.5^2 + 35^2} = 70.3 \text{ ft} \tag{10.46}$$

$$\theta_{(L2\text{-}C)} = \tan^{-1}\left(\frac{\sqrt{31^2 + 52.5^2}}{35}\right) = 60.1° \tag{10.47}$$

$$\psi_{(L2\text{-}C)} = \tan^{-1}\left(\frac{52.5}{31}\right) = 59.4° \tag{10.48}$$

Again, interpolation is used to find the desired luminous intensity $I\,(60.1°, 59.4°)$.

From Table 10.6,

$$I(60.0°, 55°) = 4385 \text{ cd} \tag{10.49}$$

and

$$I(62.5°, 55°) = 3716 \text{ cd}. \tag{10.50}$$

By interpolation,

$$I(60.1°, 55°) = 4358 \text{ cd}. \tag{10.51}$$

Again, from Table 10.6,

$$I(60.0°, 65°) = 4143 \text{ cd} \tag{10.52}$$

and

$$I(62.5°, 65°) = 3348 \text{ cd}. \tag{10.53}$$

By interpolation,

$$I(60.1°, 65°) = 4111 \text{ cd}. \tag{10.54}$$

By interpolating between the values found in Eq. (10.51) and Eq. (10.54),

$$I(60.1°, 59.4°) = 4249 \text{ cd}. \tag{10.55}$$

The illuminance at Point C due to L2 is

$$E_{(C-L2)} = \frac{4249 \cos 60.1°}{70.3^2} = 0.4286 \text{ fc}. \tag{10.56}$$

The horizontal and vertical components of $E_{(C-L2)}$ are

$$E_{H-C(L2)} = 0.4286 \cos 60.1° = 0.2137 \text{ fc} \tag{10.57}$$

$$E_{V-C(L2)} = 0.4286 \sin 60.1° = 0.3716 \text{ fc}. \tag{10.58}$$

This means the total illuminance at Point C, by component, is

$$\sum E_{H(C)} = 0.03945 + 0.2137 = 0.2532 \text{ fc} \tag{10.59}$$

and

$$\sum E_{V(C)} = 0.1066 + 0.3716 = 0.4782 \text{ fc}. \tag{10.60}$$

These values exceed the minimum illuminance values in Table 10.5, so we have a design that meets all illuminance criteria. Lastly, we must verify that the uniformity ratio is compliant with the IESNA criterion ($\leq 20:1$).

Comparing the maximum illuminance value to the minimum illuminance value for both the horizontal and vertical components gives the following ratios:

$$E_H: \frac{2.6673}{0.2532} = 10.5:1 \tag{10.61}$$

$$E_V: \frac{1.390}{0.4782} = 2.9:1 \tag{10.62}$$

Both of these ratios are below the 20:1 maximum acceptable ratio given in Table 10.5, so this design is acceptable.

10.5.2 Interior Lighting

Unlike applications of outdoor lighting, applications of lighting interior spaces require the consideration of reflected light. In addition to the light directly emitted from the source, the interior work surface is also subject to light reflected from the ceiling, walls, and floor. The IESNA-recommended method for interior lighting calculations is the *lumen method for average illuminance* based on the *zonal cavity method*.

The zonal cavity method uses the *room cavity ratio* to describe the volume of the space to be illuminated, as follows:

$$\text{RCR} = \frac{5h\,(l+w)}{l \times w}, \qquad (10.63)$$

where h is the cavity height, l is the cavity length, and w is the cavity width. If the luminaires are suspended below the ceiling, a ceiling cavity ratio (CCR) exists between the ceiling and the luminaires. Below the plane of the defined work surface is a floor cavity, described by a floor cavity ratio (FCR). Figure 10.15 shows an interior space with its cavities annotated.

This design example will be based on a large open high-bay space, perhaps a storage room. The room measures 100 ft in length by 50 ft in width. The ceiling is 20-ft high, and an illuminance of 25 fc is desired at the workplane, which is chosen to be the floor. This space to be lit is shown in Figure 10.16.

To begin the design process, a suitable luminaire must be chosen. Metal halide technology is a common choice for this type of application, but before making that choice, some drawbacks of HID lighting should be contemplated. If power is

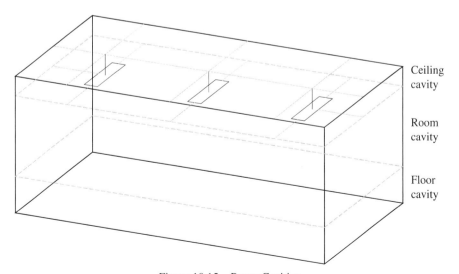

Figure 10.15   Room Cavities

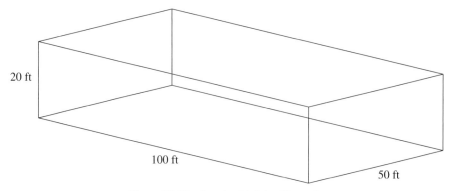

Figure 10.16   Interior Lighting Example

interrupted to a HID lighting system, a considerable time period of up to 20 minutes must elapse while the bulbs cool before they can be relit. Also, a storage room may be unoccupied for extended periods, suggesting lighting controls to conserve energy such as occupancy detectors. And since tasks requiring high illuminance levels will probably not be done in a storage area, multiple lighting levels may by useful to provide an appropriate level of lighting while conserving energy. Multiple lighting levels require multiple ballasts per luminaire, so a multiple-ballast luminaire must be selected to allow multiple lighting levels. After some careful thought, the designer may choose to forego the HID technology in favor of high-output fluorescent lighting designed especially for high-bay applications.

An appropriate high-output fluorescent luminaire is described in Figure 10.17.

Full-Body Fluorescent High-Bay Luminaire

6-T5HO lamps

90° Electronic Ballast, $\leq$ 10% total harmonic distortion, programmed rapid start

Before starting the calculation to determine the number of luminaires required to light our volume, the topic of *light loss* must be addressed. Various factors have been identified which result in a reduction of lumen production. Some of these factors are *recoverable*, meaning they can be reduced or eliminated with maintenance, while

Figure 10.17   Selected Luminaire for Example. (*Courtesy of Lithonia Lighting—an Accuity Brands Company.*)

TABLE 10.7 Light Loss Factors

Recoverable
  Lamp lumen depreciation factor
  Luminaire dirt depreciation factor
  Room surface dirt depreciation factor
  Lamp burnout factor
Nonrecoverable
  Luminaire ambient temperature factor
  Heat extraction thermal factor
  Voltage-to-luminaire factor
  Ballast factor
  Ballast-lamp photometric factor
  Equipment operating factor
  Lamp position (tilt) factor
  Luminaire surface depreciation factor

Source: *IESNA Lighting Handbook*, 9th edition.

others are inherent to the lighting system design and are said to be *nonrecoverable*. Table 10.7 shows various light loss factors.

If any of these factors can be estimated using standard practices, such as those described in the *IESNA Lighting Handbook*, these values should be used. Values of 1.0 should be used for any unknown factors. The total light loss factor is the product of all the individual factors listed in Table 10.7.

First, we will apply Eq. (10.63) to determine cavity ratio for the room cavity (between the workplane and the luminaires). Since the selected luminaires are surface-mounted, the ceiling cavity ratio is zero, and since the workplane is at the floor elevation, the floor cavity ratio is also zero.

$$\text{RCR} = \frac{5\,(20)(100+50)}{(100)(50)} = 3.0 \qquad (10.64)$$

Next, the reflectivity values of the ceiling, wall, and floor surfaces must be determined. While these values can be measured, they are often chosen based on industry practice. We will use the typical values shown below:

$$\rho_{\text{ceiling}} = 70\% \qquad (10.65)$$
$$\rho_{\text{walls}} = 60\% \qquad (10.66)$$
$$\rho_{\text{floor}} = 20\% \qquad (10.67)$$

Now we can reference photometric data from the manufacturer for the selected luminaire to determine the coefficient of utilization (CU) for the luminaire. Referencing Table 10.8, a CU of 0.64 is found by interpolating between the $\rho_{\text{walls}} = 70\%$ and $\rho_{\text{walls}} = 50\%$ values.

TABLE 10.8 Coefficients of Utilization for Selected Luminaire using the Zonal Cavity Method()

$\rho_{ceiling} \rightarrow$	80				70				50				30				10			0
$\rho_{walls} \rightarrow$	70	50	30	0	70	50	30	0	50	30	20	50	30	20	50	30	20	0
Room Cavity Ratio																			
0	0.87	0.87	0.87	0.87	0.85	0.85	0.85	0.73	0.82	0.82	0.82	0.78	0.78	0.78	0.75	0.75	0.75	0.73
1	0.81	0.78	0.76	0.73	0.79	0.77	0.74	0.65	0.73	0.71	0.70	0.71	0.69	0.68	0.68	0.67	0.66	0.64
2	0.75	0.70	0.65	0.62	0.73	0.68	0.64	0.56	0.66	0.62	0.59	0.63	0.61	0.58	0.61	0.59	0.57	0.55
3	0.69	0.62	0.57	0.53	0.67	0.61	0.56	0.49	0.59	0.55	0.51	0.57	0.53	0.51	0.55	0.52	0.50	0.48
4	0.64	0.56	0.50	0.46	0.62	0.55	0.50	0.44	0.53	0.49	0.45	0.52	0.48	0.44	0.50	0.47	0.44	0.42
5	0.59	0.51	0.45	0.40	0.58	0.50	0.44	0.39	0.48	0.44	0.40	0.47	0.43	0.39	0.46	0.42	0.39	0.38
6	0.55	0.46	0.40	0.36	0.54	0.46	0.40	0.35	0.44	0.39	0.36	0.43	0.39	0.35	0.42	0.38	0.35	0.34
7	0.52	0.43	0.37	0.33	0.50	0.42	0.36	0.32	0.41	0.36	0.32	0.40	0.35	0.32	0.39	0.35	0.32	0.30
8	0.48	0.39	0.34	0.30	0.47	0.39	0.33	0.29	0.38	0.33	0.29	0.37	0.32	0.29	0.36	0.32	0.29	0.28
9	0.45	0.36	0.31	0.27	0.44	0.36	0.31	0.27	0.35	0.30	0.27	0.34	0.30	0.27	0.34	0.30	0.27	0.25
10	0.43	0.34	0.29	0.25	0.42	0.34	0.28	0.25	0.33	0.28	0.25	0.32	0.28	0.25	0.32	0.28	0.25	0.23

Source: Acuity Brands Lighting.
$\rho_{floor} = 20\%$.

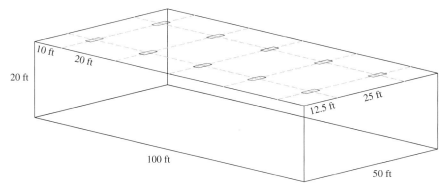

Figure 10.18 Solution to Interior Lighting Example

Next, the light loss factor is determined using the following individual factors. Guidance in determining these factors can be found in the *IESNA Lighting Handbook*.

$$\text{Lamp Lumen Depreciation Factor} = 0.95 \quad (10.68)$$

$$\text{Luminaire Dirt Depreciation Factor} = 0.894 \quad (10.69)$$

$$\text{Room Surface Dirt Depreciation Factor} = 0.95 \quad (10.70)$$

$$\text{Ballast Factor} = 0.98 \quad (10.71)$$

These factors allow us to calculate the total light loss factor.

$$\text{LLF} = (0.95)(0.894)(0.95)(0.98) = 0.79 \quad (10.72)$$

Now, the number of luminaires required can be calculated using Eq. (10.73).

$$\text{Number of luminaires} = \frac{E_{\text{maintained}} \times \text{Workplane Area}}{\text{Lamps per Luminaire} \times \text{Lamp Lumens} \times \text{CU} \times \text{LLF}} \quad (10.73)$$

Substituting into Eq. (10.73),

$$\text{Number of luminaires} = \frac{25 \times (100 \times 50)}{6 \times 4450 \times 0.64 \times 0.79} = 9.25. \quad (10.74)$$

Rounding this value up gives us a requirement of 10 luminaires to provide 25 fc of illuminance at the workplane of our volume. Spacing them uniformly gives the layout shown in Figure 10.18.

SUMMARY

Luminous intensity, measured in cd, is one of the seven basic SI measurement units. The quantity of light emitted by a source is the *luminous flux*, and is measured in lumens, where one lumen equals one candela-steradian. The total luminous flux incident on a surface per unit area is called *illuminance*, and is measured in footcandles, where one footcandle equals one lumen per square foot.

TABLE 10.9   Comparison of Luminous Efficacies

Technology	Efficacy (lm/W)	Luminous Efficiency (%)
Incandescent	15–35	2–5
Fluorescent	45–100	6.5–14.5
Mercury vapor	35–65	5–9.5
Low-pressure sodium	100–200	14.5–29
High-pressure sodium	85–150	12.5–22
Metal halide	65–115	9.5–17
Light-emitting diode	55–102	8–15

683 lm/W = 100% efficiency.

Color temperature quantifies the spectral content, or color, of the light. *Luminous efficacy* quantifies the conversion efficiency from electricity to light.

Typical efficacies for various lighting technologies are shown in Table 10.9.

Various lighting technologies have been developed and evolved over time, including *incandescent, low-pressure discharge* types such as *fluorescent* and *low-pressure sodium*, and *HID* types such as *mercury vapor, high-pressure sodium*, and *metal halide. LEDs* produce photons when electrons change energy states while propagating through semiconductor material.

Ballasting is required with discharge-type technologies to provide sufficient ignition voltage to start the lamp, to act as a constant current source when the lamp is starting, and to act as a constant power source when the lamp is operating.

Luminaires are fixtures designed to enclose lamps and provide specific dispersion patterns for the light.

Lighting systems are usually designed using illumination criteria determined by IESNA. These criteria are based on the anticipated use of the space being illuminated.

FOR FURTHER READING

IESNA Lighting Handbook, 9th edition, Illuminating Engineering Society of North America, 2000. ISBN 978-0-879-95150-4.

The Museum of Electric Lamp Technology at www.lamptech.co.uk traces the development of artificial electric light sources from their origins right up to the present day innovations. The technological developments in lamp construction, materials and manufacturing processes that have brought the electric lamp to its present popularity, convenience and high efficacy of converting electricity into light are detailed with historic examples, amply illustrated, along with references to historic catalogs and associated technical literature.

Underwriters Laboratories Standard for Safety for Luminaires, UL 1598, 2004.

QUESTIONS

1. How is the conversion efficiency from electricity to light stated?
2. How is the color of light quantified?

3. What is meant by "warm" light compared to "cool" light?
4. What three functions are performed by a ballast?
5. Describe how an incandescent lamp produces light.
6. Describe how a fluorescent lamp produces light.
7. What is the major difference between low-pressure sodium lamps and high-pressure sodium lamps?
8. Why does a high-pressure sodium lamp produce bluish-white light immediately after it lights, then gradually transitions to its characteristic orangish color?
9. Which heavy metal is used in all the discharge-type lighting technologies?
10. Describe two different ways the light color produced by LEDs can be controlled.

PROBLEMS

1. Estimate the change in expected life of an incandescent bulb if the bulb is operated at
   a. 95% of rated voltage.
   b. 105% of rated voltage.
2. Estimate the change in lumen output of an incandescent bulb if the bulb is operated at
   a. 95% of rated voltage.
   b. 105% of rated voltage.
3. Estimate the change in light output of a typical T5 fluorescent tube at an ambient temperature of
   a. 0°C.
   b. 20°C.
   c. 60°C.
4. A shopping center parking lot has a minimum measured horizontal illuminance of 0.65 fc, a minimum measured vertical illuminance of 0.35 fc, a maximum measured horizontal illuminance of 16 fc, a maximum measured vertical illuminance of 7 fc. Is this lighting design adequate per IESNA recommendations? Why or why not?
5. Using the luminaire described by Table 10.6, find the illuminance at a point on the ground 40 ft below the luminaire and 40 ft from the luminaire pole using a lateral angle of
   a. 0°.
   b. 45°.
   c. 90°.
6. Determine the room cavity ratio for a space 50-ft long, 30-ft wide, and 12-ft high.
7. If the room described in Problem 6 has luminaires suspended 2 ft below the ceiling, calculate the ceiling cavity ratio.
8. If the room described in Problem 6 has a work surface 3 ft above the floor, calculate the floor cavity ratio.

9. If an illuminance of 40 footcandles must be maintained at the work surface of the room described in Problem 6, how many 25,000-lm luminaires are needed assuming a ceiling reflectivity of 75%, a wall reflectivity of 50%, a floor reflectivity of 20%, and a light loss factor of 0.85.

10. Repeat Problem 9 for an illuminance of 35 footcandles, a ceiling reflectivity of 70%, a wall reflectivity of 60%, a floor reflectivity of 20%, and a light loss factor of 0.75.

CHAPTER 11

POWER FACTOR CORRECTION

OBJECTIVES

- Understand the importance of operating a power system at a good power factor
- Size a shunt capacitor for power factor correction
- Know the three shunt capacitor configurations for a three-phase system and the advantages and disadvantages of each
- Identify the basics of shunt capacitor protection

11.1 OVERVIEW

Shunt capacitors were first used to supply reactive power to inductive loads about 1914. These devices used mineral oil as the dielectric fluid between the capacitor plates. Until the early 1930s, widespread use of capacitors was limited due to high cost and large size and weight. In the 1930s, chlorinated aromatic hydrocarbon compounds (such as polychlorinated biphenyls, or PCBs) were routinely added to mineral oil to produce a dielectric fluid that was less flammable than pure mineral oil. This enhanced dielectric fluid also greatly reduced the size and weight of the capacitor. The failure rate also improved. By the late 1930s, shunt capacitors were used extensively to supply reactive power. World War II furthered the application of shunt capacitors, since the capacitor was the most economical means of furnishing reactive power to the factories supporting the war effort.

Capacitor dielectric fluids contained very high concentrations of PCBs. When PCBs were determined to be a carcinogen, synthetic dielectrics such as silicone oils replaced PCB-enhanced mineral oil as the dielectric fluid used in capacitors.

Today, capacitors are frequently connected to a power distribution system in shunt to supply reactive power to inductive loads. This reactive power supply counteracts the reactive power demand of induction motors, transformer magnetizing current, and other similar loads, thereby reducing the magnitude of the complex power phasor.

The effect of a shunt capacitor in parallel with an inductive load on the reactive (Q) and complex (S) power phasors can be seen in Figure 11.1.

Industrial Power Distribution, Second Edition. Ralph E. Fehr, III.
© 2016 The Institute of Electrical and Electronics Engineers, Inc. Published 2016 by John Wiley & Sons, Inc.

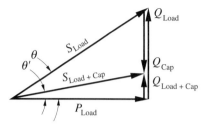

Figure 11.1 Phasor Diagram of Shunt Capacitor in Parallel with Load

The shunt capacitor also reduces the angle θ between the real power (P_{Load}) and complex power (S_{Load}) phasors. After the capacitor is paralleled with the load, the angle seen by the source is θ'. Since the cosine of this angle is the power factor, shunt capacitors can be used to raise the power factor of an inductive circuit closer to unity. This is known as power factor correction.

Operating at a power factor close to unity not only provides optimum system performance and voltage regulation, but also has economic appeal, since most industrial rate schedules penalize customers for poor power factor. Care must be taken, however, to not use too large a capacitor, as this will overcorrect the power factor, as shown in Figure 11.2. Overcorrection makes θ' a negative angle, resulting in a leading power factor. Circuits with leading power factors can experience unacceptably high voltages, especially during periods of light load.

Capacitors are usually purchased in single-phase units or *cans*. The cans are grouped in series and parallel combinations as required to produce the reactive power contribution desired. Although power capacitors have a capacitance rating expressed in microfarads, they are usually specified by a kilovar rating. The kilovar rating has a tolerance of minus zero to plus 15%, meaning that a 100 kvar capacitor will produce no less than 100 kvar nor more than 115 kvar at rated voltage. The maximum operating voltage of a power capacitor is 110% of rated voltage. Capacitors can be operated at voltages greater than 110% of their rated voltage under emergency conditions, but only for brief periods and not more than 200 to 300 times over the life of the capacitor. If the overvoltage lasts only a few cycles, the voltage magnitude can safely exceed 400% of rated voltage, but overvoltage that lasts for minutes should be limited to 135% of rated voltage to prevent damage to the capacitor. Consult the capacitor manufacturer for acceptable overvoltage limits.

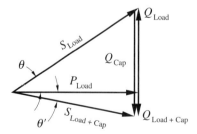

Figure 11.2 Phasor Diagram of Overcorrected Power Factor

11.2 CONFIGURATION

Three-phase capacitor banks can be connected in *delta*, *wye*, or *grounded wye*. The optimal configuration depends on whether the system is grounded or ungrounded, fusing considerations, and concerns about harmonics. Capacitors do not generate harmonic voltages, but since the impedance of a capacitor is inversely proportional to the frequency, capacitors offer a low-impedance path for high-frequency currents. This may cause harmonic currents to flow that would not flow without the presence of the capacitor. Harmonic voltages usually originate from overexcited transformers and generators.

11.2.1 Delta

Delta-connected capacitor banks are frequently used on ungrounded systems, especially if fault current is not excessive. The capacitors must be rated for line-to-line voltage. Delta-connected capacitors do not allow third harmonic currents to flow in the line currents, so interference with communication systems is not a concern. When fault currents become high, protecting a delta-connected bank can be difficult, so an ungrounded wye configuration may be preferable in these situations.

Delta-connected capacitor banks are subject to *neutral inversion* in the event that one phase becomes open-circuited. With a delta or ungrounded wye system, the system neutral (point of zero potential under balanced load conditions) is not connected to any particular reference such as ground. Consequently, the neutral is free to "float" as load imbalance varies. This phenomenon is most easily visualized on an ungrounded wye system, as shown in Figure 11.3.

When a phase of an unbalanced wye system becomes open-circuited, the neutral becomes unstable. Its position is no longer mathematically determinate, as two positions satisfy the physical constraints, as shown in Figure 11.4.

The neutral position at the intersection of the two dashed phasors, as shown in Figure 11.5, is displaced substantially from the position of the balanced system neutral. This is known as *neutral inversion*. Similarly, the neutral position of a delta system can become inverted if the delta-configured voltage phasors invert when a phase is open-circuited.

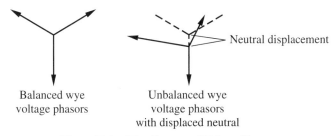

Figure 11.3 Wye-Connected Voltage Phasors

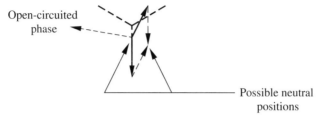

Figure 11.4 Wye-Connected Phasors with Open-Circuited Phase and Unstable Neutral

11.2.2 Wye

Ungrounded wye-connected capacitor banks can be used on either grounded or ungrounded systems. On ungrounded systems where fault current is high, ungrounded wye-connected capacitors are preferable to delta-connected capacitors because protection coordination is easier. If one phase of the three-phase bank short circuits, full line-to-line voltage will be imposed on the other two phases, so ungrounded wye-connected capacitors must be rated for line-to-line voltage. Like delta-connected banks, ungrounded wye-connected capacitors do not propagate third harmonic currents, so communication system interference does not occur. Ungrounded wye-connected capacitors are also subject to neutral inversion. High-voltage capacitor banks, such as those found in substations, are often connected ungrounded wye.

11.2.3 Grounded Wye

Grounded wye-connected capacitor banks are sometimes used on grounded systems. They are never used on ungrounded systems, as they would introduce a ground source. Grounded wye capacitors are easy to protect, but fault currents can be very high. Harmonics are also a concern with grounded wye capacitors, since zero-sequence currents can flow. Because of the grounded neutral, full line voltage can never be imposed across a capacitor unit. Consequently, the units need only be insulated for line-to-neutral voltage.

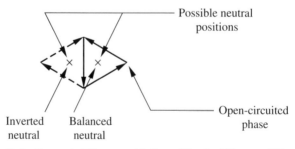

Figure 11.5 Delta-Connected Phasors with Open-Circuited Phase and Unstable Neutral

11.3 SIZING AND PLACEMENT

Capacitor banks must be properly sized to function correctly. If the capacitor is energized at all times (or *fixed*), care must be taken to not oversize the capacitor. If the reactive power furnished by the capacitor exceeds the reactive power demand of the load, the power factor will go leading. Leading power factor can cause poor voltage regulation resulting in high voltages during periods of light load. Buses with a relatively constant reactive power requirement are good candidates for fixed capacitors. To avoid very leading power factors, the capacitor should be sized such that the power factor does not exceed unity even when a predetermined percentage of the connected load is not energized. This percentage varies with the expected diversity of the load. Therefore, a good understanding of the load characteristics is essential for proper capacitor sizing.

When the reactive demand of a bus varies substantially, the capacitor may have to be *switched*. During periods of high reactive demand, the capacitor can be energized, and when the reactive demand diminishes, the capacitor can be de-energized. Two critical devices are necessary to switch capacitors: a *switching device* and a *controller*.

The switching device is usually an oil or vacuum switch for smaller capacitors or a circuit breaker or circuit switcher for larger capacitors. Particular care must be taken when selecting a switching device. Some switches and breakers are specially designed to switch capacitors. The voltage and current waveforms in a capacitor circuit are almost 90° out-of-phase. This means that when the interrupting contacts open at a current zero current crossing, the voltage will be near a maximum. This can lead to restriking or reignition of the arc. While some restriking is acceptable, excessive arcing, while the contacts part, can result in switching surges, an increased maintenance requirement, or even failure of the switching device.

The controller utilizes logic based on voltage, current, time, or a combination of these parameters to determine when the switching device should be opened and closed. A timer-based controller is simple and inexpensive, but requires that the reactive demand of the load be predictable as a function of time of day. Voltage-based controllers can be used when the capacitor is connected at a point where the circuit voltage decreases as load increases. Current-based controllers can be used when voltage alone is not a satisfactory parameter for controlling the capacitor.

Whether a capacitor is fixed or switched, its location on the distribution system is critical to its performance and effectiveness. Reactive power should be produced as close to the point of consumption as practical. Transporting reactive power through the distribution system, especially through transformers, is very inefficient due to the high inductive reactance of the system. Reactive power losses and excessive voltage drop are the primary reasons for locating capacitors close to the inductive loads to which they are intended to furnish reactive power.

Often a capacitor is installed across the terminals of an induction motor. This is an ideal location since it is located at the point of reactive power demand. The same contactor that switches the motor also switches the capacitor. This means that the capacitor is online only when the motor is running. The size of a capacitor installed at the terminals of an induction motor should be limited as to not cause excessive voltage

TABLE 11.1 Maximum Capacitor Sizes for Direct Connection to the Terminals of an Induction Motor

HP	3600 rpm[a]		1800 rpm[a]		1200 rpm[a]		900 rpm[a]		720 rpm[a]		600 rpm[a]	
	max kvar	% amp decr.	max kvar	% amp decr.	max kvar	% amp decr.	max kvar	% amp decr.	max kvar	% amp decr.	max kvar	% amp decr.
10	3	10	3	11	3.5	14	5	21	6.5	27	7.5	31
15	4	9	4	10	5	13	6.5	18	8	23	9.5	27
20	5	9	5	10	6.5	12	7.5	16	9	21	12	25
25	6	9	6	10	7.5	11	9	15	11	20	14	23
30	7	8	7	9	9	11	10	14	12	18	16	22
40	9	8	9	9	11	10	12	13	15	16	20	20
50	12	8	11	9	13	10	15	12	19	15	24	19
60	14	8	14	8	15	10	18	11	22	15	27	19
75	17	8	16	8	18	10	21	10	26	14	32.5	18
100	22	8	21	8	25	9	27	10	32.5	13	40	17
125	27	8	26	8	30	9	32.5	10	40	13	47.5	16
150	32.5	8	30	8	35	9	37.5	10	47.5	12	52.5	15
200	40	8	37.5	8	42.5	9	47.5	10	60	12	65	14

[a] Synchronous speed.

at the motor due to self-excitation when the motor contactor is opened. The NEC states maximum permissible capacitor sizes for direct connection to the terminals of an induction motor. Rule 4606 of the NEC is summarized in Table 11.1, which shows maximum capacitor size and the associated decrease in line current.

If it is desired to keep the capacitor online constantly or switch it independent of a specific motor, placement of the capacitor is often limited to medium- or low-voltage switchgear. A cubicle of the switchgear is dedicated to a capacitor breaker, any control equipment associated with the capacitor, and the capacitor units themselves.

To maximize economy, there is a tendency to connect as many kilovars as possible to a single switching device. Since the switching device represents a substantial cost, this practice may make good economic sense, but it can lead to operating problems. When capacitors are too large, it may be possible that the bus voltage is too low with the capacitor off-line, and too high with the capacitor online. This obviously makes the capacitor useless. Installing multiple capacitors in smaller units may be necessary for operational flexibility, but problems involving back-to-back switching, addressed in the next section, must be configured.

11.4 CAPACITOR SWITCHING

The out-of-phase current produced by a shunt capacitor leads to interesting behavior during switching. Energizing a shunt capacitor will be discussed by analyzing the voltage and current waveforms produced by the capacitor breaker closing operation.

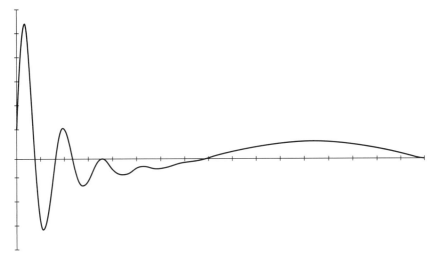

Figure 11.6 Current Waveform During Capacitor Energization

Then the de-energization process will be described by studying the same waveforms as the capacitor breaker opens.

Energizing a capacitor bank suddenly changes the voltage at the terminals of the capacitor. The capacitor appears to be a short circuit when first energized, so a sudden inrush of current flows. Since voltage across a capacitor cannot change instantaneously, the system voltage attempts to adjust to the voltage on the capacitor, which is normally zero. The time constant formed by the system inductance and the capacitance of the capacitor bank causes the inrush current to be damped until a steady-state 60-Hz current flows into the bank. This current waveform can be seen in Figure 11.6.

Since, under normal conditions, there is no trapped charge in the capacitor, and consequently no voltage present when energizing the bank, the peak inrush current can be calculated by using Eq. (11.1):

$$I_{peak} = \frac{\sqrt{2}}{\sqrt{3}} \frac{kV_{L-L}}{Z_{capacitor}} = \sqrt{\frac{2C_{capacitor}}{3L_{system}}} kV_{L-L} \quad kA \quad (11.1)$$

Alternatively, Eq. (11.1) can be rewritten in terms of more easily obtained quantities:

$$I_{peak} = \sqrt{2}\sqrt{I_{sc} \times I_{capacitor}} \quad A \quad (11.2)$$

where, I_{sc} is the available three-phase fault current, and $I_{capacitor}$ is the rated current of the capacitor bank.

The frequency of the inrush current is expressed by Eq. (11.3):

$$f_{inrush} = \frac{1}{2\pi\sqrt{L_{source}C_{capacitor}}} Hz \quad (11.3)$$

Eq. (11.3) can also be written in more convenient terms:

$$f_{inrush} = f_{system}\sqrt{\frac{I_{sc}}{I_{capacitor}}} \text{ Hz} \quad (11.4)$$

In most applications, the peak inrush current ranges between 1000 and 7000 A, while the frequency lies between 300 and 800 Hz. These parameters change significantly when there is an already energized capacitor bank nearby.

When a shunt capacitor is closed onto a system that has another energized shunt capacitor nearby, it is considered *back-to-back switching*. In this case, the inductance between the two capacitors determines the inrush current of the second bank. If the two capacitors are on the same bus, only a very small inductance, on the order of a few hundred microhenrys, exists between the capacitors, so the inrush current and frequency will be very high. The energy stored in the energized bank will discharge into the second bank, and current will oscillate between the two until the voltages on the two banks stabilize.

The peak inrush current experienced during back-to-back switching is given by Eq. (11.5):

$$I_{peak} = \sqrt{\frac{2}{3}\frac{kV_{L-L}}{Z_c}} = 1.75\sqrt{\frac{kV_{L-L} \times I_{cap1} \times I_{cap2}}{L_{eq}(I_{cap1} + I_{cap2})}} \text{ A} \quad (11.5)$$

where,

kV_{L-L} is the line-to-line voltage in kV

Z_c is the characteristic impedance formed by the two capacitor banks in series and the inductance between them (the charging current discharge path)

L_{eq} is the inductance between the two banks

I_{cap1} is the current rating of the first bank

I_{cap2} is the current rating of the second bank

The frequency of the inrush current is calculated using Eq. (11.6):

$$f = \frac{1}{2\pi\sqrt{L_{eq}C_{eq}}} = 9.5\sqrt{\frac{f_s \times kV_{L-L} \times (I_{cap1} + I_{cap2})}{L_{eq} \times I_{cap1} \times I_{cap2}}} \text{ Hz} \quad (11.6)$$

where, C_{eq} is the equivalent capacitance of the two banks in series, and f_s is the system frequency.

Damping of the high-frequency current will take place because of bus inductance and high-frequency resistance of the circuit. In practice, this damping averages about 10%.

Next, the topic of capacitor de-energization will be discussed. The solid line shown in Figure 11.7 represents the bus voltage and the dashed line represents the current drawn by a shunt capacitor bank.

The current (dashed line) leads the voltage (solid line) by nearly 90°. Now, the capacitor bank is de-energized by opening its circuit breaker. The breaker contacts

11.4 CAPACITOR SWITCHING

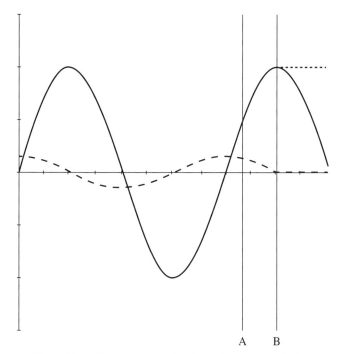

Figure 11.7 Waveforms During Capacitor De-energization

begin to part at time A, but due to arcing between the breaker contacts, current remains flowing until the current waveform crosses zero at time B. The current stops flowing, but the trapped charge in the capacitor keeps the capacitor voltage constant at a value near the system peak voltage, shown by the dotted line. Meanwhile, the voltage on the system side of the opening breaker continues its 60 Hz oscillation, decreasing toward zero. The breaker contacts are still opening, and have only been opening for a few milliseconds, so their separation is slight and the dielectric strength across the contacts is just building.

The voltage difference between the opening contacts continues to grow, and at time C in Figure 11.8, this voltage exceeds the dielectric strength between the opening breaker contacts, and a *reignition* occurs.

A reignition is the reestablishment of the arc within $1/4$ cycle of interruption. If the arc is reestablished after $1/4$ cycle of interruption, the term *restrike* is used. Reignitions and restrikes do not always occur when opening a shunt capacitor circuit, but are quite common, and must be anticipated when designing the capacitor bank.

When the reignition occurs at time C in Figure 11.7, the voltage on the capacitor cannot change abruptly, so the system voltage tries to match the capacitor voltage by rapidly increasing. As the system voltage changes abruptly, an inrush current flows into the capacitor bank. The frequency of the inrush current is given by Eq. (11.1).

The voltage recovers to its normal sinusoidal waveform by transitioning through a high-frequency oscillation period with a frequency equal to that of the inrush current. The peak value of the inrush current is defined by Eq. (11.2). This peak value

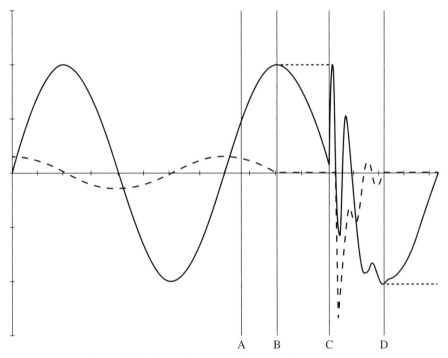

Figure 11.8 Reignition During Capacitor De-energization

can become very high when the system impedance is low, giving rise to very high voltage transients. For this reason, surge arresters are often installed with switched capacitor banks.

As time progresses to point D in Figure 11.8, the capacitor current is again interrupted, the system voltage returns to its normal sinusoid, and the trapped charge on the capacitor holds the capacitor voltage near the negative peak value of the system voltage. The breaker contacts are still parting. Keep in mind that it takes several tens of milliseconds for circuit breaker contacts to part fully.

At this point, if the dielectric strength of the gap between the breaker contacts is greater than the voltage difference across the contacts, the arc will not reestablish and the current will remain interrupted. Permanent arc extinction can occur after the first interruption, but sometimes one or two reignitions occur first. The critical issue is that the arc must be permanently extinguished by the time the contacts are fully open. If the breaker is designed properly for the application, this will always be the case, since the dielectric strength of the gap between the fully parted contacts greatly exceeds the highest voltage that will be experienced.

At time E in Figure 11.9, a second reignition occurs, resulting in another abrupt change in system voltage, another inrush of current into the capacitor, and subsequent damping until the arc is again extinguished at time F. At time F, the breaker contacts have a greater separation than they did at time D, providing a higher dielectric strength. This dielectric strength is sufficient to prevent reignitions and restrikes,

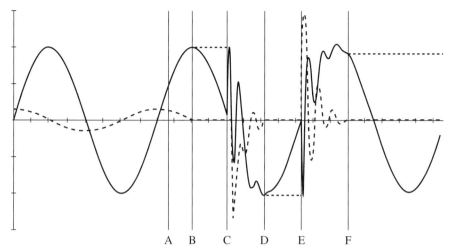

Figure 11.9   Second Reignition During Capacitor De-energization

so the capacitor remains de-energized. The trapped charge in the capacitor holds the voltage constant for a short time, but discharge resistors built into the capacitor bank bleed off this current over a period of minutes. IEEE standards 18-2012 and 1036-2010 on capacitor ratings and applications require that the residual voltage on a capacitor bank be no greater than 50 V at 5 minutes after de-energization.

11.5 HARMONICS

Harmonic currents and voltages can be a serious problem on power systems. In addition to increasing loading on system components, harmonic currents tend to cause saturation of magnetic circuits and can cause interference in nearby communication circuits.

While harmonic currents and voltages present during capacitor switching can be potentially disrupting or damaging to power system components, they are transient in nature and are quickly damped by the system impedance. Grounded capacitor banks, however, provide a path to ground which propagate harmonic currents and voltages in the steady state.

Harmonic currents and voltages are analyzed using *symmetrical components*. Positive and negative sequence harmonics are usually of little concern since both sequence currents add to zero in the three phases. Zero-sequence harmonics, however, are significant since they are in-phase in all three phases and are arithmetically additive.

Zero-sequence harmonics are of the order $(6n-3)$ where n is any integer. This means that the harmonics of primary concern are the 3rd, 9th, 15th, 21st, 27th, … Even harmonics are not present on a three-phase power system unless some device capable of half-wave rectification exists, which is usually not the case. Since zero-sequence current can only flow in a grounded wye connection, connecting the capacitor bank in delta or ungrounded wye will avoid the harmonic problem. Shunt capacitors can be beneficial on systems with substantial harmonic voltages. Because of their low impedance at high frequencies, shunt capacitors function like filters to block the flow of harmonic currents. Capacitors are sometimes installed near sources of harmonics for this reason.

11.6 RESONANCE

Shunt capacitors can create undesirable resonance conditions. Switching surges, which are not resonance conditions but have similar characteristics, can also result in transient voltages high enough to cause arrester failure and equipment damage. Selecting devices that are capable of switching the capacitor without excessive arc restriking can mitigate the switching surge problem. Resonance must be anticipated whenever capacitors are installed, and can be verified by performing a transient study. If critical frequencies correspond to frequencies that may be present on the power system, some attempt of detuning the resonant circuit should be made.

11.7 PROTECTION

Large transmission-class shunt capacitors are sometimes *fuseless*, relying on protective relays to de-energize them in the event of an internal fault or an excessive number of capacitor units shorted. Some large and mostly all smaller shunt capacitors are protected by fuses, which blow if excessive current flows from the capacitor bank. One fuse can protect a single capacitor unit (individual fusing), or a group of units (group fusing). The fusing methods are shown in Figure 11.10.

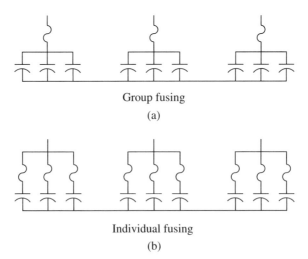

Figure 11.10 Individual and Group Capacitor Fusing

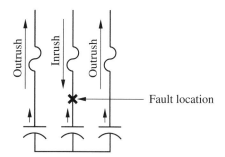

Figure 11.11 Outrush Current

Fuses must be selected that can withstand steady-state and transient currents so spurious fuse operations do not occur. The fuses must, however, effectively remove a failed or failing capacitor unit from service without detrimental effects to the rest of the system.

Fuse links are usually sized at least 125–135% of the rated capacitor current to allow for overvoltage conditions, capacitor rating tolerance, and harmonic content. Rated current for a three-phase capacitor bank is expressed by Eq. (11.7):

$$I_{\text{rated}} = \frac{\text{kvar}_{3\phi}}{\sqrt{3}\,\text{kV}_{\text{L-L}}} \qquad (11.7)$$

NEMA Type K or T tin link fuses are commonly used to protect capacitors. K-links blow faster than T-links, so a fuse coordination analysis must be done against upstream protective devices as well as against the capacitor tank rupture curve. Tank rupture curves are inverse time-current characteristic curves similar to fuse curves and are provided by the capacitor manufacturer.

Another consideration when fusing capacitors is that of *outrush*. When a capacitor unit fails, the voltage drops substantially on the faulted phase, but the other two phases also see a depression in voltage due to the capacitive coupling between the phases. Consequently, the capacitors on the unfaulted phases discharge their stored energy in an attempt to restore voltage, as shown in Figure 11.11. This discharge is called *outrush*. The fuses protecting the capacitors must be able to tolerate outrush current without blowing.

Finally, the interrupting rating of the fuse cannot be exceeded. For paper/film capacitors, energy should be limited to 10,000 J (approximately 3100 kvar) per parallel group to avoid exceeding the fuse's interrupting rating.

SUMMARY

Shunt capacitors can be used to provide reactive power to inductive loads, thereby raising or *correcting* the power factor. Care must be taken to not drive the power factor too far into the leading range by *overcorrecting* or adding too much capacitance.

Three-phase capacitor banks can be connected *delta*, *wye*, or *grounded wye*, each configuration having advantages and disadvantages. When delta or ungrounded wye banks are *single-phased*, the possibility of *neutral inversion* exists. Capacitors should be installed as close to the reactive power load as practical. Installation at the motor terminals can result in overvoltages due to self-excitation when the motor contactor is opened if the capacitor is oversized.

Switching capacitors can lead to high-transient voltages, especially if there is another energized capacitor in the vicinity of the one being switched. Inductors may have to be installed in series with the capacitor switching device to limit the overvoltages during *back-to-back switching*.

Harmonics and *resonance* are conditions that must be anticipated and avoided. Fuses in either an individual or a group arrangement protect capacitor banks. Fuses must withstand transient currents, harmonics, and *outrush* without blowing.

FOR FURTHER READING

Adams, R. A., Middlekauff, S. W., Camm, E. H., and McGee, J. A., *Solving Customer Power Quality Problems Due to Voltage Magnification*, 1998 IEEE PES Winter Meeting, Tampa, Florida, February, 1998.

Camm, E. H., *Shunt Capacitor Overvoltages and a Reduction Technique*, 1999 IEEE PES Transmission and Distribution Conference and Exposition, Overvoltages: Analysis and Protection panel session, New Orleans, Louisiana, April, 1999.

QUESTIONS

1. How does a capacitor "correct" the power factor of an inductive load?
2. What are the problems introduced by sizing a capacitor too large?
3. Why are power capacitors commonly rated in kvar instead of farads?
4. How does system voltage affect the reactive power output of a shunt capacitor?
5. Construct a chart summarizing the advantages and disadvantages of connecting a shunt capacitor as delta versus wye versus grounded wye.
6. What parameters must be considered when sizing a fuse to protect a shunt capacitor?
7. Why is switching a capacitive load more demanding on the switching equipment than switching other types of loads?
8. Why is back-to-back switching of particular concern to the engineer?
9. What is the difference between *reignition* and a *restrike*?
10. What can be done to lessen the effects of overvoltages during capacitor switching?

PROBLEMS

1. A load center feeds the following loads:
   - 150 kW lighting load
   - 2–75 hp induction motors, 85% efficiency at 88% power factor

- 1–100 hp synchronous motor, 90% efficiency, unity power factor
    a. Draw a phasor diagram showing the power components served by this load center.
    b. What is the power factor seen by the load center?
    c. Two more induction motors identical to the two existing ones are added to the load center and the synchronous motor is removed. What is the power factor now seen by the load center?
    d. If the power factor must be maintained at 98% lagging or better, how much capacitance (in kvar) must be added after the changes described in Part C are made?
    e. If the capacitor sized in Part D is connected ungrounded wye to the 480-V load center, what is the minimum current rating of the capacitor bank?

2. A 1200 rpm squirrel-cage motor rated at 50 hp, 90% efficiency, 90% power factor is installed on a 480-V bus.
    a. What is the best power factor correction that can be achieved by connecting a shunt capacitor to the motor terminals without causing self-excitation of the motor?
    b. What is the line current before and after the addition of this capacitor?

3. Calculate the frequencies of the lowest five zero-sequence harmonics on a 60-Hz system.

4. What is the rated current of a capacitor bank rated 2400 kvar at 13.8 kV?

5. The capacitor bank in Problem 4 is installed on a bus with 26 kA of fault current. What is the peak amplitude and frequency of the inrush current?

6. A second identical capacitor bank is installed on the bus described in Problem 5. Assuming an inductance of 150 microhenrys between the capacitors, what is the peak amplitude and frequency of the inrush current when the second bank is energized while the first bank is already energized?

7. How much inductance must be placed in series with the second capacitor bank in Problem 6 to limit the back-to-back inrush current to 2750 A?

8. What is the change in frequency of the inrush current when the inductor is added in Problem 7?

9. What is the largest capacitor bank (in kvar) that can be switched on a 4.16 kV bus with 42 kA of fault current without exceeding 6500 A of inrush current?

10. A 2400 kvar capacitor bank is installed on a 7.2 kV bus with 28 kA of fault current. How large can a second capacitor bank be sized (in kvar) such that the inrush current when the second bank is energized does not exceed 3600 A? Assume an inductance of 200 microhenrys between the capacitor banks.

CHAPTER *12*

POWER QUALITY

OBJECTIVES

- Understand the historical perspectives related to power quality
- Know the symptoms of power quality problems
- Be able to quantify power quality parameters
- Understand the effects of transients and harmonics on a power system

12.1 OVERVIEW

Electrical devices are designed to utilize electricity at a given voltage and frequency. If the power supplied to the device deviates from these design parameters, the performance of the device will not be optimum. When deviations from the design parameters become excessive, the device may not operate at all.

The term *power quality* describes a broad subject including continuity of service, variation in voltage magnitude, transient voltages and currents, and harmonic content in the sinusoidal waveforms. Each of these components will be investigated in detail, but first, a historical perspective of power quality is presented.

12.2 HISTORICAL PERSPECTIVE

Early in the development of power systems, the most common load elements were devices such as induction motors and discharge lighting systems. These devices were much more tolerant of voltage and frequency excursions than most of the sensitive electronic devices such as microprocessors that are commonplace today. Consequently, more problems with power quality are *perceived* today than in years past.

Ironically, many of the devices that are most sensitive to power quality problems are the very devices that create many of the voltage and frequency problems seen on today's power systems. Solid-state devices such as variable speed motor drives, switching power supplies, and rectifier/inverter circuits are major producers of *harmonics*, which distort the sinusoidal shape of the voltage and current waveforms. Arc discharge devices, such as furnaces, fluorescent and high-intensity discharge lighting systems, and welding machines, have been in use for many years, but today there

---

Industrial Power Distribution, Second Edition. Ralph E. Fehr, III.
© 2016 The Institute of Electrical and Electronics Engineers, Inc. Published 2016 by John Wiley & Sons, Inc.

are numerous sensitive loads distributed throughout the system that serve as "power quality detectors," letting people know when there is a power quality problem by malfunctioning. A common example to which we can all relate is a digital clock blinking "12:00," which indicates an interruption in service has occurred. Without the blinking clock, most power system customers would be unaware of what could have been a very brief interruption, but the clock reveals the interruption to all.

Most utility systems are significantly more reliable today than they were decades ago, yet customers are more aware of power quality issues today than ever. Improving power quality may seem like an uphill battle—even when the system improves, customers notice more problems. This does not have to be the case. Despite the fact that ever higher standards for power quality are being demanded by power system customers, significant improvements over today's reliability levels can be realized. To do so, a thorough understanding of the phenomena that cause power quality problems must be mastered. Then, mitigation measures can be designed and implemented.

One of the first concepts to understand is that power quality is *perceived* by the power system customers. Perception means everything—if a problem is perceived, it is real. Of course, we have quantitative means of measuring many components of power quality, but these numbers alone often mean little. One customer may not notice any problems when a specific event occurs, while the customer next door may experience major complications, causing not only inconvenience but also expense. This chapter explores many aspects of power quality, and starts by examining how a rather subjective concept based much on perception can be objectively measured.

12.3 QUANTIFYING POWER QUALITY

In the early days of electric power, a common power quality problem was evident by an obvious *flicker* in incandescent lights. Distortions in the voltage waveform, often caused by motors, produced a noticeable fluctuation in the intensity produced by incandescent bulbs. Individuals perceived this flicker differently. One individual may have barely noticed the fluctuations in lamp intensity, while another may have found the flicker very annoying to the point of irritation. *Flicker charts* were developed to assist the design engineer in keeping voltage flicker within tolerable limits. These charts were quite subjective, using terms such as *borderline of visibility* and *borderline of irritation* when expressing the allowable percent voltage fluctuation as a function of fluctuations per unit time. A typical flicker chart is shown in Figure 12.1.

As more sensitive electronic loads began to become common on power systems, a more quantitative measure of power quality became necessary. In 1983, Federal Information Processing Standards (FIPS) Publication 94 was issued to address power quality requirements of solid-state devices. This publication was endorsed by the Computer Business Equipment Manufacturers Association (CBEMA), and a graph indicating voltage level tolerance as a function of duration became known as the CBEMA curve, and is shown in Figure 12.2.

After CBEMA changed its name to the Information Technology Industry Council (ITIC), a working group was formed with the Electric Power Research Institute

12.3 QUANTIFYING POWER QUALITY

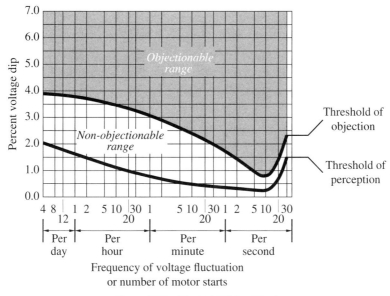

Figure 12.1   Typical Flicker Chart

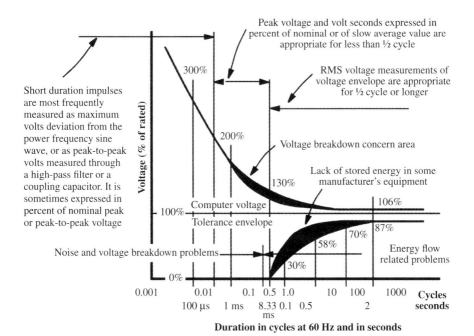

Figure 12.2   CBEMA Curve from FIPS Publication 94

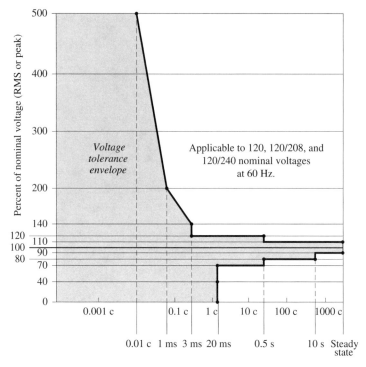

Figure 12.3   ITIC Curve

(EPRI) to revise the original CBEMA curve to accurately reflect the performance of computers and their peripherals, as well as other common information technology devices. The specific data points shown on the ITIC curve in Figure 12.3 also assist in the design of electronic equipment.

As seen in Figure 12.3, even sensitive computer equipment can tolerate a total power outage for 20 ms, or just over a cycle at 60 Hz, without malfunctioning. But for extended operation (10 seconds or more), a voltage tolerance of $\pm 10\%$ is required.

Next, some specific power quality parameters will be discussed.

12.4   CONTINUITY OF SERVICE

The most obvious power quality problem is total loss of power. Power system customers in developed countries have extraordinary expectations when it comes to continuity of service. Availability is expected to be on the order of 99.99% (less than an hour of service interruption per year), and this value has been and will continue to increase as more critical uses for electricity are implemented. Power interruptions can be *momentary*, as is the case when a circuit breaker trips to interrupt a temporary fault and then successfully recloses to restore power. Momentary interruptions last on the order of cycles. Sensitive electronic loads can tolerate a total loss of voltage for just over an electrical cycle by utilizing the energy stored in their power supply capacitors. This energy, though, is depleted in a few milliseconds—far less time than it takes for a circuit breaker to trip and reclose. As a result, momentary interruptions will disrupt the operation of most electronic circuits. If disruption of service is not acceptable, an *uninterruptable power supply*, or *UPS*, could supply the sensitive equipment. A UPS rectifies line current to charge batteries, whose output is inverted to serve the load. When input power to a UPS is interrupted, the batteries supply the load through an inverter circuit. The UPS can continue to furnish the load until the batteries are depleted. UPS batteries are typically sized to provide enough time to transfer to an alternate power source, such as a standby generator or an alternate distribution feed.

Motor loads can typically tolerate loss of power for several cycles without adverse effects. However, if power is interrupted for too long, the inertia of the load will slow the rotation of the rotor such that when power is restored, the angle between the rotor position and the corresponding position of the rotating stator field becomes very large, resulting in high transient voltages and currents. These transients produce high torques, which can be damaging to the driven load. Because of this, care must be taken when reenergizing an already rotating motor. If there is any doubt as to how much angle difference between the stator and rotor fields accumulated during the outage, reenergization of the motor should be intentionally delayed until the field in the rotor has sufficiently decayed.

Fluorescent lighting systems will tolerate momentary power interruptions quite well. At the worst, a slight blink may be detectable, but in most cases, the momentary interruption will go unnoticed. Likewise, resistive heating devices are unaffected by momentary interruptions. Arc discharge lighting systems, however, must cool down before they can reignite. This results in an extended downtime for lighting systems using high-pressure sodium, mercury vapor, and metal halide technologies.

Sometimes, a fault is not cleared during the few cycles between the tripping and reclosing of a circuit breaker. When this is the case, the breaker recloses into the fault and subsequently trips again. After the second tripping of a circuit breaker, an intentional time delay is usually added before trying to reclose again. Hopefully, during this time delay, the fault will be cleared. The service interruption during the time delay is classified as a *brief* interruption. Brief interruptions last on the order of seconds.

Running motors are perhaps most vulnerable to brief interruptions. To protect against damage caused by high torques produced when reenergizing, *undervoltage dropout* protection is often provided by motor starting circuits. With undervoltage dropout protection, when a running motor loses power, its contactor opens, preventing it from automatically restarting when power is restored. See Chapter 8 for details on this type of motor protection.

Lighting systems will shut down during brief interruptions, but resistive heating devices will tolerate brief interruptions without a perceptible degradation in performance.

Long-duration interruptions last in excess of a minute. The most likely cause of a long-duration interruption is equipment failure. Many times, equipment failure is caused by bad weather such as lightning, wind, or ice. Since long-duration outages require human intervention to resolve, the time for restoration can range from a few minutes to replace a fuse to several days or even weeks in the event of a widespread disaster such as a hurricane or earthquake.

If a long-duration interruption is not acceptable, measures must be taken when the power system is designed. To protect against equipment failure, redundant equipment can be installed. For example, if a transformer fails, provisions can be made when the system is designed so the load from the failed transformer can be fed from another transformer by operating circuit breakers or switches. Many of the source configurations presented in Chapter 1 use this technique.

In the event that the primary source of power (the utility connection) is lost, an alternate source of power is necessary. Due to the nature of many long-duration interruptions, it may be desirable to have the alternate power source on-site, since the same event that caused disruption of the primary source could affect the alternate source too. Most often, the alternate power source is a diesel generator. Any loads that must remain energized from the time that the utility source is lost until the diesel generator can pick up the load are often supplied by a UPS or similar energy storage device. To minimize the size of the standby generator, many times only critical load is fed from the diesel, and nonessential load remains de-energized until the primary source is restored. If this approach is taken, critical load and noncritical load should be supplied from different load centers to facilitate switching between the primary and backup sources.

12.5 VOLTAGE REQUIREMENTS

Voltage regulation is the ability to maintain steady-state voltage. The American National Standards Institute (ANSI) Standard C84 defines Range "A" voltage as $\pm 5\%$ of the nominal value. Range A is the voltage range at which utilities are required to deliver electricity to their customers under normal conditions. Disturbances on the utility's system, however, can cause excursions from Range A to Range B, which is $\pm 10\%$ of the nominal value. Excursions beyond Range B can be problematic to sensitive electrical equipment.

ANSI Standard 1100–1992 defines excursions outside the parameters of Range B voltage. Voltage *sags* are "a RMS reduction in the AC voltage, at the power frequency, for duration from a half a cycle to a few seconds." Voltage *swells* are similarly defined as brief increases in voltage beyond the Range B limit.

Voltage sags are caused by such events as motor starts, sudden load increases, loads that draw large intermittent currents (e.g., copiers and laser printers), loose wiring connections, and faults occurring elsewhere on the power system. Voltage swells are caused by sudden load decreases, the energization of a de-energized circuit, open neutral conductors, loose wiring connections, and the loss of a phase of a three-phase system (single-phasing).

Both sags and swells can be detrimental to equipment performance and, in extreme cases, damaging to equipment. DC power supplies are particularly susceptible to damage caused by reduced input voltage, because as the voltage input to a DC power supply declines, the power supply must draw energy from its reservoir capacitor to maintain proper DC output voltage. Electrolytic capacitors are not designed for deep discharge, so their large charge and discharge currents cause heating in the capacitor, reducing its life. Rectifiers and power transistors also draw higher currents when voltage is reduced, raising their junction temperatures and lowering the longevity of the semiconductor. Logic circuits are susceptible to logic errors if their input voltage is not adequate.

High voltage can punch through the semiconductor junctions in power electronic devices and cause a device to fail. Typically, junctions can break down when the operating voltage exceeds the design voltage of the device by more than 10% for several cycles.

12.6 TRANSIENTS

Steady-state sinusoidal voltage and current waveforms are perturbed every time a switching operation occurs on the power system. The switching operation may be a circuit breaker opening to clear a fault or to de-energize a line or transformer, or closing to energize a line or transformer. The switching operation could also be a switching device energizing or de-energizing a shunt capacitor bank. All of these switching functions introduce transient currents which give rise to transient voltages on the power system.

ANSI Standard 1100 defines *transient* as "a subcycle disturbance in the AC waveform that is evidenced by a sharp brief discontinuity of the waveform." These disturbances, although short-lived (typically in the tens of microseconds), can be of a very substantial amplitude. Voltage transients on a low-voltage system can exceed 10,000 V. Two types of transients can be found on any power system: *oscillatory* and *impulse*.

Oscillatory transients are the more common of the two types. They consist of an initial perturbation followed by a rapidly damped decay back to the steady state. Oscillatory transients are most often caused by lightning strikes, capacitor switching, and power semiconductor operation.

Impulse transients are sudden spikes that occur sporadically along the waveform. These transients are usually caused by motor load switching, loose wiring, and poor grounding.

The primary means of controlling transients and mitigating their impact on electrical equipment is the use of surge arresters. Arresters can be installed in switchgear to control the transients caused by load and capacitor switching as well as those caused by lightning strikes. Also, equipment wiring and grounding must be properly installed and maintained to limit the production of transients.

12.7 HARMONICS

Not all voltage and current waveforms on the power system are sinusoidal with a frequency equal to the operating frequency of the system. Those with a frequency

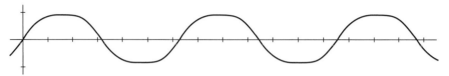

Figure 12.4 Sine Wave with 10% In-Phase Third Harmonic

equal to the power system frequency are *fundamental frequency* waveforms. Others with frequencies that are integer multiples of the fundamental frequency (e.g., 120 Hz, 180 Hz, 240 Hz, etc., on a 60 Hz system) are *harmonic* waveforms. The integer multiplier defines the *order* of the harmonic.

As harmonics are added to the fundamental frequency, the sinusoidal waveform becomes distorted. The level of distortion can vary from very slight to severe. Figure 12.4 shows a sinusoid containing a third-order harmonic with an amplitude of 10% of the fundamental. The in-phase third harmonic merely flattens the top of the sinusoid, and is hardly noticeable.

By adding equal amplitudes of fifth, seventh, ninth, eleventh, and thirteenth harmonic to the waveform shown in Figure 12.4, the level of distortion increases tremendously, as is shown in Figure 12.5.

The previous examples contained only odd-ordered harmonics. This is because only odd-ordered harmonics are typically found on a power system.

Harmonics can be classified based on their behavior in sequence networks. On a three-phase system, harmonics can be considered as symmetrical components. The fundamental frequency is a *positive sequence harmonic*. If n is a positive integer, the orders of the positive sequence harmonics is defined by Eq. (12.1):

$$h_{h1} = 3n - 2 \qquad (12.1)$$

Since even-ordered harmonics cannot exist unless half-wave rectification is present, and this is usually not the case, Eq. (12.1) can be modified to exclude the even-ordered harmonics, as shown in Eq. (12.2):

$$h_{h1} = 6n + 1 \qquad (12.2)$$

Negative sequence harmonics are those harmonics with an order defined by Eq. (12.3):

$$h_{h2} = 3n - 1 \qquad (12.3)$$

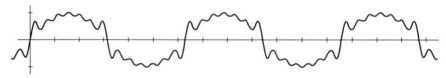

Figure 12.5 Sine Wave with 10% Third, Fifth, Seventh, Ninth, Eleventh, and Thirteenth Harmonics

Likewise, Eq. (12.3) can be modified to exclude the even-order harmonics:

$$h_{h2} = 6n - 1 \qquad (12.4)$$

Zero sequence, or *triplen harmonics* have an order defined by Eq. (12.5):

$$h_{h0} = 3n \qquad (12.5)$$

Adjusting Eq. (12.5) to exclude the even-order harmonics,

$$h_{h0} = 6n - 3 \qquad (12.6)$$

While positive and negative sequence harmonic currents tend to cancel in the line currents of a three-phase system, triplen harmonics are additive and tend to be of most concern. IEEE Standard 519–2014 establishes steady-state waveform distortion goals for designing electrical systems to supply nonlinear loads.

The harmonic content of any periodic waveform can be analyzed using *Fourier analysis*.

12.7.1 Fourier Analysis

While studying heat transfer in 1822, Joseph Fourier determined that all continuous periodic functions can be resolved into a series of sinusoidal waves of different frequencies. This brilliant breakthrough, known as Fourier analysis, has been applied in many branches of mathematics, science, and engineering. The sinusoid with the lowest frequency defines the *fundamental frequency*. The higher frequency sinusoids have frequencies that are integer multiples of the fundamental frequency. The higher frequency sinusoids are called *harmonics* of the fundamental frequency.

Fourier analysis is used to decompose a periodic waveform into its fundamental frequency sinusoid and harmonics. The Fourier series of a general periodic waveform is expressed as

$$f(t) = a_0 + \sum_{n=1}^{\infty} a_n \cos \frac{2n\pi}{T} t + b_n \sin \frac{2n\pi}{T} t, \qquad (12.7)$$

where the coefficients a_n and b_n are defined as

$$a_n = \frac{2}{T} \int_{-T/2}^{T/2} f(t) \cos \left(\frac{2n\pi t}{T} \right) dt \qquad (12.8)$$

$$b_n = \frac{2}{T} \int_{-T/2}^{T/2} f(t) \sin \left(\frac{2n\pi t}{T} \right) dt \qquad (12.9)$$

The fundamental frequency can also be called the *first harmonic*. The variable n indicates the *order* of the harmonic.

The constant coefficient a_0 represents a DC offset in the periodic waveform. It can be determined by calculating the area under one cycle of the periodic waveform and dividing it by the period:

$$a_0 = \frac{1}{T} \int_{-T/2}^{T/2} f(t)dt \qquad (12.10)$$

Another way to visualize the meaning of a_0 is as the average amplitude of the periodic waveform over an entire period. Using this interpretation of a_0, it is apparent that if the periodic waveform is symmetric about the horizontal axis, a_0 will have a value of zero.

The sine and cosine terms for a single frequency simply adjust the magnitude and phase of that harmonic. For a single frequency,

$$a \cos \theta + b \sin \theta = \sqrt{a^2 + b^2} \cos \left(\theta - \tan^{-1} \frac{b}{a} \right) \qquad (12.11)$$

Applying Eq. (12.11) to each frequency of the Fourier series would produce a series containing cosine terms only.

Using the Fourier series to synthesize a periodic waveform can be shown graphically. Using a Fourier series consisting of only sine terms, a symmetric square wave can be produced by adding only odd-ordered harmonics. The components are all in-phase ($t = 0$ is at a positive-going x-axis crossing). The first 11 harmonics are shown in Figure 12.6.

The waveform resulting from the first 11 harmonics is just an approximation of a square wave, thus the nonzero rise and fall times, the ripple, and the overshoot. To produce "perfect" square waves, we would have to add harmonic orders 13 through infinity.

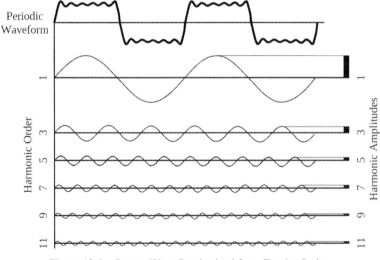

Figure 12.6 Square Wave Synthesized from Fourier Series

Figure 12.7 Fourier Transform of the Periodic Waveform in Figure 12.6

On the right side of Figure 12.6, bars representing the magnitudes of each harmonic are shown. These bars can be arranged as in Figure 12.7, producing a graphical representation of the Fourier transform of the original waveform.

A graphical representation like that shown in Figure 12.7 is displayed by a power quality analyzer, which uses a fast Fourier transform algorithm to process the input waveform in real time.

A periodic waveform may possess *symmetry* characteristics. If the negative lobe of the periodic waveform is an exact reproduction of the positive lobe with only the sign of the amplitude negated, the periodic waveform is said to have *half-wave symmetry*. Mathematically, for a periodic waveform to have half-wave symmetry,

$$f(x) = -f(x+180°) \tag{12.12}$$

A method of verifying half-wave symmetry is to slide the negative lobe of the wave to the left until its zero crossing coincides with the zero crossing of the positive lobe. If the wave possesses half-wave symmetry, the positive and negative lobes will be mirror images of each other, as can be seen in Figure 12.8.

A combination of odd-order harmonics with the fundamental will always produce a periodic wave with half-wave symmetry. The introduction of even-order harmonics destroys the half-wave symmetry of the periodic wave.

Harmonics on a power system are common, and are produced in a variety of ways. Iron-core devices such as transformers are common harmonic sources. When an iron-core transformer is energized with the secondary terminals open-circuited, a current path for excitation current (I_E) exists to neutral through the conductance and susceptance of the core, as shown in Figure 12.9.

The impedance of this circuit is very inductive, so the excitation current I_E lags the applied voltage V_A by almost 90°. As the current reaches its peak, the voltage is near a zero crossing. Also as the current peaks, the magnetic flux in the core peaks, and the iron approaches saturation. The core is intentionally designed to approach saturation when the magnetic flux peaks for economic reasons. The

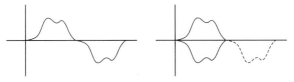

Figure 12.8 Graphical Test for Half-Wave Symmetry

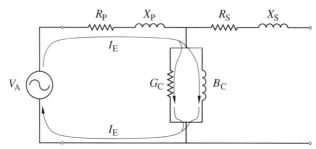

Figure 12.9   Transformer Circuit Model and Excitation Current

saturation characteristics of the core give rise to a "B–H curve" or graph of magnetic flux density as a function of magnetic field, commonly called a *hysteresis loop*.

The nonlinear excitation current I_E is rich in second and third harmonic content, and to a lesser degree, other odd-ordered harmonics. The predominant third harmonic component is approximately 90° out of phase (lagging) with the fundamental, producing a waveform as shown in Figure 12.10.

Iron-core devices are not the only source of harmonics on a power system. Furnaces and welders produce a great deal of harmonics, due to the arcing that is inherent to their operation. Even synchronous generators produce some fifth and seventh harmonic content due to the winding pitch and nonlinearities caused by the air gap between the rotor and stator circuits.

12.7.2   Effects of Harmonics

Harmonics can have many effects on a power system—and none of them are good. One of the more serious impacts of harmonics is the overheating of power system components, particularly circuit neutral conductors and transformers. Since harmonic currents are at a higher frequency than the fundamental frequency, skin effect is more pronounced than at the fundamental frequency. This increases the resistance significantly to currents at the higher frequencies, which in turn increases the power dissipated by these currents.

The increased power dissipation caused by the higher frequencies of current produce more eddy current losses than the fundamental frequency current acting alone. Additional eddy current losses increase the temperature rise of the core steel, resulting in a reduction of transformer life expectancy. Special transformers can be specified for applications with harmonic-rich loads. In fact, the NEC requires that transformers supplying loads with high harmonic content be of this variety, commonly known as *k-factor transformers*.

K-factor transformers are typically delta-wye transformers designed to operate at a reduced flux density. They have oversized delta windings to accommodate circulating triplen harmonic currents, and smaller-than-usual winding conductors paralleled to reduce skin effect and eddy current losses. The neutral has a current rating twice that of the secondary phase conductors at a basic impulse level (BIL) rating of 10 kV. The *k*-factor of the load must be calculated to specify the proper *k*-factor transformer.

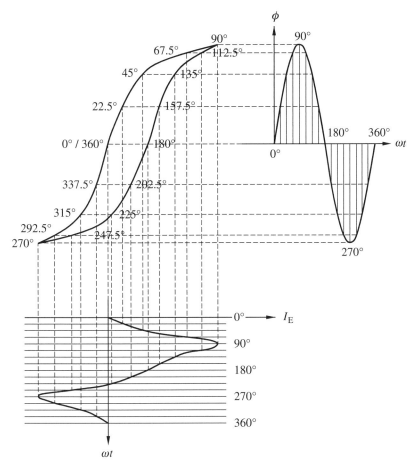

Figure 12.10 Nonlinear Transformer Excitation Current

The first step in calculating the *k*-factor is to perform a harmonic analysis using a power quality meter. This provides an RMS magnitude for each frequency component I_n of the current waveform, where n represents the harmonic order. The RMS magnitude for the effective current can be determined using Eq. (12.13):

$$I_{\text{eff}} = \sqrt{\sum_{n=1}^{\infty} I_n^2} \qquad (12.13)$$

In practical applications, the magnitude of I_n decreases rapidly as n increases, so considering only the first several harmonics is usually adequate. When the magnitude of I_n is less than 1% of I_1, its contribution to the effective current becomes insignificant.

TABLE 12.1   Harmonic Analysis of Load Current

Harmonic Order	Magnitude (A)
1	385
3	120
5	85
7	40
9	15
11	12
13	8
15	6
17	4
19	3

After the effective current is calculated, the magnitude of each harmonic current must be squared, multiplied by the square of its harmonic order, and summed. This quantity divided by the square of the effective current is the *k*-factor:

$$K = \frac{\sum_{n=1}^{\infty} n^2 I_n^2}{I_{eff}^2} \tag{12.14}$$

Consider the following example. A load draws a current containing the harmonic content shown in Table 12.1.

The effective current is calculated to be

$$I_{eff} = \sqrt{385^2 + 120^2 + 85^2 + 40^2 + 15^2 + 12^2 + 8^2 + 6^2 + 4^2 + 3^2} = 415 \text{ A} \tag{12.15}$$

Next, the *k*-factor is calculated:

$$K = \frac{1^2 385^2 + 3^2 120^2 + 5^2 85^2 + 7^2 40^2 + 9^2 15^2 + 11^2 12^2 + 13^2 8^2 + 15^2 6^2 + 17^2 4^2 + 19^2 3^2}{415^2} = 3.48 \tag{12.16}$$

A transformer with a *k*-rating of at least 3.48 must be specified to supply this load.

Another term that is used to express the magnitude of harmonic content in a periodic waveform is *total harmonic distortion* or *THD*. THD is the ratio of the effective magnitude of the harmonic components to the magnitude of the fundamental and is expressed in percent:

$$\text{THD} = \frac{\sqrt{\sum_{n=2}^{\infty} X_n^2}}{X_1} \times 100\% \tag{12.17}$$

THD can be calculated for both the voltage and current waveforms, so the *X* in Eq. (12.17) can represent either voltage or current. If the waveform is a perfect sinusoid, THD equals zero. As the harmonic content increases, so does the THD.

TABLE 12.2 Transmission and Distribution System Voltage Distortion Limits

Nominal Voltage (V)	Individual Harmonic Order Voltage Distortion Limit	THD
V ≤ 69 kV	3.0%	5.0%
69 kV < V < 161 kV	1.5%	2.5%
V ≥ 161 kV	1.0%	1.5%

Source: Reprinted from ANSI/IEEE 141, Table 9-6, with permission.

In addition to overheating iron-core devices, harmonic currents result in larger neutral currents on 4-wire systems. Theoretically, the neutral current can have a magnitude of three times the average phase current. Realistically, neutral currents in circuits with high harmonic content can be upwards of twice the average phase current.

IEEE Standard 141 states distortion limits for both voltage and current waveforms on transmission and distribution systems. Voltage distortion limits are based on nominal operating voltage, while current distortion limits are based on the ratio of short circuit availability to load current (SCA/I_L) at the point of delivery from the utility. The voltage distortion standard also limits the distortion caused by any one harmonic frequency. The current distortion standard limits the distortion caused by specific harmonic frequencies, based on their order. The voltage and current distortion limits are summarized in Tables 12.2 and 12.3, respectively.

12.7.3 Harmonic Filters

Harmonic currents can be prevented from flowing in a power system by using *harmonic filters*. Two types of filters can be used: a high-impedance series filter, or a low-impedance shunt filter. Since the series filter must accommodate full-load current and must be insulated for line-to-line voltage, it is usually a very expensive option. The shunt filter is not only less expensive, but also provides reactive power at the fundamental frequency. Because of their dual functionality of removing harmonics

TABLE 12.3 Current Distortion Limits

SCA/I_L	Individual Harmonic Order (h) Current Distortion Limit					THD
	h < 11	11 ≤ h < 17	17 ≤ h < 23	23 ≤ h < 35	h ≥ 35	
<20[a]	4.0%	2.0%	1.5%	0.6%	0.3%	5.0%
20–50	7.0%	3.5%	2.5%	1.0%	0.5%	8.0%
50–100	10.0%	4.5%	4.0%	1.5%	0.7%	12.0%
100–1000	12.0%	5.5%	5.0%	2.0%	1.0%	15.0%
>1000	15.0%	7.0%	6.0%	2.5%	1.4%	20.0%

Source: Reprinted from ANSI/IEEE 141, Table 9-5, with permission.
[a]Including all power generation equipment.

and correcting power factor, shunt filters are the more common type of harmonic filter, and the remainder of this section will deal exclusively with them.

Before a harmonic filter can be designed, a harmonic analysis must be done to determine the magnitudes and frequencies of the harmonic currents to be filtered. This analysis can be done with a power quality meter.

The harmonic filter will consist of a capacitor in series with a reactor, connected line-to-neutral. The capacitor should be sized first since it will also be a source of reactive power. Sizing it correctly can optimize its use for both filtering and power factor correction.

Perhaps an existing capacitor bank can be utilized as part of the harmonic filter. If a new capacitor is required, several considerations should be made. It is advisable to specify a capacitor with a higher voltage rating than the nominal line-to-neutral voltage of the system to account for fundamental frequency overvoltages. Also, the current rating of the capacitor must be selected wisely. In addition to the harmonic currents identified by the harmonic analysis, the filter will undoubtedly experience slightly higher current magnitudes because it will tend to act as a sink for harmonic currents produced elsewhere on the system. If the filter must handle large magnitudes of high-order harmonics, particular attention must be paid to the current rating of the capacitor, as high-frequency currents tend to substantially increase loading on the capacitor. Finally, capacitor fusing must be considered. It is common to fuse capacitors as low as 125% of rated current. When applied as a harmonic filter, the fuse rating may be more of a limiting factor than the capacitor itself. Increasing the fuse size may be necessary.

After a capacitor is selected, a reactor must be sized to achieve the desired tuning. The current rating of the reactor must be based on the total effective current that will flow through the reactor. The inductance value is selected so that the filter is resonant at the desired frequency. If several harmonic orders are problematic, filtering the lowest frequency first will also attenuate the higher order harmonics somewhat. If this attenuation is not sufficient, multiple filters can be connected in parallel. The reactor impedance at the fundamental frequency to achieve resonance at the nth harmonic frequency is given by Eq. (12.18).

$$X_L = \frac{kV_{LL}^2}{(Mvar)_{3\phi} n^2} = \frac{X_C}{n^2} \qquad (12.18)$$

A common practice is to tune the harmonic filter slightly below (3–10%) the harmonic of concern. Doing so reduces the duty on the filter, since the path to neutral will not be a short circuit at the harmonic frequency. This allows other system components to absorb a portion of the harmonic currents while still providing a low-impedance path to neutral for the harmonic of concern. Tuning slightly below the harmonic frequency also allows for the tolerance in the kvar rating of the capacitor, which is 0 to +10%.

After the reactor is sized, the interaction of the filter with the rest of the system should be simulated. The THD should be checked throughout the range of capacitor and reactor tolerances. An analysis for unwanted resonant conditions should also be made. These analyses are complex and require computer modeling.

12.7 HARMONICS

Consider the following example, which converts an existing 600 kvar capacitor bank to a fifth harmonic filter. The harmonic analysis indicated that the fifth harmonic current comes from a 3000 kVA load with a 5% fifth harmonic component. The line-to-line voltage is 12.47 kV, so the impedance of the reactor at 60 Hz needed for the harmonic filter is

$$X_L = \frac{12.47^2}{(0.600)5^2} = 10.4\,\Omega \tag{12.19}$$

The rated current of the capacitor at the fundamental frequency is

$$I_C = \frac{600}{\sqrt{3}\,12.47} = 27.8\text{ A} \tag{12.20}$$

The impedance of the capacitor at the fundamental frequency must then be

$$X_C = \frac{12{,}470/\sqrt{3}}{27.8} = 259\,\Omega \tag{12.21}$$

Since the algebraic signs of X_L and X_C are opposite, their impedances in series are subtractive. Therefore, the shunt impedance at 60 Hz of the harmonic filter is

$$X_f = X_C - X_L = 259 - 10.4 = 248.6\,\Omega \tag{12.22}$$

The current through the harmonic filter at the fundamental frequency is

$$I_{f1} = \frac{12{,}470/\sqrt{3}}{248.6} = 29.8\text{ A} \tag{12.23}$$

Note that this current is slightly higher than the current drawn by the shunt capacitor alone.

The fifth harmonic current to which the harmonic filter will be subjected is

$$I_{f5} = 0.05\left(\frac{3000}{12.47\sqrt{3}}\right) = 6.9\text{ A} \tag{12.24}$$

Therefore, the effective current flowing through the harmonic filter is

$$I_\text{eff} = \sqrt{I_{f1}^2 + I_{f5}^2} = 29.8\text{ A} \tag{12.25}$$

The next step is to verify that the voltage across the capacitor does not exceed the capacitor's maximum voltage rating. The fundamental frequency voltage across the capacitor is

$$V_{C1} = (29)(259) = 7511\text{ V} \tag{12.26}$$

This means that the fundamental frequency voltage across the reactor is

$$V_{R1} = 7200 - 7511 = -311\text{ V} \tag{12.27}$$

The fifth harmonic voltage across the capacitor is

$$V_{C5} = (6.9)\left(\frac{259}{5}\right) = 357\text{ V} \tag{12.28}$$

Assuming that V_{C1} and V_{C5} are in-phase, which produces a worst-case scenario,

$$V_{C(\text{total})} = \sqrt{7511^2 + 357^2} = 7519 \text{ V} \qquad (12.29)$$

This is less than 110% of the nominal voltage, so is within the capacitor rating.

The last step is to verify that the total kvar production of the capacitor is less than 135% of nominal, which is an ANSI-defined limit:

$$\text{kvar}_{1\Phi} = V_{\text{total}} \cdot I_{\text{eff}} = 224.1 \text{ kvar}/\Phi \qquad (12.30)$$

The three-phase kVAR production is

$$\text{kvar}_{3\Phi} = 3 \cdot \text{kvar}_{1\Phi} = 672.2 \text{ kvar.} \qquad (12.31)$$

This value exceeds the nominal kvar rating by only 12%, so is within the acceptable range. Therefore, simply placing a 10.4 Ω (at 60 Hz) reactor in series with the existing 600 kvar shunt capacitor creates an acceptable fifth harmonic filter.

12.8 POWER FACTOR

Power factor is a measurement of electricity's ability to do useful work. If the current sinusoid does not peak at the same time as the voltage, the ability to do useful work will be reduced, much like if the members of a crew team do not row in unison. The scull will still move forward, but less effectively than when the coxswain keeps the rowers synchronized.

The ability to do useful work is also reduced as the current sinusoid becomes distorted. As seen earlier, a distorted sinusoid contains harmonic components, but these harmonics do not produce usable power since the product of a fundamental frequency voltage and a harmonic current is zero. Only the fundamental frequency component of the current contributes to usable power.

Thus, power factor has two components: the angular separation (displacement) of the voltage and current sinusoids, called the *displacement power factor*, and the amount of distortion in the current sinusoid, or the *distortion power factor*.

Displacement power factor, the familiar cos θ, is used extensively in numerous formulas.

$$\text{pf}_{\text{displacement}} = \cos \theta \qquad (12.32)$$

where θ is the angle between the voltage and current phasors,
or the angle between P and S in the power triangle,
or the angle between R and Z in the impedance triangle.

Distortion power factor is the ratio of the fundamental frequency current to the effective current, which is calculated using Eq. (12.13):

$$\text{pf}_{\text{distortion}} = \frac{I_1}{I_{\text{eff}}} \qquad (12.33)$$

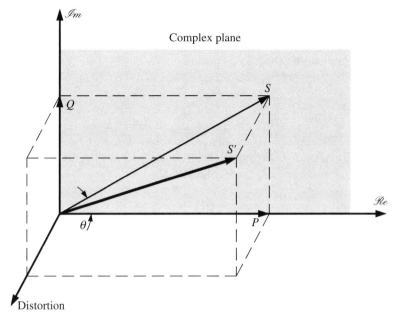

Figure 12.11 Apparent Power S' Considering Distortion

A true reflection of electricity's ability to do useful work can be shown graphically by adding a third dimension to the familiar Argand diagram. The z-axis represents distortion of the current sinusoid. Projecting the complex power phasor S to the plane parallel to the complex plane that represents the amount of current distortion present gives the S' phasor, or the apparent power. The ratio of real power (the only component of power that can do useful work) to the apparent power S' is an accurate depiction of the electricity's ability to do useful work. The P phasor is the only component that can do useful work, but S' represents the amount of power that must be handled by the system. The relationships between P, S, S', and θ are shown in Figure 12.11.

We can now define total power factor, which is the ratio of P to S':

$$\text{Total Power Factor} = \frac{P}{S'} = \frac{\text{Real power}}{\text{Apparent power}} = \frac{\sqrt{3}\, V I_1 \cos \theta}{\sqrt{3}\, V I_{\text{eff}}} = \frac{I_1}{I_{\text{eff}}} \cos \theta \quad (12.34)$$

From Eq. (12.34), it can be seen that total power factor (or simply *power factor*) is the product of displacement power factor and distortion power factor:

$$\text{pf} = (\text{pf}_{\text{displacement}})(\text{pf}_{\text{distortion}}) \quad (12.35)$$

SUMMARY

Problems with power quality have existed since the energization of the first electrical system, but in recent years, more emphasis has been placed on power quality because the increasing use of voltage-sensitive loads such as computers make it easier to perceive problems with power quality. It is important to be able to quantify power quality problems so measurements can be made and improvements can be realized. *Flicker charts* were the first attempt to quantify power quality. Later, the CBEMA and ITI curves were developed to guide manufactures to produce equipment that would perform satisfactorily during routine voltage perturbations.

Continuity of service is simply the ability to provide electrical service. If service is lost, no other power quality concerns exist, so continuity of service should be viewed as the most basic power quality issue. Power systems have incredibly high availability expectations, exceeding 99.99% and this figure will undoubtedly grow as society becomes even more dependent of electricity.

Voltage regulation is the ability to maintain steady-state voltage. In general, steady-state voltage must be maintained within $\pm 5\%$ of nominal ($\pm 10\%$ during contingencies) at the point of power delivery to comply with ANSI C84. Sags or swells that exceed the ANSI limits are power quality issues.

Oscillatory transients consist of an initial perturbation followed by a rapidly damped decay back to the steady state. Oscillatory transients are most often caused by lightning strikes, capacitor switching, and power semiconductor operation. *Impulse transients* are sudden spikes that occur sporadically along the waveform. These transients are usually caused by motor load switching, loose wiring, and poor grounding.

Harmonics are sinusoids with frequencies equal to integer multiples of the fundamental frequency. When harmonics are added to the fundamental frequency, the waveform becomes distorted.

Harmonic filters are essentially low-pass filters, often formed by adding a reactor in series with an existing power factor correction (shunt) capacitor, designed to attenuate a troublesome harmonic to a tolerable level by shunting some of the harmonic to ground. Harmonic filters are usually slightly detuned to reduce the amount of current they must handle.

Power factor is a measurement of electricity's ability to do useful work. Power factor has two components: *displacement power factor* and *distortion power factor*. Power factor is the ratio of real power (P) to apparent power (S').

FOR FURTHER READING

Dugan, R. C., McGranaghan, M. F., Santoso, S., and Beaty, H. W., *Electrical Power Systems Quality*, 3rd edition, McGraw-Hill, 2012. ISBN 978-0-07-176155-0.

FIPS Publication 94, Guideline on Electrical Power for ADP Installations, Federal Information Processing Standards, National Institute of Standards and Technology, September 1983.

IEEE Recommended Practice and Requirements for Harmonic Control in Electric Power Systems, IEEE Standard 519, 2014.

Joffe, E. B. and Lock, K., *Grounds for Grounding: A Circuit to System Handbook*, Wiley-IEEE Press, 2010. ISBN 978-0-471-66008-8.

Kusko, A. and Thompson, M. T., *Power Quality in Electrical Systems*, McGraw Hill, 2007. ISBN 0-07-147075-1.

QUESTIONS

1. Why is power quality more of an issue today than in years past?
2. How did business equipment manufacturers quantify their power quality requirements?
3. Different types of loads are tolerant of various levels of power quality degradation. Give examples of several categories of loads and their basic power quality requirements.
4. Propose a strategy that could be used to improve the power quality at a highly automated manufacturing plant, where downtime is extremely costly.
5. What measures can be taken to minimize the risk of equipment failure due to transients?
6. Why are harmonics problematic to many electrical devices?
7. How can transformers be used to mitigate the effects of some harmonics?
8. If a voltage waveform is nonsinusoidal, but exhibits half-wave symmetry, what conclusion can be drawn about the harmonic content of the waveform?
9. How can low power factor and harmonic current content both be corrected with a single device?
10. Describe the *displacement power factor* and *distortion power factor*, and cite methods of improving each.

PROBLEMS

1. A machine that complies with the original CBEMA curve from FIPS Publication 94 is installed in a manufacturing plant. A solid-state transfer switch is proposed to swap the source of the machine to an alternative power source upon the failure of the primary source. How fast must the transfer switch be so that the machine continues to operate properly through the source transfer?
2. A switchgear bus is subject to voltage variations of $\pm 15\%$ of the nominal value lasting for up to one-fourth of a second due to an automatic welding machine.
   a. Can a device that is compliant with the ITIC curve tolerate these voltage deviations?
   b. What is the impact if the duration of the voltage swings lasts for 2 seconds?
3. A harmonic analysis shows substantial 11th and 15th harmonic content in the current feeding a solid-state load through a grounded-wye–grounded-wye transformer. How would replacing the transformer with a delta–grounded-wye transformer change the situation?
4. Compute the Fourier series of the full-wave rectified sinusoid
$$V(t) = |V_0 \cos \omega t|$$
5. Compute the Fourier series of a square wave having a maximum amplitude of V_0, a minimum amplitude of $-V_0$, and a period of $2\pi/\omega$.
6. Calculate the effective current magnitude for the waveform
$$I(t) = 70\cos 120\pi t + 15\cos 360\pi t + 3\cos 600\pi t$$

7. Determine the minimum k-factor rating of a transformer that can adequately serve the load with the current requirement calculated in Problem 6.

8. A load draws a fundamental frequency current of 100 A and a third harmonic current of unknown magnitude. Assuming that no other harmonic orders are present, what is the maximum third harmonic current amplitude that can be adequately supplied by a transformer having a k-factor of 2.6?

9. Calculate the THD of the voltage waveform:

$$V(t) = 480\cos 120\pi t + 36\cos 360\pi t + 24\cos 600\pi t + 16\cos 840\pi t$$

10. Design a 13.8 kV harmonic filter by converting a 1200 kvar capacitor bank to a seventh harmonic filter adequate for a 5000 kVA load with a 7% seventh harmonic current component.

APPENDIX A

UNITS OF MEASUREMENT

All quantities can be measured using a set of seven metric (Système International, or SI) units referred to as *base units*. The seven base units are dimensionally independent and are defined in Table A.1 for the MKS (meter–kilogram–second) system.

Table A.2 lists *supplementary units*, which along with the base units, can be combined to form any *derived unit*.

Table A.3 lists the commonly derived electrical, magnetic, and illumination units along with their equivalence in terms of other electrical units and SI base units.

THE AMERICAN WIRE GAUGE (AWG)

The American Wire Gauge, also known as the Brown and Sharpe wire gauge, has been used since 1857, primarily in the United States, as a system for sizing round electrically conductive wires. Steel wires are not measured using the AWG system—their diameters are usually specified in inches. Most countries outside the United States measure wire diameters in millimeters and cross sections in square millimeters. Table A.4 shows the American Wire Gauge from size 4/0 to 40, along with the diameters of each wire size, both in inches and in millimeters.

By definition, AWG 4/0 (pronounced "four aught") is exactly 0.46 inches in diameter and AWG 36 is exactly 0.005 inch in diameter. The larger of these two standard sizes has a diameter that is exactly 92 times the diameter of the smaller. There are 38 AWG sizes between these two standard sizes, so the diameter of an n AWG wire is

$$d_n = (0.005) \times 92^{\frac{36-n}{39}} \text{ inch}$$

Note that larger AWG numbers indicate smaller wire sizes. The even-numbered AWG sizes are more commonly used than the odd-numbered sizes, which are often reserved for special uses. For example, AWG 9 is sometimes used for current transformer secondary circuits.

Conductors larger than AWG 4/0 are measured in *circular mils*, a measure of cross-sectional area equal to that of a circle with a diameter of a *mil*, or 0.001 inch. A circular mil is approximately equal to 0.5067 mm^2. Since the circular mil is very

---

Industrial Power Distribution, Second Edition. Ralph E. Fehr, III.
© 2016 The Institute of Electrical and Electronics Engineers, Inc. Published 2016 by John Wiley & Sons, Inc.

358 APPENDIX A UNITS OF MEASUREMENT

TABLE A.1   SI Base Units (MKS System)

Unit	Quantity
candela (cd)	Luminous intensity
coulomb (C)	Electric charge
kelvin (K)	Thermodynamic temperature
kilogram (kg)	Mass
meter (m)	Length
mole (mol)	Amount of a substance
second (s)	Time

TABLE A.2   SI Supplementary Units

Unit	Quantity
radian (rad)	Plane angle
steradian (sr)	Solid angle

TABLE A.3   Common Electrical, Magnetic, and Illumination Units

Unit	Quantity	Equivalent Units	SI Base Units
ampere (A)	Electric current	W/V	C/s
ampere-hour (A-h)	Electric charge	—	3600 C
farad (F)	Capacitance	C/V	$C^2 \cdot s^2/kg\ m^2$
henry (H)	Inductance	Wb/A	$kg\ m^2/C^2$
hertz (Hz)	Frequency	—	s^{-1}
horsepower (hp)	Power	746 W = 550 ft-lb/s	746 $kg\ m^2/s^3$
joule (J)	Energy	N m[a]	$kg\ m^2/s^2$
ohm (Ω)	Resistance	V/A	$kg\ m^2/C^2\ s$
siemens (S)[b]	Conductance	A/V	$C^2\ s/kg\ m^2$
volt (V)	Electric potential	W/A	$kg\ m^2/C\ s^2$
watt (W)[c]	Power	J/s	$kg\ m^2/s^3$
ampere-turn (A-t)[d]	Magnetomotive Force (MMF)	$4\pi/10$ Gi	C/s
gauss (G) or oersted (Oe)	Magnetic flux Density	Mx/cm^2	10^{-4} kg/C s
gilbert (Gi)	MMF	$10/4\pi$ A-turns	$10/4\pi$ C/s
maxwell (Mx)[e]	Magnetic flux	10^{-8} Wb	$10^{-8}\ kg\ m^2/C\ s$
tesla (T)	Magnetic flux density	Wb/m^2	kg/C s
weber (Wb)	Magnetic flux	V s	$kg\ m^2/C\ s$
footcandle (fc)	Illuminance	10.764 lx	10.764 $cd\ sr/m^2$
lumen (lm)	Luminous flux	—	cd sr
lux (lx)	Illuminance	lm/m^2	$cd\ sr/m^2$

[a] "N" designates the *Newton*, the SI unit of force.
[b] The traditional unit of conductance is the *mho* (℧), where 1 mho = 1 siemens. Note that *mho* is *ohm* spelled backward.
[c] The volt-ampere (VA) and the volt-ampere reactive (var) are dimensionally equivalent to the watt.
[d] To distinguish MMF from electric current, the unit *ampere-turn* is commonly used, although a *turn* is not a true dimension.
[e] The maxwell was previously called a *line* of flux.

TABLE A.4 The American Wire Gauge

AWG	Diameter (Inches)	Diameter (mm)	AWG	Diameter (Inches)	Diameter (mm)
4/0	**0.4600**	**11.684**	19	0.0359	0.912
3/0	0.4096	10.404	20	0.0320	0.812
2/0	0.3648	9.266	21	0.0285	0.723
1/0	0.3249	8.252	22	0.0253	0.644
1	0.2893	7.348	23	0.0226	0.573
2	0.2576	6.544	24	0.0201	0.511
3	0.2294	5.827	25	0.0179	0.455
4	0.2043	5.189	26	0.0159	0.405
5	0.1819	4.621	27	0.0142	0.361
6	0.1620	4.115	28	0.0126	0.321
7	0.1443	3.665	29	0.0113	0.286
8	0.1285	3.264	30	0.0100	0.255
9	0.1144	2.906	31	0.00893	0.227
10	0.1019	2.588	32	0.00795	0.202
11	0.0907	2.305	33	0.00708	0.180
12	0.0808	2.053	34	0.00630	0.160
13	0.0720	1.828	35	0.00561	0.143
14	0.0641	1.628	**36**	**0.00500**	**0.127**
15	0.0571	1.450	37	0.00445	0.113
16	0.0508	1.291	38	0.00397	0.101
17	0.0453	1.150	39	0.00353	0.0897
18	0.0403	1.024	40	0.00314	0.0799

small, the metric multiplier *kilo-* is often associated with the circular mil, leading to *kcmil*, or *thousands of circular mils*. This measure is equivalent to the traditional *MCM* (no longer recommended for use), where the first *M* represents the Roman numeral for 1000, and the *CM* stands for *circular mils*.

APPENDIX B

CIRCUIT ANALYSIS TECHNIQUES

The following methods can be used to combine impedances to simplify the topology of an electric circuit. Also, formulae are given for voltage and current division across/through impedances.

SERIES IMPEDANCES

$$X \circ\!\!-\!\!\boxed{Z_A}\!\!-\!\!\boxed{Z_B}\!\!-\!\!\circ Y \qquad X \circ\!\!-\!\!\boxed{Z_A + Z_B}\!\!-\!\!\circ Y$$

PARALLEL IMPEDANCES

$$X \circ\!\!-\!\!\begin{array}{c}\boxed{Z_A}\\ \boxed{Z_B}\end{array}\!\!-\!\!\circ Y \qquad X \circ\!\!-\!\!\boxed{\dfrac{Z_A Z_B}{Z_A + Z_B}}\!\!-\!\!\circ Y$$

DELTA-TO-WYE TRANSFORMATION

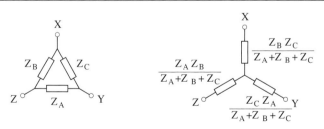

Industrial Power Distribution, Second Edition. Ralph E. Fehr, III.
© 2016 The Institute of Electrical and Electronics Engineers, Inc. Published 2016 by John Wiley & Sons, Inc.

WYE-TO-DELTA TRANSFORMATION

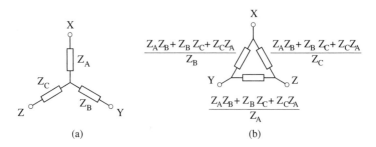

VOLTAGE DIVIDER

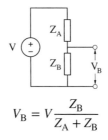

$$V_B = V \frac{Z_B}{Z_A + Z_B}$$

CURRENT DIVIDER

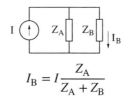

$$I_B = I \frac{Z_A}{Z_A + Z_B}$$

MESH-CURRENT ANALYSIS

A *mesh* is defined as any closed path through a planar circuit that contains no other closed paths. A planar circuit is one where no conductors cross over each other. To apply mesh-current analysis to a nonplanar circuit, it must first be redrawn as a planar circuit.

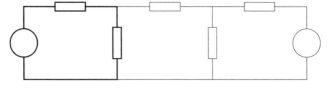

Example of a mesh

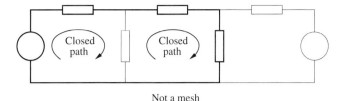

Not a mesh

Per Kirchhoff's voltage law (KVL), the sum of the voltage *rises* around any mesh must equal the sum of the voltage *drops* around the same mesh. This is an extension of the law of conservation of energy.

Mesh-current analysis leads to a system of n equations with n unknowns, where n is the number of meshes in the circuit. Matrix methods, such as Cramer's rule, are helpful to solve these systems of equations.

Variables are assigned to each of the mesh currents. While the direction of the mesh current is arbitrary, clockwise currents are often assumed for uniformity.

Mesh-current analysis can be applied to the following circuit to produce the equations shown.

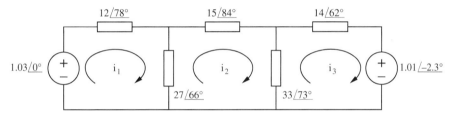

Writing the equation for the mesh defined by mesh current i_1,

$$1.03 \underline{/0°} = (12 \underline{/78°})i_1 + (27 \underline{/66°})(i_1 - i_2)$$

The algebraic sign of the voltage source is determined by the terminal from which one *leaves* the source while traversing the mesh. Traversing the i_1 mesh clockwise, one leaves the *positive* terminal of the source, so $1.03 \underline{/0°}$ is positive.

Distributing the $27 \underline{/66°}$ and simplifying algebraically by combining like terms,

$$(38.8 \underline{/70°})i_1 + (-27 \underline{/66°})i_2 = 1.03 \underline{/0°}$$

Similarly the second mesh gives

$$0 = (15 \underline{/84°})i_2 + (33 \underline{/73°})(i_2 - i_3) + (27 \underline{/66°})(i_2 - i_1)$$

which simplifies to

$$(-27 \underline{/66°})i_1 + (75 \underline{/73°})i_2 + (-33 \underline{/73°})i_3 = 0$$

The third mesh yields

$$-1.01 \underline{/-2.3°} = (14 \underline{/62°})i_3 + (33 \underline{/73°})(i_3 - i_2)$$

which simplifies to

$$(-33 \underline{/73°})i_2 + (47 \underline{/70°})i_3 = -1.01 \underline{/-2.3°}.$$

Note that while traversing the third mesh clockwise, the *negative* terminal of the source is exited, so $-1.01 \underline{/-2.3°}$ is used in the equation.

These three equations can be written as a single matrix equation to facilitate implementation of a linear algebra solution method such as Cramer's Rule:

$$\left. \begin{array}{l} (38.8 \underline{/70°})i_1 + (-27 \underline{/66°})i_2 = 1.03 \underline{/0°} \\ (-27 \underline{/66°})i_1 + (75 \underline{/73°})i_2 + (-33 \underline{/73°})i_3 = 0 \\ (-33 \underline{/73°})i_2 + (47 \underline{/70°})i_3 = -1.01 \underline{/-2.3°} \end{array} \right\} \rightarrow$$

$$\begin{bmatrix} 38.8 \underline{/70°} & -27 \underline{/66°} & 0 \\ -27 \underline{/66°} & 42 \underline{/72°} & 14 \underline{/62°} \\ 0 & -1 \underline{/0°} & 1 \underline{/0°} \end{bmatrix} \cdot \begin{bmatrix} i_1 \\ i_2 \\ i_3 \end{bmatrix} = \begin{bmatrix} 1.03 \underline{/0°} \\ -1.01 \underline{/-2.3°} \\ 3.6 \underline{/-31.7°} \end{bmatrix}$$

When a current source is present, an expression for the voltage across the current source cannot be written. Defining a *supermesh* that avoids the current source avoids the problem of not being able to express the voltage across the current source. A supermesh is not a mesh since it contains at least one closed path, but KVL applies to all closed paths through a circuit, not just meshes. An additional equation must be written addressing the current source (and any other circuit elements bypassed by the supermesh).

The following circuit illustrates the use of a supermesh.

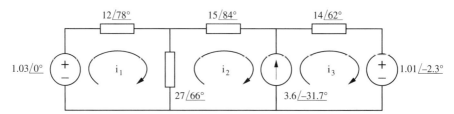

The equation for the first mesh can be written as usual:

$$1.03 \underline{/0°} = (12 \underline{/78°})i_1 + (27 \underline{/66°})(i_1 - i_2)$$
$$(38.8 \underline{/70°})i_1 + (-27 \underline{/66°})i_2 = 1.03 \underline{/0°}$$

Writing mesh equations for the second and third meshes would involve the current source, so the second two meshes are combined into a supermesh as follows:

$$-1.01 \underline{/-2.3°} = (15 \underline{/84°})i_2 + (14 \underline{/62°})i_3 + (27 \underline{/66°})(i_2 - i_1)$$
$$(-27 \underline{/66°})i_1 + (42 \underline{/72°})i_2 + (14 \underline{/62°})i_3 = -1.01 \underline{/-2.3°}$$

The third equation is developed from the branch avoided by the supermesh—the branch containing the current source. By examining that branch, we can see by inspection that

$$i_3 - i_2 = 3.6 \underline{/-31.7°}$$

These three equations can be written as a single matrix equation as follows:

$$(38.8\,\underline{/70°})i_1 + (-27\,\underline{/66°})i_2 = 1.03\,\underline{/0°}$$
$$(-27\,\underline{/66°})i_1 + (42\,\underline{/72°})i_2 + (14\,\underline{/62°})i_3 = -1.01\,\underline{/-2.3°} \quad \rightarrow$$
$$-i_2 + i_3 = 3.6\,\underline{/-31.7°}$$

$$\begin{bmatrix} 38.8\,\underline{/70°} & -27\,\underline{/66°} & 0 \\ -27\,\underline{/66°} & 42\,\underline{/72°} & 14\,\underline{/62°} \\ 0 & -1\,\underline{/0°} & 1\,\underline{/0°} \end{bmatrix} \begin{bmatrix} i_1 \\ i_2 \\ i_3 \end{bmatrix} \begin{bmatrix} 1.03\,\underline{/0°} \\ -1.01\,\underline{/-2.3°} \\ 3.6\,\underline{/-31.7°} \end{bmatrix}$$

NODE-VOLTAGE ANALYSIS

A *node* is defined as any closed path enclosing part of a circuit. When the size of a node approaches zero, it becomes a single point.

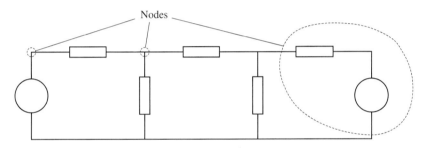

Examples of nodes

Per Kirchhoff's current law (KCL), the sum of the currents *entering* a node must equal the sum of the currents *exiting* that node. This is an extension of the law of conservation of charge.

Node-voltage analysis leads to a system of *n* equations with *n* unknowns, where $n + 1$ is the number of nodes in the circuit. The equation for the last node (the reference node) is not linearly independent of the first *n* equations, so it is not necessary. Matrix methods, such as Cramer's rule, are helpful to solve these systems of equations.

Variables are assigned to represent the node voltages. Then, KCL is applied to each labeled node, assuming a zero voltage reference at the reference node at the bottom of the circuit.

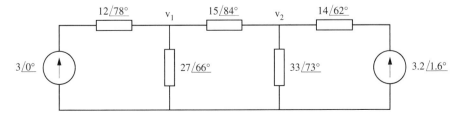

At the node labeled v_1, a current of $3 \underline{/0°}$ enters from the left. Assuming the other two currents exit the node, the following node equation can be written:

$$3 \underline{/0°} = \frac{v_1}{27 \underline{/66°}} + \frac{v_1 - v_2}{15 \underline{/84°}}.$$

Combining the v_1 terms and taking the reciprocals of the denominators,

$$0.103 \underline{/-77.6°} v_1 + 0.0667 \underline{/96°} v_2 = 3 \underline{/0°}.$$

A current of $3.2 \underline{/1.6°}$ enters node v_2 from the right. Assuming the other two currents exit the node, the following node equation can be written:

$$3.2 \underline{/1.6°} = \frac{v_2}{33 \underline{/73°}} + \frac{v_2 - v_1}{15 \underline{/84°}}.$$

Combining the v_2 terms and taking the reciprocals of the denominators,

$$0.0667 \underline{/-84°} v_1 + 0.0966 \underline{/-81°} v_2 = 3.2 \underline{/1.6°}.$$

These two equations can be written as a single matrix equation:

$$0.103 \underline{/-77.6°} v_1 + 0.0667 \underline{/96°} v_2 = 3 \underline{/0°}$$

$$0.0667 \underline{/-84°} v_1 + 0.0966 \underline{/-81°} v_2 = 3.2 \underline{/1.6°}$$

$$\begin{bmatrix} 0.103 \underline{/-77.6°} & 0.0667 \underline{/96°} \\ 0.0667 \underline{/-84°} & 0.0966 \underline{/-81°} \end{bmatrix} \begin{bmatrix} v_1 \\ v_2 \end{bmatrix} = \begin{bmatrix} 3 \underline{/0°} \\ 3.2 \underline{/1.6°} \end{bmatrix}$$

When a voltage source is encountered during node-voltage analysis, the current through the voltage source cannot be expressed.

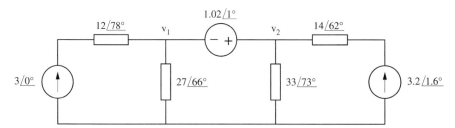

Creating a *supernode* that encompasses the voltage source will allow the node-voltage process to be used.

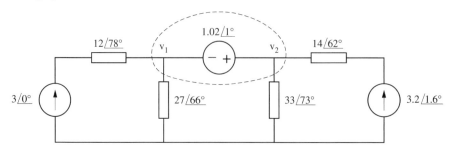

Now, a single node equation can be written for the supernode, assuming the currents through the $27\underline{/66°}$ and $33\underline{/73°}$ impedances leave the supernode.

$$3\underline{/0°} + 3.2\underline{/1.6°} = \frac{v_1}{27\underline{/66°}} + \frac{v_2}{33\underline{/73°}}$$

Simplifying the equation above,

$$0.0370\underline{/-66°}\,v_1 + 0.0303\underline{/-73°}\,v_2 = 6.2\underline{/0.8°}.$$

Since the supernode equation contains two unknowns, a second equation must be written. By examining the supernode, it can be seen that

$$v_2 - v_1 = 1.02\underline{/1°}$$

These two equations can be written as a single matrix equation:

$$0.0370\underline{/-66°}\,v_1 + 0.0303\underline{/-73°}\,v_2 = 6.2\underline{/0.8°}$$

$$-v_1 + v_2 = 1.02\underline{/1°}$$

$$\begin{bmatrix} 0.0370\underline{/-66°} & 0.0303\underline{/-73°} \\ -1\underline{/0°} & 1\underline{/0°} \end{bmatrix} \begin{bmatrix} v_1 \\ v_2 \end{bmatrix} = \begin{bmatrix} 6.2\underline{/0.8°} \\ 1.02\underline{/1°} \end{bmatrix}.$$

EVALUATING DETERMINANTS

The determinant of any square matrix can be evaluated in a number of ways. The determinant of a 2×2 matrix is defined as

$$\det \begin{bmatrix} a & b \\ c & d \end{bmatrix} = \begin{vmatrix} a & b \\ c & d \end{vmatrix} = ad - bc.$$

For a 3×3 matrix, the pattern that defines the 2×2 determinant can be expanded as follows:

$$\begin{vmatrix} a & b & c \\ d & e & f \\ g & h & k \end{vmatrix} = (aek + bfg + cdh) - (ceg + bdk + afh).$$

An alternative method is to decompose the larger determinant to 2×2 determinants using a process called *minoring*. Minoring must be used to evaluate determinants larger than 3×3. One way to apply minoring to find a 3×3 determinant is

$$\begin{vmatrix} a & b & c \\ d & e & f \\ g & h & k \end{vmatrix} = a\begin{vmatrix} e & f \\ h & k \end{vmatrix} - b\begin{vmatrix} d & f \\ g & k \end{vmatrix} + c\begin{vmatrix} d & e \\ g & h \end{vmatrix}.$$

Minoring can be done on any row or column. In the example above, minoring was done on the first row. It is advantageous to minor on a row or column that contains zeroes, since a zero entry will eliminate a term in the expansion.

A minor has a dimension one less than the original determinant. Minoring can be repeated until every determinant is a 2×2 determinant. In the example above, when the "a" element is minored, the resulting minor is the 2×2 determinant remaining when the row and column containing the "a" element is eliminated. The minor must be multiplied by $(-1)^{i+j}$, where i and j are the row and column of the minored element in the original determinant. This term causes the algebraic sign of each term in the minor expansion to alternate.

CRAMER'S RULE

This powerful linear algebra technique is useful for solving systems of equations, particularly when the coefficients are complex.

For example, a 3×3 system of equations can be written as a matrix equation:

$$\left.\begin{array}{c} ax + by + cz = m \\ dx + ey + fz = n \\ gx + hy + kz = p \end{array}\right\} \rightarrow \begin{bmatrix} a & b & c \\ d & e & f \\ g & h & k \end{bmatrix} \cdot \begin{bmatrix} x \\ y \\ x \end{bmatrix} = \begin{bmatrix} m \\ n \\ p \end{bmatrix}$$

Four determinants can be defined: the first being the determinant of the coefficient matrix, and the next three being the determinant of the coefficient matrix with the constant vector substituted for one of the columns:

$$D = \begin{vmatrix} a & b & c \\ d & e & f \\ g & h & k \end{vmatrix} \quad A = \begin{vmatrix} m & b & c \\ n & e & f \\ p & h & k \end{vmatrix} \quad B = \begin{vmatrix} a & m & c \\ d & n & f \\ g & p & k \end{vmatrix} \quad C = \begin{vmatrix} a & b & m \\ d & e & n \\ g & h & p \end{vmatrix}$$

The solution for the system of equations is

$$x = \frac{A}{D}, \quad y = \frac{B}{D}, \quad z = \frac{C}{D}.$$

Cramer's rule can be extended to handle any size system of equations, and evaluation of the determinants can be done easily with software.

APPENDIX C

PHASORS AND COMPLEX NUMBER MATHEMATICS

Phasors are time-varying vectors that exist in the complex plane. The complex plane consists of a horizontal axis (the *real* axis) and a vertical axis (the *imaginary* axis). A graphical representation of the complex plane is called an *Argand diagram*. Time variation produces a *counterclockwise rotation* in the complex plane. The rate of rotation is the *radian frequency* of the phasor, ω ($\omega = 2\pi f$, where f is the frequency of the phasor in hertz). Figure C.1 shows a phasor plotted on an Argand diagram.

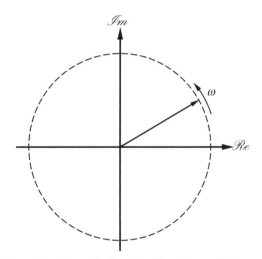

Figure C.1   Phasor in Complex Plane (Argand Diagram)

Industrial Power Distribution, Second Edition. Ralph E. Fehr, III.
© 2016 The Institute of Electrical and Electronics Engineers, Inc. Published 2016 by John Wiley & Sons, Inc.

COMPLEX OPERATORS

$$j \equiv \sqrt{-1} = 1\,\underline{/90°}$$

(multiplying by j rotates the multiplier counterclockwise in the complex plane by 90°)

$$a \equiv \frac{-1}{2} + j\frac{\sqrt{3}}{2} = 1\,\underline{/120°}$$

(multiplying by a rotates the multiplier counterclockwise in the complex plane by 120°)

NOTATIONS FOR COMPLEX NUMBERS

Rectangular (Cartesian) coordinates: $A + jB$
Polar coordinates: $R\,\underline{/\theta}$
Rectangular to polar conversion

$$R = \sqrt{A^2 + B^2}$$

$$\theta = \tan^{-1} \frac{B}{A}$$

Polar to rectangular conversion

$$A = R\cos\theta$$
$$B = R\sin\theta$$

Figure C.2 shows the relationship between rectangular and polar coordinates.

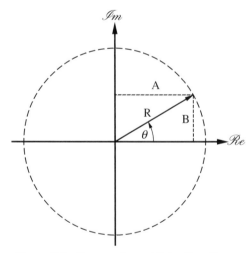

Figure C.2 Rectangular and Polar Coordinates

Arithmetic operations for complex numbers are shown in Table C.1.

TABLE C.1 Complex Arithmetic Operations

Operation	Rectangular Coordinates	Polar Coordinates
Negation	$-(A + jB) = -A - jB$	$-(R\,\underline{/\theta}) = R\,\underline{/\theta \pm 180°}$
Conjugation	$(A + jB)^* = A - jB$	$(R\,\underline{/\theta})^* = R\,\underline{/-\theta}$
Addition and Subtraction	$(A + jB) \pm (C + jD) =$ $(A \pm C) + j(B \pm D)$	$(W\,\underline{/X}) \pm (Y\,\underline{/Z}) =$ $\sqrt{W^2 \pm 2WY\cos(X - Z) + Y^2}$ $\Big/\tan^{-1}\left(\dfrac{W\sin X \pm Y\sin Z}{W\cos X \pm Y\cos Z}\right)$
Multiplication	$(A + jB) \times (C + jD) =$ $(AC - BD) + j(BC + AD)$	$(W\,\underline{/X}) \times (Y\,\underline{/Z}) =$ $(W \times Y)\,\underline{/X + Z}$
Division	$(A + jB) \div (C + jD) =$ $[(AC + BD) \div (C^2 + D^2)] +$ $j[(BC - AD) \div (C^2 + D^2)]$	$(W\,\underline{/X}) \div (Y\,\underline{/Z}) =$ $(W \div Y)\,\underline{/X - Z}$ for $Y \neq 0$
Exponentiation	$(A + jB)^n =$ $(A^2 + B^2)^{n/2} \cos[n \tan^{-1}(B/A)]$ $+ j(A^2 + B^2)^{n/2} \sin[n \tan^{-1}(B/A)]$	$(R\,\underline{/\theta})^n = R^n\,\underline{/\theta \times n}$
Roots	$\sqrt[n]{A + jB} =$ $(A^2 + B^2)^{1/2n} \cos[(1/n) \tan^{-1}(B/A)]$ $+ j(A^2 + B^2)^{1/2n} \sin[(1/n) \tan^{-1}(B/A)]$	$\sqrt[n]{R\,\underline{/\theta}} = \sqrt[n]{R}\,\underline{/\theta \div n}$

PROPERTIES OF CONJUGATION

Conjugation is distributive over the four basic binary arithmetic operations. If X and Y are complex numbers, then

$$(X + Y)^* = X^* + Y^*$$
$$(X - Y)^* = X^* - Y^*$$
$$(X \times Y)^* = X^* \times Y^*$$
$$(X \div Y)^* = X^* \div Y^*$$

The real and imaginary components of a complex number can be expressed using conjugation. If Z is a complex number, then

$$\text{Re}\{Z\} = \frac{1}{2}(Z + Z^*)$$

and

$$\text{Im}\{Z\} = \frac{-j}{2}(Z - Z^*)$$

The reciprocal of a complex number can also be expressed using conjugation. If Z is a complex number, then

$$\frac{1}{Z} = \frac{Z^*}{[\text{Re}\{Z\}]^2 + [\text{Im}\{Z\}]^2}.$$

EULER'S FORMULA

$$e^{j\theta} = \cos\theta + j\sin\theta$$

Letting $\theta = \pi$ gives Euler's Identity: $e^{j\pi} + 1 = 0$.

Euler's Formula can be used to develop exponential and trigonometric forms to express a phasor, based on polar coordinates:

$$R\,\underline{/\theta} = Re^{j\theta} = R\cos\theta + jR\sin\theta$$

APPENDIX D

IMPEDANCE DATA

TABLE D.1 Typical Subtransient and Transient Reactance Values for Rotating Machines

Machine Type	Direct-Axis Subtransient Reactance (X_d'')	Direct-Axis Transient Reactance (X_d')
Turbine-generator (2 poles)	0.09	0.15
Turbine-generator (4 poles)	0.15	0.23
Salient-pole generators w/damper windings (12 poles or less)	0.16	0.33
Salient-pole generators w/damper windings (14 poles or more)	0.21	0.33
Synchronous motors (6 poles)	0.15	0.23
Synchronous motors (8–14 poles)	0.20	0.30
Synchronous motors (16 poles or more)	0.28	0.40
Synchronous condensers	0.24	0.37
Large induction motors (above 600 V)	0.17	—
Induction motors (600 V and below, 50 hp and above)	0.20	—
Induction motors (600 V and below, smaller than 50 hp)	0.28	—

Industrial Power Distribution, Second Edition. Ralph E. Fehr, III.
© 2016 The Institute of Electrical and Electronics Engineers, Inc. Published 2016 by John Wiley & Sons, Inc.

TABLE D.2 Constants of Medium-Voltage ACSR/Copper Conductors at One-Foot Delta Spacing

kcmil or AWG	Aluminum Conductor Steel Reinforced (ACSR)		Copper	
	Resistance (R) at 25°C, 60 Hz (mΩ/conductor/100 ft)	Reactance (X_A) at 1 ft spacing, 60 Hz (mΩ/conductor/100 ft)	Resistance (R) at 25°C, 60 Hz (mΩ/conductor/100 ft)	Reactance (X_A) at 1 ft spacing, 60 Hz (mΩ/conductor/100 ft)
1590.0	1.18	6.79	—	—
1272.0	1.47	7.04	—	—
1000.0	—	—	1.19	7.58
954.0	1.94	7.38	—	—
795.0	2.22	7.44	—	—
750.0	—	—	1.53	7.90
556.5	3.21	7.86	—	—
500.0	—	—	2.24	8.39
477.0	3.38	8.02	—	—
350.0	—	—	3.17	8.83
336.4	4.80	8.43	—	—
250.0	—	—	4.44	9.22
4/0	7.62	10.99	5.24	9.53
2/0	12.13	12.12	8.31	10.10
2	24.35	12.15	16.51	10.80
4	38.67	12.40	26.27	11.30
6	61.47	12.73	41.31	12.10

TABLE D.3 60-Hz Reactance Spacing Correction Factor (X_B) in Milliohms per Conductor per 100 ft

(feet)	Separation (inches)											
	0	1	2	3	4	5	6	7	8	9	10	11
0	—	−5.71	−4.12	−3.19	−2.52	−2.01	−1.59	−1.24	−0.93	−0.66	−0.42	−0.20
1	0.00	0.18	0.35	0.51	0.61	0.80	0.93	1.06	1.17	1.29	1.39	1.49
2	1.59	1.69	1.78	1.86	1.95	2.03	2.11	2.18	2.55	2.32	2.39	2.46
3	2.52	2.59	2.65	2.71	2.77	2.82	2.88	2.93	2.99	3.04	3.09	3.14
4	3.19	3.23	3.28	3.33	3.37	3.41	3.46	3.50	3.54	3.58	3.62	3.66
5	3.71	3.74	3.77	3.81	3.85	3.88	3.92	3.95	3.99	4.02	4.05	4.09
6	4.12	4.15	4.18	4.21	4.24	4.27	4.30	4.33	4.36	4.39	4.42	4.45
7	4.47	4.50	4.53	4.55	4.58	4.60	4.63	4.66	4.68	4.71	4.73	4.76
8	4.78											

TABLE D.4 Medium-Voltage Cable in Conduit Reactance Factor (M) for Various Constructions and Installations

	Multiple-Conductor Cable in Conduit or Armor			
	Nonshielded		Shielded	
kcmil	Nonmagnetic	Magnetic	Nonmagnetic	Magnetic
≤250	1.000	1.149	0.945	1.086
300	1.000	1.146	0.946	1.084
350	1.000	1.140	0.947	1.080
400	1.000	1.134	0.949	1.076
450	1.000	1.128	0.951	1.073
500	1.000	1.122	0.952	1.068
600	1.000	1.111	0.955	1.061
700	1.000	1.100	0.957	1.053
750	1.000	1.095	0.959	1.050
1000	1.000	1.070	0.962	1.029
	Three Single-Conductor Cables in Conduit with Random Lay			
	Nonshielded		Shielded	
kcmil	Nonmagnetic	Magnetic	Nonmagnetic	Magnetic
≤250	1.200	1.500	1.320	1.650
300	1.200	1.500	1.302	1.628
350	1.200	1.500	1.298	1.623
400	1.200	1.500	1.295	1.619
450	1.200	1.500	1.292	1.616
500	1.200	1.500	1.290	1.613
600	1.200	1.500	1.285	1.607
700	1.200	1.500	1.282	1.602
750	1.200	1.500	1.278	1.598
1000	1.200	1.500	1.271	1.589

TABLE D.5 Impedance of 15-kV, Three-Phase, 175-Mil Cross-Linked Polyethylene, Concentric Neutral Underground Cable in Ohms/1000 ft at 60 Hz

Conductor Temperatures
Phase: 90°C
Neutral: 70°C
Earth Resistivity = 100 Ω-m

AWG or kcmil	$R_1 = R_2$	$X_1 = X_2$	R_0	X_0
Aluminum Phase Conductor				
1/0	0.2182	0.0955	0.5215	0.2906
2/0	0.1782	0.0926	0.4697	0.2463
3/0	0.1433	0.0893	0.4049	0.1825
4/0	0.1181	0.0858	0.3497	0.1402
250	0.1038	0.0827	0.3085	0.1114
350	0.0837	0.0761	0.2315	0.0691
500	0.0680	0.0674	0.1653	0.0428
750	0.0550	0.0581	0.1188	0.0305
1000	0.0493	0.0495	0.0905	0.0235
Copper Phase Conductor				
1/0	0.1451	0.0944	0.4066	0.1852
2/0	0.1181	0.0908	0.3492	0.1428
3/0	0.0989	0.0867	0.2907	0.1033
4/0	0.0854	0.0813	0.2318	0.0718
250	0.0785	0.0770	0.2008	0.0578
350	0.0657	0.0685	0.1495	0.0408
500	0.0554	0.0574	0.1060	0.0289
750	0.0463	0.0446	0.0724	0.0216
1000	0.0404	0.0358	0.0554	0.0181

TABLE D.6 Impedance of 15-kV, Three-Phase, 220-Mil Cross-Linked Polyethylene, Concentric Neutral Underground Cable in Ohms/1000 ft at 60 Hz

AWG or kcmil	$R_1 = R_2$	$X_1 = X_2$	R_0	X_0
	Conductor Temperatures			
	Phase: 90°C			
	Neutral: 70°C			
	Earth Resistivity = 100 Ω-m			
	Aluminum Phase Conductor			
1/0	0.2177	0.0956	0.5205	0.2927
2/0	0.1777	0.0927	0.4688	0.2484
3/0	0.1427	0.0894	0.4043	0.1846
4/0	0.1174	0.0860	0.3493	0.1423
250	0.1031	0.0829	0.3082	0.1134
350	0.0828	0.0765	0.2314	0.0709
500	0.0671	0.0681	0.1653	0.0444
750	0.0542	0.0589	0.1188	0.0319
1000	0.0486	0.0504	0.0905	0.0247
	Copper Phase Conductor			
1/0	0.1444	0.0946	0.4060	0.1876
2/0	0.1173	0.0911	0.3488	0.1451
3/0	0.0980	0.0870	0.2904	0.1055
4/0	0.0844	0.0818	0.2316	0.0738
250	0.0774	0.0777	0.2007	0.0597
350	0.0647	0.0694	0.1494	0.0425
500	0.0545	0.0585	0.1059	0.0304
750	0.0456	0.0459	0.0724	0.0229
1000	0.0400	0.0370	0.0554	0.0193

TABLE D.7 Impedance of 15-kV, Three-Phase, 175-Mil Cross-Linked Polyethylene, Conventional Cable with Full-Size Neutral in Ohms/1000 ft at 60 Hz

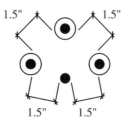

Conductor Temperatures
Phase: 90°C
Neutral: 70°C
Earth Resistivity = 100 Ω-m

AWG or kcmil	$R_1 = R_2$	$X_1 = X_2$	R_0	X_0
\multicolumn{5}{c}{Aluminum Phase Conductor}				
4	0.5350	0.0662	0.8580	0.5888
2	0.3360	0.0609	0.7191	0.4375
1	0.2680	0.0569	0.6558	0.3443
1/0	0.2100	0.0543	0.5864	0.2709
2/0	0.1690	0.0516	0.5043	0.1728
3/0	0.1320	0.0490	0.4317	0.1209
4/0	0.1050	0.0463	0.3635	0.0779
250	0.0890	0.0440	0.3071	0.0461
300	0.0750	0.0419	0.2647	0.0307
350	0.0650	0.0402	0.2269	0.0170
\multicolumn{5}{c}{Copper Phase Conductor}				
4	0.3260	0.0662	0.7122	0.4387
2	0.2050	0.0609	0.5846	0.2692
1	0.1630	0.0569	0.5001	0.1710
1/0	0.1260	0.0543	0.4271	0.1186
2/0	0.1010	0.0516	0.3604	0.0752
3/0	0.0810	0.0490	0.2997	0.0437
4/0	0.0640	0.0463	0.2417	0.0200

TABLE D.8 Impedance of 15-kV, Single-Phase, 220-Mil Low-Density Thermoplastic Polyethylene, Conventional Cable with Full-Size Neutral in Ohms/1000 ft at 60 Hz

Conductor Temperatures
Phase: 75°C
Neutral: 50°C
Earth Resistivity = 100 Ω-m

AWG or kcmil	$R_1 = R_2$	$X_1 = X_2$	R_0	X_0
Aluminum Phase Conductor				
4	0.5100	0.0662	0.8410	0.5734
2	0.3200	0.0609	0.7040	0.4158
1	0.2550	0.0569	0.6384	0.3216
1/0	0.2000	0.0543	0.5677	0.2498
2/0	0.1600	0.0516	0.4911	0.1717
3/0	0.1250	0.0490	0.4204	0.1208
4/0	0.1000	0.0463	0.3544	0.0785
250	0.0850	0.0440	0.2993	0.0486
300	0.0710	0.0419	0.2522	0.0306
350	0.0610	0.0402	0.2154	0.0180
Copper Phase Conductor				
4	0.3100	0.0662	0.6967	0.4171
2	0.1950	0.0609	0.5652	0.2493
1	0.1550	0.0569	0.4877	0.1703
1/0	0.1200	0.0543	0.4166	0.1189
2/0	0.0970	0.0516	0.3522	0.0762
3/0	0.0770	0.0490	0.2919	0.0455
4/0	0.0610	0.0463	0.2355	0.0223

APPENDIX E
AMPACITY DATA

The National Electrical Code (NEC) defines allowable ampacities for insulated conductors in Article 310. The tables in Article 310, some of which are reprinted in this Appendix, do not take voltage drop into consideration. It may be necessary to increase the size of a conductor above that specified by the NEC to avoid excessive voltage drop.

Ampacities for cables rated 0–2000 V are stated in NEC Tables 310–16 through 310–19, and are reprinted as Tables E.1 through E.4. Ampacities for solid dielectric cables rated 2001 through 35,000 V are stated in NEC Tables 310–67 through 310–86, and are reprinted as Tables E.5 through E.24.

Ampacities are based on the temperature rating of the conductor, which is a function of the type of cable.

The following cable types are rated for 60°C (140°F): TW, UF.

The following cable types are rated for 75°C (167°F): FEPW, RH, RHW, THHW, THW, THWN, XHHW, USE, ZW.

The following cable types are rated for 90°C (194°F): TBS, SA, SIS, FEP, FEPB, MI, RHH, RHW-2, THHN, THHW, THW-2, THWN-2, USE-2, XHH, XHHW, XHHW-2, ZW-2, MV-90.

The following cable type is rated for 105°C (221°F): MV-105.

The following cable type is rated for 150°C (302°F): Z.

The following cable types are rated for 200°C (392°F): FEP, FEPB, PFA.

The following cable types are rated for 250°C (482°F): PFAH, TFE.

Industrial Power Distribution, Second Edition. Ralph E. Fehr, III.
© 2016 The Institute of Electrical and Electronics Engineers, Inc. Published 2016 by John Wiley & Sons, Inc.

TABLE E.1 NEC Table 310–16 Ampacities of Insulated Conductors (0–2000 V) for Not More Than Three Current-Carrying Conductors in Raceway, Cable, or Direct-Buried in Earth (30°C Ambient)

Size	Temperature Rating of Conductor						Size
	Copper			Aluminum or Copper-Clad Aluminum			
AWG or kcmil	60°C (140°F)	75°C (167°F)	90°C (194°F)	60°C (140°F)	75°C (167°F)	90°C (194°F)	AWG or kcmil
14	20[a]	20[a]	25[a]	—	—	—	14
12	25[a]	25[a]	30[a]	20[a]	20[a]	25[a]	12
10	30	35[a]	40[a]	25	30[a]	35[a]	10
8	40	50	55	30	40	45	8
6	55	65	75	40	50	60	6
4	70	85	95	55	65	75	4
2	95	115	130	75	90	100	2
1/0	125	150	170	100	120	135	1/0
2/0	145	175	195	115	135	150	2/0
3/0	165	200	225	130	155	175	3/0
4/0	195	230	260	150	180	205	4/0
250	215	255	290	170	205	230	250
300	240	285	320	190	230	255	300
350	260	310	350	210	250	280	350
400	280	335	380	225	270	305	400
500	320	380	430	260	310	350	500
600	355	420	475	285	340	385	600
700	385	460	520	310	375	420	700
750	400	475	535	320	385	435	750
800	410	490	555	330	395	450	800
900	435	520	585	355	425	480	900
1000	455	545	615	375	445	500	1000
1250	495	590	665	405	485	545	1250
1500	520	625	705	435	520	585	1500
1750	545	650	735	455	545	615	1750
2000	560	665	750	470	560	630	2000

Correction Factors

Ambient °C	Multiply ampacities shown above by appropriate correction factor shown below						Ambient °F
21–25	1.08	1.05	1.04	1.08	1.05	1.04	70–77
26–30	1.00	1.00	1.00	1.00	1.00	1.00	78–86
31–35	0.91	0.94	0.96	0.91	0.94	0.96	87–95
36–40	0.82	0.88	0.91	0.82	0.88	0.91	96–104
41–45	0.71	0.82	0.87	0.71	0.82	0.87	105–113
46–50	0.58	0.75	0.82	0.58	0.75	0.82	114–122
51–55	0.41	0.67	0.76	0.41	0.67	0.76	123–131
56–60	—	0.58	0.71	—	0.58	0.71	132–140
61–70	—	0.33	0.58	—	0.33	0.58	141–158
71–80	—	—	0.41	—	—	0.41	159–176

[a] Unless specifically permitted elsewhere in this *Code*, the overcurrent protection for conductors shall not exceed 15 A for No. 14, 20 A for No. 12, and 30 A for No. 10 copper; or 15 A for No. 12 and 25 A for No. 10 aluminum after any correction factors for ambient temperature and number of conductors have been applied.

TABLE E.2 NEC Table 310–17 Ampacities of Single Insulated Conductors (0–2000 V) in Free Air (30°C Ambient)

Size	Temperature Rating of Conductor						Size
	Copper			Aluminum or Copper-Clad Aluminum			
AWG or kcmil	60°C (140°F)	75°C (167°F)	90°C (194°F)	60°C (140°F)	75°C (167°F)	90°C (194°F)	AWG or kcmil
14	25^a	30^a	35^a	—	—	—	14
12	30^a	35^a	40^a	25^a	30^a	35^a	12
10	40^a	50^a	55^a	35^a	40^a	40^a	10
8	60	70	80	45	55	60	8
6	80	95	105	60	75	80	6
4	105	125	140	80	100	110	4
2	140	170	190	110	135	150	2
1/0	195	230	260	150	180	205	1/0
2/0	225	265	300	175	210	235	2/0
3/0	260	310	350	200	240	275	3/0
4/0	300	360	405	235	280	315	4/0
250	340	405	455	265	315	355	250
300	375	445	505	290	350	395	300
350	420	505	570	330	395	445	350
400	455	545	615	355	425	480	400
500	515	620	700	405	485	545	500
600	575	690	780	455	540	615	600
700	630	755	855	500	595	675	700
750	655	785	885	515	620	700	750
800	680	815	920	535	645	725	800
900	730	870	985	580	700	785	900
1000	780	935	1055	625	750	845	1000
1250	890	1065	1200	710	855	960	1250
1500	980	1175	1325	795	950	1075	1500
1750	1070	1280	1445	875	1050	1185	1750
2000	1155	1385	1560	960	1150	1335	2000

Correction Factors

Ambient °C	Multiply ampacities shown above by appropriate correction factor shown below						Ambient °F
21–25	1.08	1.05	1.04	1.08	1.05	1.04	70–77
26–30	1.00	1.00	1.00	1.00	1.00	1.00	78–86
31–35	0.91	0.94	0.96	0.91	0.94	0.96	87–95
36–40	0.82	0.88	0.91	0.82	0.88	0.91	96–104
41–45	0.71	0.82	0.87	0.71	0.82	0.87	105–113
46–50	0.58	0.75	0.82	0.58	0.75	0.82	114–122
51–55	0.41	0.67	0.76	0.41	0.67	0.76	123–131
56–60	—	0.58	0.71	—	0.58	0.71	132–140
61–70	—	0.33	0.58	—	0.33	0.58	141–158
71–80	—	—	0.41	—	—	0.41	159–176

^{a}Unless specifically permitted elsewhere in this *Code,* the overcurrent protection for conductors shall not exceed 15 A for No. 14, 20 A for No. 12, and 30 A for No. 10 copper; or 15 A for No. 12 and 25 A for No. 10 aluminum after any correction factors for ambient temperature and number of conductors have been applied.

TABLE E.3 NEC Table 310–18 Ampacities of Three Single Insulated Conductors (0–2000 V) in Raceway or Cable (40°C Ambient)

Size	Temperature Rating of Conductor				Size
	Copper		Nickel or Nickel-Coated Copper	Aluminum or Copper-Clad Aluminum	
AWG or kcmil	150°C (302°F)	200°C (392°F)	250°C (482°F)	150°C (302°F)	AWG or kcmil
14	34	36	39	—	14
12	43	45	54	30	12
10	55	60	73	44	10
8	76	83	93	57	8
6	96	110	117	75	6
4	120	125	148	94	4
2	160	171	191	124	2
1/0	215	229	244	169	1/0
2/0	251	260	273	198	2/0
3/0	288	297	308	227	3/0
4/0	332	346	361	260	4/0

Correction Factors

Ambient °C	Multiply ampacities shown above by appropriate correction factor shown below				Ambient °F
41–50	0.95	0.97	0.98	0.95	105–122
51–60	0.90	0.94	0.95	0.90	123–140
61–70	0.85	0.90	0.93	0.85	141–158
71–80	0.80	0.87	0.90	0.80	159–176
81–90	0.74	0.83	0.87	0.74	177–194
91–100	0.67	0.79	0.85	0.67	195–212
101–120	0.52	0.71	0.79	0.52	213–248
121–140	0.30	0.61	0.72	0.30	249–284
141–160	—	0.50	0.65	—	285–320
161–180	—	0.35	0.58	—	321–356
181–200	—	—	0.49	—	357–392
201–225	—	—	0.35	—	393–437

TABLE E.4   NEC Table 310–19 Ampacities of Single Insulated Conductors (0–2000 V) in Free Air (40°C Ambient)

Size	Temperature Rating of Conductor					Size
	Copper			Nickel or Nickel- Coated Copper	Aluminum or Copper- Clad Aluminum	
AWG or kcmil	150°C (302°F)	200°C (392°F)	Bare or Covered Conductors	250°C (482°F)	150°C (302°F)	AWG or kcmil
14	46	54	30	59	—	14
12	60	68	35	78	47	12
10	80	90	50	107	63	10
8	106	124	70	142	83	8
6	155	165	95	205	112	6
4	190	220	125	278	148	4
2	255	293	175	381	198	2
1/0	339	399	235	532	263	1/0
2/0	390	467	275	591	305	2/0
3/0	451	546	320	708	351	3/0
4/0	529	629	370	830	411	4/0

Correction Factors

Ambient °C	Multiply ampacities shown above by appropriate correction factor shown below					Ambient °F
41–50	0.95	0.97	—	0.98	0.95	105–122
51–60	0.90	0.94	—	0.95	0.90	123–140
61–70	0.85	0.90	—	0.93	0.85	141–158
71–80	0.80	0.87	—	0.90	0.80	159–176
81–90	0.74	0.83	—	0.87	0.74	177–194
91–100	0.67	0.79	—	0.85	0.67	195–212
101–120	0.52	0.71	—	0.79	0.52	213–248
121–140	0.30	0.61	—	0.72	0.30	249–284
141–160	—	0.50	—	0.65	—	285–320
161–180	—	0.35	—	0.58	—	321–356
181–200	—	—	—	0.49	—	357–392
201–225	—	—	—	0.35	—	393–437

TABLE E.5 NEC Table 310–67 Ampacities of Insulated Single Copper-Conductor Cables Triplexed in Air (40°C Ambient)

Size	Temperature Rating of Conductor			
	2001–5000 V		5001–35,000 V	
AWG or kcmil	90°C (194°F)	105°C (221°F)	90°C (194°F)	105°C (221°F)
8	65	74	—	—
6	90	99	100	110
4	120	130	130	140
2	160	175	170	195
1/0	215	240	225	255
2/0	250	275	260	295
3/0	290	320	300	340
4/0	335	375	345	390
250	375	415	380	430
350	465	515	470	525
500	580	645	580	650
750	750	835	730	820
1000	880	980	850	950

TABLE E.6 NEC Table 310–68 Ampacities of Insulated Single Aluminum-Conductor Cables Triplexed in Air (40°C Ambient)

Size	Temperature Rating of Conductor			
	2001–5000 V		5001–35,000 V	
AWG or kcmil	90°C (194°F)	105°C (221°F)	90°C (194°F)	105°C (221°F)
8	50	57	—	—
6	70	77	75	84
4	90	100	100	110
2	125	135	130	150
1/0	170	185	175	200
2/0	195	215	200	230
3/0	225	250	230	265
4/0	265	290	270	305
250	295	325	300	335
350	365	405	370	415
500	460	510	460	515
750	600	665	590	660
1000	715	800	700	780

TABLE E.7 NEC Table 310–69 Ampacities of Insulated Single Copper Conductor Isolated in Air (40°C Ambient)

Size	Temperature Rating of Conductor					
	2001–5000 V		5001–15,000 V		15,001–35,000 V	
AWG or kcmil	90°C (194°F)	105°C (221°F)	90°C (194°F)	105°C (221°F)	90°C (194°F)	105°C (221°F)
8	83	93	—	—	—	—
6	110	120	110	125	—	—
4	145	160	150	165	—	—
2	190	215	195	215	—	—
1/0	260	290	260	290	260	290
2/0	300	330	300	335	300	330
3/0	345	385	345	385	345	380
4/0	400	445	400	445	395	445
250	445	495	445	495	440	490
350	550	615	550	610	545	605
500	695	775	685	765	680	755
750	900	1000	885	990	870	970
1000	1075	1200	1060	1185	1040	1160
1250	1230	1370	1210	1350	1185	1320
1500	1365	1525	1345	1500	1315	1465
1750	1495	1665	1470	1640	1430	1595
2000	1605	1790	1575	1755	1535	1710

TABLE E.8 NEC Table 310–70 Ampacities of Insulated Single Aluminum Conductor Isolated in Air (40°C Ambient)

Size	Temperature Rating of Conductor					
	2001–5000 V		5001–15,000 V		15,001–35,000 V	
AWG or kcmil	90°C (194°F)	105°C (221°F)	90°C (194°F)	105°C (221°F)	90°C (194°F)	105°C (221°F)
8	64	71	—	—	—	—
6	85	95	87	97	—	—
4	115	125	115	130	—	—
2	150	165	150	170	—	—
1/0	200	225	200	225	200	225
2/0	230	260	235	260	230	260
3/0	270	300	270	300	270	300
4/0	310	350	310	350	310	345
250	345	385	345	385	345	380
350	430	480	430	480	430	475
500	545	605	535	600	530	590
750	710	790	700	780	685	765
1000	855	950	840	940	825	920
1250	980	1095	970	1080	950	1055
1500	1105	1230	1085	1215	1060	1180
1750	1215	1355	1195	1335	1165	1300
2000	1320	1475	1295	1445	1265	1410

TABLE E.9 NEC Table 310–71 Ampacities of an Insulated Three-Conductor Copper Cable Isolated in Air (40°C Ambient)

Size	Temperature Rating of Conductor			
	2001–5000 V		5001–35,000 V	
AWG or kcmil	90°C (194°F)	105°C (221°F)	90°C (194°F)	105°C (221°F)
8	59	66	—	—
6	79	88	93	105
4	105	115	120	135
2	140	154	165	185
1	160	180	185	210
1/0	185	205	215	240
2/0	215	240	245	275
3/0	250	280	285	315
4/0	285	320	325	360
250	320	355	360	400
350	395	440	435	490
500	485	545	535	600
750	615	685	670	745
1000	705	790	770	860

TABLE E.10 NEC Table 310–72 Ampacities of an Insulated Three-Conductor Aluminum Cable Isolated in Air (40°C Ambient)

Size	Temperature Rating of Conductor			
	2001–5000 V		5001–35000 Volts	
AWG or kcmil	90°C (194°F)	105°C (221°F)	90°C (194°F)	105°C (221°F)
8	46	51	—	—
6	61	68	72	80
4	81	90	95	105
2	110	120	125	145
1	125	140	145	165
1/0	145	160	170	185
2/0	170	185	190	215
3/0	195	215	220	245
4/0	225	250	255	285
250	250	280	280	315
350	310	345	345	385
500	385	430	425	475
750	495	550	540	600
1000	585	650	635	705

TABLE E.11 NEC Table 310–73 Ampacities of an Insulated Triplexed or Three Single-Conductor Copper Cables in Isolated Conduit in Air (40°C Ambient)

Size AWG or kcmil	Temperature Rating of Conductor			
	2001–5000 V		5001–35,000 V	
	90°C (194°F)	105°C (221°F)	90°C (194°F)	105°C (221°F)
8	55	61	—	—
6	75	84	83	93
4	97	110	110	120
2	130	145	150	165
1	155	175	170	190
1/0	180	200	195	215
2/0	205	225	225	255
3/0	240	270	260	290
4/0	280	305	295	330
250	315	355	330	365
350	385	430	395	440
500	475	530	480	535
750	600	665	585	655
1000	690	770	675	755

TABLE E.12 NEC Table 310–74 Ampacities of an Insulated Triplexed or Three Single-Conductor Aluminum Cables in Isolated Conduit in Air (40°C Ambient)

Size AWG or kcmil	Temperature Rating of Conductor			
	2001–5000 V		5001–35,000 V	
	90°C (194°F)	105°C (221°F)	90°C (194°F)	105°C (221°F)
8	43	48	—	—
6	58	65	65	72
4	76	85	84	94
2	100	115	115	130
1	120	135	130	150
1/0	140	155	150	170
2/0	160	175	175	200
3/0	190	210	200	225
4/0	215	240	230	260
250	250	280	255	290
350	305	340	310	350
500	380	425	385	430
750	490	545	485	540
1000	580	645	565	640

TABLE E.13 NEC Table 310–75 Ampacities of Insulated Three-Conductor Copper Cable in Isolated Conduit in Air (40°C Ambient)

Size	Temperature Rating of Conductor			
	2001–5000 V		5001–35,000 V	
AWG or kcmil	90°C (194°F)	105°C (221°F)	90°C (194°F)	105°C (221°F)
8	52	58	—	—
6	69	77	83	92
4	91	100	105	120
2	125	135	145	165
1/0	165	185	195	215
2/0	190	210	220	245
3/0	220	245	250	280
4/0	255	285	290	320
250	280	315	315	350
350	350	390	385	430
500	425	475	470	525
750	525	585	570	635
1000	590	660	650	725

TABLE E.14 NEC Table 310–76 Ampacities of Insulated Three-Conductor Aluminum Cable in Isolated Conduit in Air (40°C Ambient)

Size	Temperature Rating of Conductor			
	2001–5000 V		5001–35,000 V	
AWG or kcmil	90°C (194°F)	105°C (221°F)	90°C (194°F)	105°C (221°F)
8	41	46	—	—
6	53	59	64	71
4	71	79	84	94
2	96	105	115	125
1/0	130	145	150	170
2/0	150	165	170	190
3/0	170	190	195	220
4/0	200	225	225	255
250	220	245	250	280
350	275	305	305	340
500	340	380	380	425
750	430	480	470	520
1000	505	560	550	615

TABLE E.15 NEC Table 310–77 Ampacities of Three Single Insulated Copper Conductors in Underground Electrical Ducts (3 Conductors Per Duct, 20°C Earth Ambient, Thermal Resistance (Rho) of 90)

		Temperature Rating of Conductor			
	Size	2001–5000 V		5001–35,000 V	
Duct Bank Layout	AWG or kcmil	90°C (194°F)	105°C (221°F)	90°C (194°F)	105°C (221°F)
One Circuit 1 × 1	8	64	69	—	—
(11.5″ × 11.5″)	6	85	92	90	97
	4	110	120	115	125
	2	145	155	155	165
	1	170	180	175	185
	1/0	195	210	200	215
	2/0	220	235	230	245
	3/0	250	270	260	275
	4/0	290	310	295	315
	250	320	345	325	345
	350	385	415	390	415
	500	470	505	465	500
	750	585	630	565	610
	1000	670	720	640	690
Three Circuits 2 × 2	8	56	60	—	—
(19″ × 19″) or 1 × 3	6	73	79	77	83
(27″w × 11.5″d)	4	95	100	99	105
	2	125	130	130	135
	1	140	150	145	155
	1/0	160	175	165	175
	2/0	185	195	185	200
	3/0	210	225	210	225
	4/0	235	255	240	255
	250	260	280	260	280
	350	315	335	310	330
	500	375	405	370	395
	750	460	495	440	475
	1000	525	565	495	535
Six Circuits 3 × 2	8	48	52	—	—
(19″w × 27″d) or	6	62	67	64	68
2 × 3 (27″w × 19″d)	4	80	86	82	88
	2	105	110	105	115
	1	115	125	120	125
	1/0	135	145	135	145
	2/0	150	160	150	165
	3/0	170	185	170	185
	4/0	195	210	190	205
	250	210	225	210	225
	350	250	270	245	265
	500	300	325	290	310
	750	365	395	350	375
	1000	410	445	390	415

TABLE E.16 NEC Table 310–78 Ampacities of Three Single Insulated Aluminum Conductors in Underground Electrical Ducts (3 Conductors Per Duct, 20°C Earth Ambient, Thermal Resistance (Rho) of 90)

		Temperature Rating of Conductor			
	Size	2001–5000 V		5001–35,000 V	
Duct Bank Layout	AWG or kcmil	90°C (194°F)	105°C (221°F)	90°C (194°F)	105°C (221°F)
One Circuit 1 × 1	8	50	54	—	—
(11.5″ × 11.5″)	6	66	71	70	75
	4	86	93	91	98
	2	115	125	120	130
	1	130	140	135	145
	1/0	150	160	155	165
	2/0	170	185	175	190
	3/0	195	210	200	215
	4/0	225	245	230	245
	250	250	270	250	270
	350	305	325	305	330
	500	370	400	370	400
	750	470	505	455	490
	1000	545	590	525	565
Three Circuits 2 × 2	8	44	47	—	—
(19″ × 19″) or 1 × 3	6	57	61	60	65
(27″w × 11.5″d)	4	74	80	77	83
	2	96	105	100	105
	1	110	120	110	120
	1/0	125	135	125	140
	2/0	145	155	145	155
	3/0	160	175	165	175
	4/0	185	200	185	200
	250	205	220	200	220
	350	245	265	245	260
	500	295	320	290	315
	750	370	395	355	385
	1000	425	460	405	440
Six Circuits 3 × 2	8	38	41	—	—
(19″w × 27″d) or	6	48	52	50	54
2 × 3 (27″w × 19″d)	4	62	67	64	69
	2	80	86	80	88
	1	91	98	90	99
	1/0	105	110	105	110
	2/0	115	125	115	125
	3/0	135	145	130	145
	4/0	150	165	150	160
	250	165	180	165	175
	350	195	210	195	210
	500	240	255	230	250
	750	290	315	280	305
	1000	335	360	320	345

TABLE E.17 NEC Table 310–79 Ampacities of Three-Conductor Copper Cable in Underground Electrical Ducts (1 Cable Per Duct, 20°C Earth Ambient, Thermal Resistance (Rho) of 90)

		Temperature Rating of Conductor			
	Size	2001–5000 V		5001–35,000 V	
Duct Bank Layout	AWG or kcmil	90°C (194°F)	105°C (221°F)	90°C (194°F)	105°C (221°F)
One Circuit 1 × 1	8	59	64	—	—
(11.5″ × 11.5″)	6	78	84	88	95
	4	100	110	115	125
	2	135	145	150	160
	1	155	165	170	185
	1/0	175	190	195	210
	2/0	200	220	220	235
	3/0	230	250	250	270
	4/0	265	285	285	305
	250	290	315	310	335
	350	355	380	375	400
	500	430	460	450	485
	750	530	570	545	585
	1000	600	645	615	660
Three Circuits 2 × 2	8	53	57	—	—
(19″ × 19″) or 1 ×	6	69	74	75	81
3 (27″w × 11.5″d)	4	89	96	97	105
	2	115	125	125	135
	1	135	145	140	155
	1/0	150	165	160	175
	2/0	170	185	185	195
	3/0	195	210	205	220
	4/0	225	240	230	250
	250	245	265	255	270
	350	295	315	305	325
	500	355	380	360	385
	750	430	465	430	465
	1000	485	520	485	515
Six Circuits 3 × 2	8	46	50	—	—
(19″w × 27″d) or 2	6	60	65	63	68
× 3 (27″w × 19″d)	4	77	83	81	87
	2	98	105	105	110
	1	110	120	115	125
	1/0	125	135	130	145
	2/0	145	155	150	160
	3/0	165	175	170	180
	4/0	185	200	190	200
	250	200	220	205	220
	350	240	270	245	275
	500	290	310	290	305
	750	350	375	340	365
	1000	390	420	380	405

TABLE E.18 NEC Table 310–80 Ampacities of Three-Conductor Aluminum Cable in Underground Electrical Ducts (1 Cable Per Duct, 20°C Earth Ambient, Thermal Resistance (Rho) of 90)

		Temperature Rating of Conductor			
	Size	2001–5000 V		5001–35,000 V	
Duct Bank Layout	AWG or kcmil	90°C (194°F)	105°C (221°F)	90°C (194°F)	105°C (221°F)
One Circuit 1 × 1	8	46	50	—	—
(11.5″× 11.5″)	6	61	66	69	74
	4	80	86	89	96
	2	105	110	115	125
	1	120	130	135	145
	1/0	140	150	150	165
	2/0	160	170	170	185
	3/0	180	195	195	210
	4/0	205	220	220	240
	250	230	245	245	265
	350	280	310	295	315
	500	340	365	355	385
	750	425	460	440	475
	1000	495	535	510	545
Three Circuits 2 × 2	8	41	44	—	—
(19″ × 19″) or 1 ×	6	54	58	59	64
3 (27″w × 11.5″d)	4	70	75	75	81
	2	90	97	100	105
	1	105	110	110	120
	1/0	120	125	125	135
	2/0	135	145	140	155
	3/0	155	165	160	175
	4/0	175	185	180	195
	250	190	205	200	215
	350	230	250	240	255
	500	280	300	285	305
	750	345	375	350	375
	1000	400	430	400	430
Six Circuits 3 × 2	8	36	39	—	—
(19″w × 27″d) or 2	6	46	50	49	53
× 3 (27″w × 19″d)	4	60	65	63	68
	2	77	83	80	86
	1	87	94	90	98
	1/0	99	105	105	110
	2/0	110	120	15	125
	3/0	130	140	130	140
	4/0	145	155	150	160
	250	160	170	160	170
	350	190	205	190	205
	500	230	245	230	245
	750	280	305	275	295
	1000	320	345	315	335

TABLE E.19 NEC Table 310–81 Ampacities of Single Insulated Copper Conductors Directly Buried in Earth (20°C Earth Ambient, Thermal Resistance (Rho) of 90)

		Temperature Rating of Conductor			
	Size	2001–5000 V		5001–35,000 V	
Circuit Layout	AWG or kcmil	90°C (194°F)	105°C (221°F)	90°C (194°F)	105°C (221°F)
One Circuit – Three Conductors Flat Horizontal Spacing 7.5″ Between Adjacent Cables	8	110	115	—	—
	6	140	150	130	140
	4	180	195	170	180
	2	230	250	210	225
	1	260	280	240	260
	1/0	295	320	275	295
	2/0	335	365	310	335
	3/0	385	415	355	380
	4/0	435	465	405	435
	250	470	510	440	475
	350	570	615	535	575
	500	690	745	650	700
	750	845	910	805	865
	1000	980	1055	930	1005
Two Circuits – Six Conductors Flat Horizontal Spacing 7.5″ Between Adjacent Cables in Circuit, 24″ Between Circuits	8	100	110	—	—
	6	130	140	120	130
	4	165	180	160	170
	2	215	230	195	210
	1	240	260	225	240
	1/0	275	295	255	275
	2/0	310	335	290	315
	3/0	355	380	330	355
	4/0	400	430	375	405
	250	435	470	410	440
	350	520	560	495	530
	500	630	680	600	645
	750	775	835	740	795
	1000	890	960	855	920

TABLE E.20 NEC Table 310–82 Ampacities of Single Insulated Aluminum Conductors Directly Buried in Earth (20°C Earth Ambient, Thermal Resistance (Rho) of 90)

		Temperature Rating of Conductor			
	Size	2001–5000 V		5001–35,000 V	
Circuit Layout	AWG or kcmil	90°C (194°F)	105°C (221°F)	90°C (194°F)	105°C (221°F)
One Circuit – Three Conductors Flat Horizontal Spacing 7.5″ Between Adjacent Cables	8	85	90	—	—
	6	110	115	100	110
	4	140	150	130	140
	2	180	195	165	175
	1	205	220	185	200
	1/0	230	250	215	230
	2/0	265	285	245	260
	3/0	300	320	275	295
	4/0	340	365	315	340
	250	370	395	345	370
	350	445	480	415	450
	500	540	580	510	545
	750	665	720	635	680
	1000	780	840	740	795
Two Circuits – Six Conductors Flat Horizontal Spacing 7.5″ Between Adjacent Cables in Circuit, 24″ Between Circuits	8	80	85	—	—
	6	100	110	95	100
	4	130	140	125	130
	2	165	180	155	165
	1	190	200	175	190
	1/0	215	230	200	215
	2/0	245	260	225	245
	3/0	275	295	255	275
	4/0	310	335	290	315
	250	340	365	320	345
	350	410	440	385	415
	500	495	530	470	505
	750	610	655	580	625
	1000	710	765	680	730

TABLE E.21 NEC Table 310–83 Ampacities of Three-Conductor Copper Cable Directly Buried in Earth (20°C Earth Ambient, Thermal Resistance (Rho) of 90)

		Temperature Rating of Conductor			
	Size	2001–5000 V		5001–35,000 V	
Circuit Layout	AWG or kcmil	90°C (194°F)	105°C (221°F)	90°C (194°F)	105°C (221°F)
One Cable	8	85	89	—	—
	6	105	115	115	120
	4	135	150	145	155
	2	180	190	185	200
	1	200	215	210	225
	1/0	230	245	240	255
	2/0	260	280	270	290
	3/0	295	320	305	330
	4/0	335	360	350	375
	250	365	395	380	410
	350	440	475	460	495
	500	530	570	550	590
	750	650	700	665	720
	1000	730	785	750	810
Two Cables 24″ Flat Horizontal Spacing	8	80	84	—	—
	6	100	105	105	115
	4	130	140	135	145
	2	165	180	170	185
	1	185	200	195	210
	1/0	215	230	220	235
	2/0	240	260	250	270
	3/0	275	295	280	305
	4/0	310	335	320	345
	250	340	365	350	375
	350	410	440	420	450
	500	490	525	500	535
	750	595	640	605	650
	1000	665	715	675	730

TABLE E.22 NEC Table 310–84 Ampacities of Three-Conductor Aluminum Cable Directly Buried in Earth (20°C Earth Ambient, Thermal Resistance (Rho) of 90)

		Temperature Rating of Conductor			
	Size	2001–5000 V		5001–35,000 V	
Circuit Layout	AWG or kcmil	90°C (194°F)	105°C (221°F)	90°C (194°F)	105°C (221°F)
One Cable	8	65	70	—	—
	6	80	88	90	95
	4	105	115	115	125
	2	140	150	145	155
	1	155	170	165	175
	1/0	180	190	185	200
	2/0	205	220	210	225
	3/0	230	250	240	260
	4/0	260	280	270	295
	250	285	310	300	320
	350	345	375	360	390
	500	420	450	435	470
	750	520	560	540	580
	1000	600	650	620	665
Two Cables 24″ Flat Horizontal Spacing	8	60	66	—	—
	6	75	83	80	95
	4	100	110	105	115
	2	130	140	135	145
	1	145	155	150	165
	1/0	165	180	170	185
	2/0	190	205	195	210
	3/0	215	230	220	240
	4/0	245	260	250	270
	250	265	285	275	295
	350	320	345	330	355
	500	385	415	395	425
	750	480	515	485	525
	1000	550	590	560	600

TABLE E.23 NEC Table 310–85 Ampacities of Three Triplexed Single-Conductor Copper Cables Directly Buried in Earth (20°C Earth Ambient, Thermal Resistance (Rho) of 90, 90°C Conductor Temperature)

Circuit Layout	Size AWG or kcmil	2001–5000 V	5001–35,000 V
One Circuit Three Triplexed Cables	8	90	—
	6	120	115
	4	150	150
	2	195	190
	1	225	215
	1/0	255	245
	2/0	290	275
	3/0	330	315
	4/0	375	360
	250	410	390
	350	490	470
	500	590	565
	750	725	685
	1000	825	770
Two Circuits Three Triplexed Cables Each 24″ Between Circuits	8	85	—
	6	110	105
	4	140	140
	2	180	175
	1	205	200
	1/0	235	225
	2/0	265	255
	3/0	300	290
	4/0	340	325
	250	370	355
	350	445	426
	500	535	510
	750	650	615
	1000	740	690

TABLE E.24 NEC Table 310–86 Ampacities of Three Triplexed Single-Conductor Aluminum Cables Directly Buried in Earth (20°C Earth Ambient, Thermal Resistance (Rho) of 90, 90°C Conductor Temperature)

	Size		
Circuit Layout	AWG or kcmil	2001–5000 V	5001–35,000 V
One Circuit Three Triplexed Cables	8	70	—
	6	90	90
	4	120	115
	2	155	145
	1	175	165
	1/0	200	190
	2/0	225	215
	3/0	255	245
	4/0	290	280
	250	320	305
	350	385	370
	500	465	445
	750	580	550
	1000	670	635
Two Circuits Three Triplexed Cables Each 24″ Between Circuits	8	65	—
	6	85	85
	4	110	105
	2	140	135
	1	160	155
	1/0	180	175
	2/0	205	200
	3/0	235	225
	4/0	265	255
	250	290	280
	350	350	335
	500	420	405
	750	520	485
	1000	600	565

APPENDIX F

CONDUIT DATA

TABLE F.1   Rigid Steel Conduit Technical Data

Nominal Size	Pipe					Elbows		
	UL Min. Weight (lb/100 ft) incl. couplings	O.D. (inch)	I.D. (inch)	Average Wall (inch)	Threads per Inch	Weight (lb/100 ft)	Radius (inch)	Offset (inch)
$\frac{1}{2}''$	79	0.84	0.632	0.104	14	82	4	6.5
$\frac{3}{4}''$	105	1.05	0.836	0.107	14	109	4.5	7.25
$1''$	153	1.315	1.063	0.126	$11\frac{1}{2}$	201	5.75	8.75
$1\frac{1}{4}''$	201	1.66	1.394	0.133	$11\frac{1}{2}$	313	7.25	10.5
$1\frac{1}{2}''$	249	1.9	1.624	0.138	$11\frac{1}{2}$	441	8.25	11.75
$2''$	332	2.375	2.083	0.146	$11\frac{1}{2}$	707	9.5	13.5
$2\frac{1}{2}''$	527	2.875	2.489	0.193	8	1711	10.5	15
$3''$	682	3.5	3.09	0.205	8	1850	13	18
$3\frac{1}{2}''$	831	4	3.57	0.215	8	2979	15	20.5
$4''$	972	4.5	4.055	0.225	8	3528	16	22
$5''$	1314	5.563	5.073	0.245	8	6575	24	31
$6''$	1745	6.625	6.093	0.266	8	9645	30	37.5

Source: Data provided by Western Tube and Conduit Corporation.

Industrial Power Distribution, Second Edition. Ralph E. Fehr, III.
© 2016 The Institute of Electrical and Electronics Engineers, Inc. Published 2016 by John Wiley & Sons, Inc.

TABLE F.2 Intermediate Metal Conduit Technical Data

Nominal Size	O.D. (inch)	I.D. (inch)	Average Wall (inch)	Nominal Weight (lb/100 ft) incl. couplings	Threads per Inch	Length of Thread (inch)
$1/2''$	0.815	0.675	0.07	65	14	0.78
$3/4''$	1.029	0.879	0.075	92	14	0.79
$1''$	1.29	1.12	0.085	120	$11 1/2$	0.98
$1 1/4''$	1.638	1.468	0.085	153	$11 1/2$	1.01
$1 1/2''$	1.883	1.703	0.09	188	$11 1/2$	1.03
$2''$	2.36	2.17	0.095	251	$11 1/2$	1.06
$2 1/2''$	2.857	2.597	0.14	422	8	1.57
$3''$	3.476	3.216	0.14	520	8	1.63
$3 1/2''$	3.971	3.711	0.14	605	8	1.68
$4''$	4.466	4.206	0.14	674	8	1.73

Source: Data provided by Western Tube and Conduit Corporation.

TABLE F.3 Electrical Metallic Tubing Technical Data

Nominal Size	Nominal Weight (lb/100 ft)	O.D. (inch)	I.D. (inch)	Average Wall (inch)
$1/2''$	29.5	0.706	0.622	0.042
$3/4''$	44.5	0.922	0.824	0.049
$1''$	65.0	1.163	1.049	0.057
$1 1/4''$	96.0	1.51	1.38	0.065
$1 1/2''$	111.0	1.74	1.61	0.065
$2''$	141.0	2.197	2.067	0.065
$2 1/2''$	230.0	2.875	2.731	0.072
$3''$	270.0	3.5	3.356	0.072
$3 1/2''$	326.0	4	3.834	0.083
$4''$	400.0	4.5	4.334	0.083

Source: Data provided by Western Tube and Conduit Corporation.

TABLE F.4 Schedule 40 PVC Pipe Data

Nominal Pipe Size	O.D. (inch)	Average I.D. (inch)	Minimum Wall (inch)	Nominal Weight (lb/ft)
1"	1.315	1.029	0.133	0.333
$1\frac{1}{4}"$	1.660	1.360	0.140	0.450
$1\frac{1}{2}"$	1.900	1.590	0.145	0.537
2"	2.375	2.047	0.154	0.720
$2\frac{1}{2}"$	2.875	2.445	0.203	1.136
3"	3.500	3.042	0.216	1.488
$3\frac{1}{2}"$	4.000	3.521	0.226	1.789
4"	4.500	3.998	0.237	2.118
5"	5.563	5.016	0.258	2.874
6"	6.625	6.031	0.280	3.733
8"	8.625	7.942	0.322	5.619

Source: Data provided by Harvel Plastics, Inc.

INDEX

a operator, **118**, 370
AC power, 18
accelerating torque, 263
acetylene, 28
acoustic resonance, 303
across-the-line, 267
additive polarity, 71, **72**
advanced metering infrastructure, 62
air-magnetic circuit breaker, 218
altitude derating, 217
ambient temperature derating, 172, 217
ambient temperature
　effect on cable ampacity, 26
　effect on fluorescent lamps, 293
　effect on fuses, 172
American Wire Gauge, **357**
AMI, 62
ammeter switch, 55, **57**
AMR, **37**, 62
analysis of open delta transformer, 82
ANSI accuracy classes, 49
arc extinction, 156, 218, 328
arc flash, **226**
arc flash warning label, 228
arc-resistant switchgear, 208, 211
Argand diagram, 353, **369**
Askarel, 27
ASTM D-635, 196
asymmetry factor, 157
automatic meter reading, **37**, 62
automatic source transfer, 11, 214
autotransformer starting, 267, **273**
auxiliary relay, 239, **240**
AWG, **357**

back-to-back switching, 324, 326
ballast, 292, **301**, 313
base quantities, 15, **16**

basic impulse level, 27, 216
Becquerel, Alexandre, 290
BIL, 27, 216
Blondel, Andre, 52
Blondel's theorem, 52
bonding jumper, 198
breakdown torque, 260
breaker failure, 7, **8**
Brown and Sharpe wire gauge, **357**
burden, **37**, 42, 49
burden calculations, 47
burnthrough, 211

cable sizing, 23, 264
cable tray, 179, **194**
Cable Tray Institute, 194
candela, 287, 358
capacitor controller, 323
capacitor de-energization, 326
capacitor energization, 325
capacitor switching, 324
carbon dioxide, 28
carbon monoxide, 28
Carson's equations, 127
cast coil transformers, 29
cataphoretic segregation, 303
CBEMA curve, 336, **337**
ceiling cavity ratio, 311
center hanger support, 196
channel cable tray, 180, 195
Chlorinol, 27
chopped wave test, 216
CIE, 300
circuit breaker, 5, **217**
circuit protection, 225, 266
circular mil, 357
closed transfer, 11, 269
closing and latching, 159

Industrial Power Distribution, Second Edition. Ralph E. Fehr, III.
© 2016 The Institute of Electrical and Electronics Engineers, Inc. Published 2016 by John Wiley & Sons, Inc.

coefficient of friction, 186, 188
color temperature, 288
combination instrument transformer, 44
combination motor starter, 222
compensating metering, 59
components of AC power, 18
concentric neutral, 182, **185**
conductor sizing, 23, 264
conduit, 179, 181, **401**
contact-parting fault current, 112
contact-parting impedance diagram, 113
contact-parting symmetrical current, 158
contact parting time, 159
contactor, 224
continuity of service, 338
continuous current rating, 37, 45, 158, 215, 224
control power transformer, 235
cool color, 288
coordination, 169
core laminations, 69
counter EMF, 224, 255
coupling capacitor voltage transformers (CCVTs), 39
cradled configuration, 187
Cramer's rule, **368**
cross-connection, 94
cross-phasing, **58**
CT, 5, 42, 57, 357
CT rating factor (RF), 45
CT switch, 42
current base, 15, **16**
current distortion, 349
current metering, 55
current transformer, 5, 42, 57, 357
current transformer ratio (CTR), 38
current-limiting fuse, 160, **166**

delta-delta, **74**, 82
delta-double delta, 92
delta-double wye, 92
delta-wye transformer current analysis, 80
delta-wye transformer zero-sequence model, 129
delta-wye, **78**, 93
demand metering, 52
digital metering, 61
displacement power factor, 352
dissolved gas analysis, 28
distortion, 353

distortion power factor, 352
diversity of load, 26, 323
dot convention, 41, 74
double line-to-ground fault circuit model, **139**
double line-to-ground fault, 138
double-headed switchgear, 213
drawout, 5, 208, 214, 221, 238
drop out, 236
dry-type transformers, 29
duct bank, 179, 187

eddy current, 69, 179, 346
Edison, Thomas, 288
effective current, **347**
efficacy, 288, **316**
efficiency, 108, 259
electroluminescence, 297
elevated strut support, 196
enclosed non-ventilated transformers, 29
ethane, 29
Euler's formula, 20, **372**
Euler's identity, 372
excitation current, 345

Faraday's law, 66
fast transfer, 11, 214
fault impedance, 134, 137, 141
ferroresonance, 97
first-cycle symmetrical current, **105**, 113, 157
first-cycle impedance diagram, 106
fixed capacitor, 323
flash protection boundary, 226
flicker, 336
flicker chart, 336, **337**
floor cavity ratio, 311
flow switch, 238
fluorescent lighting, 288, **290**
footcandle, 287, 358
Fortescue, Charles, 113
Fourier analysis, 343
Fourier series, 343
Fourier transform, 345
Fourier, Joseph, 343
full voltage reversing starter, 224
full voltage non-reversing starter, 224
full wave test, 216
fundamental frequency, 342
fuse tolerances, 172

fused low-voltage circuit breaker, 160
fusible disconnect switch, 225
FVNR starter, 224
FVR starter, 224

getter, 293
grounding transformer, 95
group fusing, 330

half-wave symmetry, 345
hanger rod clamp, 196
harmonic filter, 349
harmonic order, 342
harmonics, 76, 280, 283, 322, 329, **341**
heater, 225
HID lighting, 288, **294**
high-intensity discharge lighting, 288, **294**
high-leg delta, 3
high-pressure sodium lighting, 288, **295**
hi-pot test, 220
Holonyak, Nick, 297
horizontal illuminance, 303
hydrogen, 28
hydrogen sulfide, 213
hysteresis, 69, 71, 346

ideal transformer, 66
IEC, 223
IEEE C37.010, 103, 110, 157
IEEE C37.04, 215
IEEE C37.06, 215
IEEE C37.09, 215
IEEE C37.13, 161, 162
IEEE C37.20.1, 215
IEEE C37.20.2, 215
IEEE C37.20.3, 208
IEEE C57.12.00, 31, 72, 81
IEEE C57.13, 46, 47, 49
IEEE 141, 103, 349
IEEE 242, 105
IEEE 1584, 226, 232
IESNA, 287, 303
illuminance, 287, 303, 311, 358
Illuminating Engineering Society of North America, 287, 303
impedance base, 15, **16**
impulse transient, 341
incandescent lighting, 288
indicator lamp, 212, 235, 238, 245
individual fusing, 330

infinite bus, 22
instrument transformers, 39
interlock, 222, 224, 238, **243**, 265
internal pressure, 212
International Commission on Illumination, 300
International Electrotechnical Commission, 223
interrupter, 217, 220
interrupting rating, 158
interrupting time, 158
inverter-duty motors, 283
ITIC curve, 338

j operator, 119, 370
jam ratio, 182
jogging, 225, **248**

kelvin, 288, 358
k-factor, 348
k-factor transformer, 346
K_h, **38**, 50
kilovar demand, 53
kilowatt demand, 53
kVA demand, 53

ladder logic, 235
ladder-type cable tray, 180, 195
laminations, 69
leading power factor, 59, 320, 323
LED driver circuit, 299
LED lighting, **297**
Lenz's law, 66
"let go" threshold, 1
level switch, 238
light load (metering), 38
light loss, 312
light loss factors, 313
light-emitting diode lighting, **297**
limit switch, 238
limited approach boundary, 226
line-to-ground fault circuit model, **137**
line-to-ground fault, 136
line-to-line fault circuit model, **141**
line-to-line fault, **141**, 247
liquid-immersed transformer cooling class, 30
liquid-immersed transformers, 27
load center, 207, 340
locked rotor, 256

locked rotor torque, 260
low-voltage fuse, 168
low-pressure discharge lighting, 288, **290**
low-pressure sodium lighting, **292**
lumen, 287, 290, 313, 358
lumen method, 311
luminaire, 296, 299, 312
luminous efficacy, 288
luminous flux, 287
luminous intensity, 287

magnetic ballast, 292
main transformer, 26
main transformer (T-connection), 89
manhole, 187
medium-voltage fuse, 163
mercury vapor lighting, 288, **294**
metal halide lighting, 288, **296**
metering class instrument transformers, 45
metering class, 45
metering correction factor, 46
metering element, 38
metering form letter, 38
metering form number, 38
metering units, 44
methane, 29
mineral oil, 27, 319
minimum melt characteristic, 170
molded-case circuit breaker, 162
moment of inertia, 263
motor circuit protector, 225
motor control center, 207, **222**
motor nameplate voltage, 258
motor protection, 265
motor starting methods, 267
motor starting time, 263
multiratio CT, 56
multispeed starter, 224
multi-stator watthour metering, 52
Museum of Electric Lamp Technology, **316**
mutual inductance, 66

negative sequence, 117
negative-sequence component, 118, 119
negative-sequence harmonics, 342
negative-sequence network, 126
negative-sequence reactance, 126
negative-sequence reference bus, 126
NEMA 250, 208
NEMA code letter, 256

NEMA design class, 261
NEMA enclosure, 208
NEMA FG1, 194
NEMA field angle, 301
NEMA fuse links, 331
NEMA MG 1-10.37.2, 256
NEMA motor starters, 223
NEMA VE1, 194
NEMA VE2, 194
network protector, 14
neutral inversion, 321
NFPA 70E, 226
nonrecoverable light loss, 313
non-standard delta-wye, 94

one-line diagrams, 4
one-line-open circuit model, **144**
one-line-open fault, 143
open (source) transfer, 11
open delta-open delta, 82
open wye-open delta, 86
open-circuit faults, 102, **143**
open-circuiting CTs, 42, 56
oscillatory transient, 341
outrush current, 331
overcurrent device coordination, 169
overlapping zones, 6
overload protection, 223, 225, 238

PACF, 47
parallel transfer, 214
part-winding starting, 278
PCBs, **27**, 292, 319
percent transformer impedance, 27
personal protective equipment, 2, 226
per-unit, 14
per-unit base conversion, 16
per-unit quantity, 16
phantom leg, 74
phase angle correction factor, 47
phase rotation, 117
phase sequencing, 117
phase shift, 58, 78, **93**, 95
phase-displaced metering, 53
phaseformer, 57
phasor, 369
phosphor, 291, 294, 296, 297
phosphor-converted, 297
pick up, 236
PLC, 239

plugging, 225, **250**
polarity, 41, 43, 66, 71, 73
polyalpha olefins, 27
polychlorinated biphenyls, **27**, 292, 319
positive sequence, 81, 93, 117
positive-sequence component, 118
positive-sequence harmonics, 342
positive-sequence network, 124
positive-sequence reference bus, 124
potential transformers, 39
power base, 15
power circuit breakers, 158
power factor, 20, 46, 49, 59, 86, 88, 108, 288, **319**, 350, **352**
power factor correction, 319
power quality, 335
PPE, 2, 226
predamage, 172
preloading, 172
pressure switch, 238
primary loop, 11
primary selective, 9
primary winding, 27, 43, 67, 74
programmable logic controller, 239
protection, 5, 225, 265, 266, 330
protection boundary, 226
protective devices, 155
PT, 39
puffer, 218
pullbox, 183, 185, 186
pulling tension, 181, 186, **187**, 200
pull-up torque, 260
pulse initiator, 54, 60
pulse recorder, 60
pulse-operated meters, 54
pulse-width modulation, 282
PWM, 282
Pyranol, 27

Q metering, 57

raceway, 179
radial bus, 6
rating factor (RF), 45
ratio correction factor (RCF), 46
reactance factor (M), 105, 375
reactiformer, 57
reactor starting, 267, **271**
recoverable light loss, 312, **313**
reduced voltage starter, 267

reduced voltage starting, 267
register ratio, 52
reignition, 327
relaying class instrument transformers, 45
resistivity, 24, 25, 69
resistor starting, 267, **269**
resonance, 76, 97, 330, 350
restricted approach boundary, 226
restrike, 327
restrike time, 297
reversing starter, 224, 245, **246**, 249
ring bus, 7
room cavity ratio, 311
rotor bar design, 261
R_r, 52

SCA, 9, 14, **23**, 349
SCADA, 61
Scott connection, 89
sealed gas-filled transformers, 29
seal-in, 240
secondary selective, 13
secondary spot network, 14
secondary winding, 39, 42, 67, 81
self-contained (SC) metering, **38**, 39, 50
self-inductance, 66, 68, 71
sequence currents, **121**
sequence network, **124**
service voltage, 38
service wiring, 38
75% method, 172
shielded cable, 182
short-circuit availability, 9, 14, **23**, 349
short-circuit current rating, 215
short-circuit fault, 102, **134**, 169
shunt capacitor, 319
sidewall pressure, 181, 186, **188**
silicone oils, 27
silver sulfide, 213
single rail cable tray, 180, 195
single-phase transformer connections, 71
single-stator watthour metering, 50
six-phase transformer connections, 92
skin depth, 24
skin effect, 24, 128, 346
slip, 256
slow transfer, 214
smart grid, 62
smart meters, 61
soft starter, 278

solid-bottom cable tray, 180, 195
solid-state ballast, 292
solid-state lighting, 297
solid-state motor starting, 278
source transfer, 7, 11, **213**
sparing transformer, 14
speed-torque curve, **260**, 262
SSR, 97
standard burdens, **49**
standard delta-wye, 94
START pushbutton, 245
starter, 222, **223**, 224
stators (metering), 39
STOP pushbutton, 245
stray flux, 66
subsynchronous resonance, 97
subtractive polarity, 72, 73
subtransient impedance, 106
sulfur dioxide, 213
sulfur trioxide, 213
Swan, Sir Joseph, 288
switched capacitor, 323
switchgear, 207, **208**
symmetrical components, 101, **116**, 329
symmetry, 4, 85, 88, 101, 114, 118, 308, 345
synchronous speed, 256
system control and data acquisition, 61
system impedance, 22, **23**, 103, 328

T-connection, 89
tape shield, 182, **184**
teaser transformer, 89
temperature coefficient of resistivity, 104, **105**
temperature switch, 238
test amps (TA), 39
tetrachromatic LED, 298
THD, 348
thermal-magnetic circuit breaker, 225
Thévenin impedance, 22, 112, 143
third harmonic currents, 128, 321, 346
three-conductor cable, 188
three-phase fault, 134
three-phase fault circuit model, **135**
three-phase transformer connections, 73
three-wire control, 242
tie breaker, 13
time-current characteristic (TCC), 169
time-of-use (TOU) metering, 54

toroidal CT, 43
total clear characteristic, 171
total harmonic distortion, 348
totalizing metering, 60
totem pole, 282
Toxic Substances Control Act (TSCA), 27
transformer circuit model, 71
transformer impedance, 27
transformer-rated (TR) metering, 39, 50, 56
transformer sizing, 26
transient impedance, 113
transients, 11, 220, 269, 273, 277, 289, 328, **341**
trapeze support, 196
triangular configuration, 187
trichromatic LED, 297
triplen harmonic, **343**, 346
trough cable tray, 180, 195
turns ratio, 14, 21, 39, 42, 48, **67**, 80
two-lines-open circuit model, **148**
two-lines-open fault, 147
two-phase power, 88
two-phase transformer connections, 88
two-wire control, 242
Type A contacts, 236
Type B contacts, 236

undervoltage dropout, 105, 236, 241
uniformity ratio, 303
uninterruptable power supply, 339
unit substations, 4
UPS, 339
utilization voltage, 3

vacuum circuit breaker, 219
vacuum pressure impregnated transformers, 29
var metering, 57
variable-frequency drive, 280
vector components, 115
vegetable oil, 28
ventilated transformers, 29
vertical illuminance, 303
VFD, 280
voltage base, 15
voltage distortion, 349
voltage drop, 20
voltage metering, 55
voltage ratio, 67
voltage regulation, 27, 320, 323, 340

voltage sag, 340
voltage swell, 340
voltage transformer, 5, 39, 41, 49, 55, 56
voltage transformer ratio (VTR), 39
voltmeter switch, 55, 56
VT, 5, 39, 41, 49, 55, 56

wall support, 196
warm color, 288
watthour constant, 38, 50
watthour metering, 50
weight correction factor, 187
winding protection, 266
wire mesh cable tray, 180, 195
WR^2, 263
wye-delta, 82, 96
wye-delta motor starting, 276

wye-double delta, 92
wye-double wye, 92
wye-wye, 76

X/R ratio, 22, **103**, **104**, 107, 110, 157, **175**

zero-sequence component, 118, 121
zero-sequence harmonics, 329, 343
zero-sequence impedance, 126
zero-sequence network alteration rules, **128**
zero-sequence network, 126
zero-sequence reference bus, 126
zero-sequence topology, 129
zig-zag connection, 96
zonal cavity method, 311
zones of protection, 5

IEEE Press Series on Power Engineering

Series Editor: M. E. El-Hawary, Dalhousie University, Halifax, Nova Scotia, Canada

The mission of IEEE Press Series on Power Engineering is to publish leading-edge books that cover the broad spectrum of current and forward-looking technologies in this fast-moving area. The series attracts highly acclaimed authors from industry/academia to provide accessible coverage of current and emerging topics in power engineering and allied fields. Our target audience includes the power engineering professional who is interested in enhancing their knowledge and perspective in their areas of interest.

1. *Principles of Electric Machines with Power Electronic Applications, Second Edition*
   M. E. El-Hawary
2. *Pulse Width Modulation for Power Converters: Principles and Practice*
   D. Grahame Holmes and Thomas Lipo
3. *Analysis of Electric Machinery and Drive Systems, Second Edition*
   Paul C. Krause, Oleg Wasynczuk, and Scott D. Sudhoff
4. *Risk Assessment for Power Systems: Models, Methods, and Applications*
   Wenyuan Li
5. *Optimization Principles: Practical Applications to the Operations of Markets of the Electric Power Industry*
   Narayan S. Rau
6. *Electric Economics: Regulation and Deregulation*
   Geoffrey Rothwell and Tomas Gomez
7. *Electric Power Systems: Analysis and Control*
   Fabio Saccomanno
8. *Electrical Insulation for Rotating Machines: Design, Evaluation, Aging, Testing, and Repair, Second Edition*
   Greg Stone, Edward A. Boulter, Ian Culbert, and Hussein Dhirani
9. *Signal Processing of Power Quality Disturbances*
   Math H. J. Bollen and Irene Y. H. Gu
10. *Instantaneous Power Theory and Applications to Power Conditioning*
    Hirofumi Akagi, Edson H. Watanabe, and Mauricio Aredes
11. *Maintaining Mission Critical Systems in a 24/7 Environment*
    Peter M. Curtis
12. *Elements of Tidal-Electric Engineering*
    Robert H. Clark
13. *Handbook of Large Turbo-Generator Operation and Maintenance, Second Edition*
    Geoff Klempner and Isidor Kerszenbaum

14. *Introduction to Electrical Power Systems*
    Mohamed E. El-Hawary
15. *Modeling and Control of Fuel Cells: Distributed Generation Applications*
    M. Hashem Nehrir and Caisheng Wang
16. *Power Distribution System Reliability: Practical Methods and Applications*
    Ali A. Chowdhury and Don O. Koval
17. *Introduction to FACTS Controllers: Theory, Modeling, and Applications*
    Kalyan K. Sen and Mey Ling Sen
18. *Economic Market Design and Planning for Electric Power Systems*
    James Momoh and Lamine Mili
19. *Operation and Control of Electric Energy Processing Systems*
    James Momoh and Lamine Mili
20. *Restructured Electric Power Systems: Analysis of Electricity Markets with Equilibrium Models*
    Xiao-Ping Zhang
21. *An Introduction to Wavelet Modulated Inverters*
    S. A. Saleh and M. A. Rahman
22. *Control of Electric Machine Drive Systems*
    Seung-Ki Sul
23. *Probabilistic Transmission System Planning*
    Wenyuan Li
24. *Electricity Power Generation: The Changing Dimensions*
    Digambar M. Tigare
25. *Electric Distribution Systems*
    Abdelhay A. Sallam and Om P. Malik
26. *Practical Lighting Design with LEDs*
    Ron Lenk and Carol Lenk
27. *High Voltage and Electrical Insulation Engineering*
    Ravindra Arora and Wolfgang Mosch
28. *Maintaining Mission Critical Systems in a 24/7 Environment, Second Edition*
    Peter Curtis
29. *Power Conversion and Control of Wind Energy Systems*
    Bin Wu, Yongqiang Lang, Navid Zargari, and Samir Kouro
30. *Integration of Distributed Generation in the Power System*
    Math H. Bollen and Fainan Hassan
31. *Doubly Fed Induction Machine: Modeling and Control for Wind Energy Generation Applications*
    Gonzalo Abad, Jesus Lopez, Miguel Rodrigues, Luis Marroyo, and Grzegorz Iwanski
32. *High Voltage Protection for Telecommunications*
    Steven W. Blume

33. *Smart Grid: Fundamentals of Design and Analysis*
    James Momoh
34. *Electromechanical Motion Devices, Second Edition*
    Paul C. Krause, Oleg Wasynczuk, and Steven D. Pekarek
35. *Electrical Energy Conversion and Transport: An Interactive Computer-Based Approach, Second Edition*
    George G. Karady and Keith E. Holbert
36. *ARC Flash Hazard and Analysis and Mitigation*
    J. C. Das
37. *Handbook of Electrical Power System Dynamics: Modeling, Stability, and Control*
    Mircea Eremia and Mohammad Shahidehpour
38. *Analysis of Electric Machinery and Drive Systems, Third Edition*
    Paul Krause, Oleg Wasynczuk, S. D. Sudhoff, and Steven D. Pekarek
39. *Extruded Cables for High-Voltage Direct-Current Transmission: Advances in Research and Development*
    Giovanni Mazzanti and Massimo Marzinotto
40. *Power Magnetic Devices: A Multi-Objective Design Approach*
    S. D. Sudhoff
41. *Risk Assessment of Power Systems: Models, Methods, and Applications, Second Edition*
    Wenyuan Li
42. *Practical Power System Operation*
    Ebrahim Vaahedi
43. *The Selection Process of Biomass Materials for the Production of Bio-Fuels and Co-Firing*
    Najib Altawell
44. *Electrical Insulation for Rotating Machines: Design, Evaluation, Aging, Testing, and Repair, Second Edition*
    Greg C. Stone, Ian Culbert, Edward A. Boulter, and Hussein Dhirani
45. *Principles of Electrical Safety*
    Peter E. Sutherland
46. *Advanced Power Electronics Converters: PWM Converters Processing AC Voltages*
    Euzeli Cipriano dos Santos Jr. and Edison Roberto Cabral da Silva
47. *Optimization of Power System Operation, Second Edition*
    Jizhong Zhu
48. *Digital Control of High-Frequency Switched-Mode Power Converters*
    Luca Corradini, Dragan Maksimovic, Paolo Mattavelli, and Regan Zane
49. *Power System Harmonics and Passive Filter Designs*
    J. C. Das
50. *Industrial Power Distribution, Second Edition*
    Ralph E. Fehr, III